W9-CTY-141

3/2010

Soils and Their Environment

JOHN J. HASSETT

The University of Illinois

WAYNE L. BANWART

The University of Illinois

JOHN TAGGART HINKLEY LIBRARY
NORTHWEST COLLEGE
POWELL 82435

WITHDRAWN

Prentice Hall
Englewood Cliffs, New Jersey

Library of Congress Cataloging-in-Publication Data

Hassett, John J., (date)
   Soils and their environment / John J. Hassett, Wayne L. Banwart.
      p.   cm.
   Includes index.
   ISBN 0-13-484049-6
   1. Soils. t. Soil ecology. I. Banwart, W. L. (Wayne L.)
   II. Title.
   S591.H32   1992
   574.5′26404—dc20   91–31888
                                            CIP

Acquisition Editor: ROBIN BALISZEWSKI
Editorial Assistant: ROSE MARY FLORIO
Cover Designer: BEN SANTORA
Prepress Buyer: ILENE LEVY
Manufacturing Buyer: ED O'DOUGHERTY
Production Editor: PATRICK WALSH

 © 1992 by Prentice-Hall, Inc.
A Simon & Schuster Company
Englewood Cliffs, New Jersey 07632

All rights reserved. No part of this book may be
reproduced, in any form or by any means,
without permission in writing from the publisher.

Printed in the United States of America
10 9 8 7 6 5 4 3 2 1

ISBN 0-13-484049-6

Prentice-Hall International (UK) Limited, *London*
Prentice-Hall of Australia Pty. Limited, *Sydney*
Prentice-Hall Canada Inc., *Toronto*
Prentice-Hall Hispanoamericana, S.A., *Mexico*
Prentice-Hall of India Private Limited, *New Delhi*
Prentice-Hall of Japan, Inc., *Tokyo*
Simon & Schuster Asia Pte. Ltd., *Singapore*
Editora Prentice-Hall do Brasil, Ltda., *Rio de Janeiro*

574.526404
H 355s

3/2010

Contents

Preface

Soils and Their Environment by John J. Hassett and Wayne L. Banwart is the product of over twenty years of teaching introductory soils courses. The book offers a comprehensive coverage of soils for students in soil science, horticulture, conservation, forestry or environmental science and those peripherally interested in soils. The book is designed to provide a solid foundation for advanced level study in soil science.

The study of soils is not restricted to agriculture, but is of importance to a wide variety of disciplines. This book provides a comprehensive treatment of the basic principles of soils as they exist and interact in the environment. It is written at a higher technical level than many introductory texts by incorporating basic components of chemistry, physics, geology, and plant physiology in an examination of soil properties and soil–plant relationships. Many existing books render limited use of the hard sciences and offer limited explanations of scientific concepts. This book is designed so that students can better grasp difficult concepts such as mechanisms controlling soil pH without having taken advanced courses in soil science. Whenever necessary students are introduced to the relevant science and then shown how to apply these concepts to soils.

This book emphasizes soil as a natural body and soil–plant relationships that affect plant growth. It is the authors' intent to produce a book that is equally useful to nonagricultural students in plant science, ecology, and environmental studies and to students in curricula such as agronomy, soil science, horticulture, and forestry.

The book is well illustrated with many original computer generated line drawings and photographs, obtained from the Soil Conservation Service photo library and other sources, to help the student visualize and to solidify the principles and concepts discussed.

John J. Hassett

Wayne L. Banwart

1

Introduction

Soil is the upper portion of the earth's crust, or more precisely, soil is the upper portion of the regolith. *Regolith* is the unconsolidated material that occurs above consolidated rock. In areas where the regolith is thin, soil may include the entire regolith; in other regions where the regolith is much thicker, the soil represents only a small portion of the regolith. The soil, which varies in thickness from less than 1 m to a maximum of approximately 50 m, with an average of approximately 2 m, represents only a small fraction of the earth's crust, yet this thin fragile layer of earth, consisting of weathered rocks, humus (soil organic materials), and organisms, is the foundation on which civilization is built.

1.1 SOIL AS A SLOWLY RENEWABLE RESOURCE

Soil is a very slowly renewable natural resource. A *resource* is any material that benefits humanity. Soil and the plant life that it supports provide food, fiber, and wood products for humanity. In addition, soil is an important component in all natural cycles. For example, as part of the hydrologic cycle, soil and associated vegetative cover mediate the flow of precipitation into streams, lakes, and other surface waterways, as well as controlling groundwater recharge. Soil is also an important part of the global cycles for carbon, nitrogen, sulfur, and other plant nutrients. Soil is an open system: materials are constantly added to soil by natural and anthropogenic activities, and great quantities of materials have been lost from soil due to the constant leaching of precipitation through soil. *Anthropogenic activities* are activities related to people.

1

Since soil is a very slowly renewable resource, it is important that our usage of soil does not lessen its value to future generations or irreversibly alter its value for other purposes. To achieve this goal, we must understand the physical, chemical, and biological processes and their interactions that make soil resources unique. It is the purpose of this book to initiate this understanding and to explore the many varied processes that occur in soil. In this book we provide an introduction to the many specialized areas of soil science, such as soil physics, soil chemistry, pedology, soil fertility, soil microbiology, and environmental soil science. In the first portion of the book we emphasize soil as a natural body: its description, formation, and relationship to other soils (pedology); in the second portion we emphasize the relationship of soil to the plant life that it supports (edaphology).

1.2 THE SOIL PROFILE

Soil is a natural product. It is formed by the actions and interactions of climate, organisms, and topography acting on a parent material over long periods of time. These factors differentiate the upper portion of the regolith from the underlying materials of the earth's crust to produce soil. Soil can be considered the zone of interaction, that is, the interface between the atmosphere and the rock, or the unconsolidated materials of the earth's crust. Because of its position, soil is the more chemically, physically, and biologically altered portion of the regolith.

The soil is differentiated into layers (horizons) that form somewhat parallel to the earth's surface (Figure 1-1). In some horizons the dominant process is organic matter accumulation. In others the dominant process is the accumulation of clay or iron oxides, or the excessive loss of materials due to leaching. Collectively, the horizons make up the soil profile. Individual soils can be distinguished from other soils and nonsoil material by the type and arrangement of their horizons. Many important soil properties can be determined from the texture, color, structure, and arrangement of soil horizons.

Individual soils are products of nature and hence occur in unique ecological and landscape positions. In a given region, soils formed under native grasses will develop soil profiles and occupy landscape positions that differ from soils formed under trees. Within any region, well-drained soils will probably occur on slopes and uplands, while poorly drained soils will probably occur in bottomlands and depressions or where characteristics of the parent material inhibit water flow through the soil profile. The internal drainage of a soil, its suitability for many uses, as well as many other important soil properties can be determined from profile characteristics. Any potential homeowner would be well advised to understand these relationships to avoid potential flooding, wet basements, and failed septic systems.

Some of the most important activities of soil scientists include the mapping and classification of soils. Soils are classified so that relationships between soils and the factors that control soil formation and weathering can be determined. Soils are also classified so that information about one specific soil can be extended to

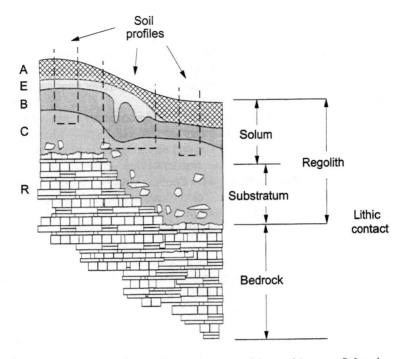

Figure 1-1 Relationships between portions of the earth's crust. Solum is the combination of all pedogenically derived horizons (i.e., A + E + B). Substratum is the portion of the regolith below the solum. Lithic contact is boundary between the regolith and hard rock. The soil profile is a vertical section through the soil. (After D. S. Fanning and M. C. B. Fanning, *Soil Morphology, Genesis, and Classification*, John Wiley & Sons, New York, 1989.)

similar soils in other states, regions, or countries. Soil maps, such as those provided by the U. S. Department of Agriculture (USDA), are available for many counties in most states. These maps and their interpretive text and tables are a wealth of information that is available to a person with some knowledge of soil.

1.3 SOIL–PLANT RELATIONSHIPS

Our interest in soil has developed primarily from the need to produce food, fiber, and forest crops. Early plant scientists were of the opinion that plants were composed entirely of materials that were obtained from the atmosphere and water. Jan Baptiste van Helmont (1577–1644) grew a willow shoot for 5 years in a container containing 90.8 kg (200 lb) of soil. At the end of the 5 years the tree weighed 76.81 kg. He could account for all but 56.7 g (2 oz) of the original 90.8 kg of soil. He erroneously concluded that the loss of the 56.7 g of soil was due to experimental error and that the plant had obtained all its nutrition from water and the atmosphere. It was not until the research of Justus von Liebig (1803–1873) and others that it was demonstrated that the soil provides plants with various mineral nutrients. We now understand that the 56.7 g of material lost from the soil in van Helmont's experiment was not experimental error but represented nutrients removed from the soil by the plant.

Much later soil and plant scientists demonstrated that plants cannot utilize all the various chemical forms of nutrients found in soil. Most nutrients are tightly bound inside the crystal lattices of soil minerals or are an integral part of soil organic matter. Plants can only utilize nutrients that are dissolved in the soil solution, weakly held to the inorganic and organic colloids or recently released by decay processes from organic matter. Colloids are very small particles that because of their very high surface areas per gram of material are responsible for many of the properties of the soil. Much of the work of soil scientists has centered around developing "tests" that measure the level of plant-available nutrients in the soil or that predict the yield increases of crops due to added nutrients (fertilizers). Soil scientists have been responsible for determining the forms of nutrients that can be added to the soil to raise soil fertility, as well as understanding the interactions of these fertilizers with different types of soils.

The interaction of fertilizers with the soil can be very complex and may differ dramatically from soil to soil. For example, the addition of limestone ($CaCO_3$) not only supplies needed calcium, but also increases soil pH. Increasing the pH of an acid soil will decrease the level of aluminum and manganese in the soil solution, and these elements can be toxic to plants. It is important not only from the viewpoint of soil fertility but also from the viewpoint of minimizing environmental pollution to be able to distinguish between the total amount of nutrients or other chemicals in the soil and the amount that is available to plants or other organisms. It is also important to be able to distinguish between chemicals that are free to move with the soil solution, potentially into the groundwater, and chemicals that are tightly bound to the soil particles and hence are immobile.

The soil also plays an important role in supplying the plant with water. Water does not simply fill the void spaces (pores) in the soil, but rather, is bound to some degree to soil particles. This means that not all the water in the soil is available for use by plants and other organisms. Much of the water in soils is bound too tightly to soil particles to be easily removed by plants. In addition, there is a critical balance between water, air, and solids in the soil. When the soil is excessively wet, the soil atmosphere is displaced by water and root respiration is adversely affected. When soil contains a high percentage of large pores that facilitate rapid gas exchange and good soil aeration, it is often at the expense of plant-available water. Much of the work of soil scientists has been concerned with the factors and processes that control the retention or movement of water in and through the soil. These factors and processes are also related to the movement of dissolved substances, heat, and even gases through the soil.

1.4 PROBLEMS RELATED TO LONG-TERM SOIL USE

The long-term use of soil demands an understanding of the several major problems associated with its use. Erosion of the surface soil represents loss of the most productive portion of the soil. This material can often be replaced, but only over very long periods of time and at great expense in terms of lost production and potential downstream pollution. Eroded soil can affect water quality when deposited as sediment in streams, lakes, and reservoirs. Sediment reduces the water-holding or water-carrying capacity of the waterway, and also represents the major pathway by which agricultural chemicals are transported to surface waters. The control of water erosion protects the soil and minimizes downstream pollution.

Large expanses of arid land are often irrigated to increase crop production. Irrigated agriculture must be based on an understanding of proper water management including an understanding not only of the water needs of plants, but also of the salinity and sodium hazards of the irrigation water. The productivity of soils can be reduced dramatically by the accumulation of soluble salts or sodium in the soil profile resulting from improper water management or use of poor-quality water for irrigation. In arid regions, the major emphasis of soil science is often related to these problems. Large areas of soils have been damaged because they were irrigated without adequate knowledge of these phenomena.

Modern agriculture utilizes a variety of materials to provide nutrients, control weeds and pests, or regulate plant growth. The addition of these chemicals, whether synthetic or natural, to provide nutrients or to control weeds or pests must be based on an understanding of the processes that control the fate of these chemicals in the environment. Rates must be high enough to compensate for adsorption to soil minerals or organic matter or for degradation by soil microbes, but not so high as to result in plant injury or the movement of these materials into surface or groundwater supplies. Soil conditions dictate the amount of a material that must be added to an area of soil to achieve the desired effect. The same amount of material added to different soils may be adequate for one soil, whereas for a different soil the addition may not be sufficient to achieve the desired effect. Addition of the same amount of material to a third soil may result in plant injury or groundwater pollution. We must develop our understanding of soil processes that control the fate of these materials, not only to prevent pollution but also to allow sustainable low-cost production of food and fiber.

We cannot provide answers for all of these issues in this book; they are far too complex and often beyond the scope of soil science alone. But it is our hope that this book will provide insight and give the student a firm base for advanced studies that will resolve many of these issues.

2

Physical Properties of Soils

The physical properties of a soil include texture, structure, density, porosity, consistency, and color. These properties are concerned with the size and content of the particles that make up the soil, how the particles are arranged into larger units or aggregates, and how the units and individual particles affect other soil properties. Soil physical properties also include soil water, temperature, and aeration. The latter properties are subjects of separate chapters.

Soil consists of three major classes of components: solids, liquids, and gases. The *solid component* is composed of inorganic minerals and organic matter. The minerals were either constituents of the original rock material (primary minerals) or were produced during the course of soil formation (secondary minerals). Soil organic matter is the result of decay processes acting on plant, animal, and microbial remains. The *liquid component* is composed of water, dissolved ions, molecules, and gases and is collectively known as the *soil solution*. The soil's *gaseous component*, that is, the *soil atmosphere*, is composed of gases similar to those in the atmosphere above the soil, but often in very different proportions.

The soil solution and soil atmosphere occupy the void or pore space between the solid particles (Figure 2-1). The total volume (V_t) of the soil is equal to the pore or void volume (V_p) plus the solid volume (V_s). For nonswelling soils V_t can be considered a constant. For soils that contain swelling clays the volume of the soil is not constant, but is dependent on the soil's water content.

An ideal soil would contain about 50% pore and 50% solid space (Figure 2-2). The solid space would be subdivided into 45% inorganic minerals and 5% organic matter, while the pore space would be equally divided between larger pores (macropores) that drain free of water and contain the soil atmosphere and smaller pores (micropores) that retain water against the pull of gravity.

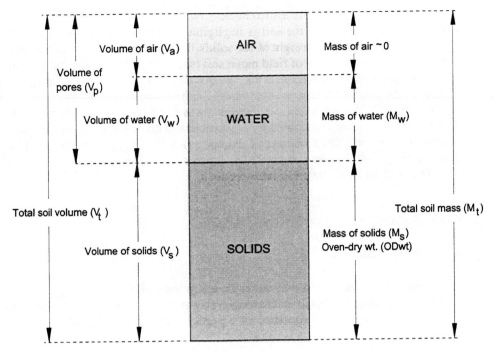

Figure 2-1 Mass and volume relationships of soil components.

Figure 2-2 Composition of an ideal soil by volume.

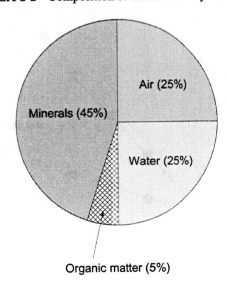

The weight of a soil is approximately equal to the weight of solids plus water since the weight of air in the soil is negligible. For analytical purposes the weight of the soil is taken as the weight of the solids. The weight of the solids is reasonably constant, while the weight of field moist soil (water plus solids) would increase after a rain or irrigation and decrease with evaporation, transpiration, or drainage. The weight of the solids, on an oven-dry basis, can be determined by placing a soil in an oven at 105°C for 48 hours to remove the water, cooling and weighing the soil. The weight of solids determined by this procedure is known as the oven-dry weight (ODwt) of the soil. The amount of calcium, potassium, organic matter, and even water in the soil is usually expressed as a percentage of the oven-dry weight of the soil. This will be illustrated in later sections.

2.1 SOIL TEXTURE

2.1.1 Size Separates

The mineral particles in a soil occur in a continuum of sizes ranging from large rocks and cobbles to small microscopic clay particles. Many chemical, physical, and even mineralogical properties of a particle are functions of particle size. For

Figure 2-3 Effect of particle size on surface area of a constant mass of material.

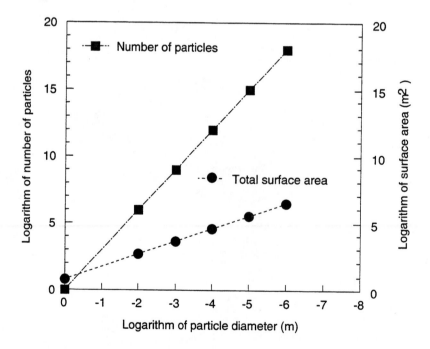

example, Figure 2-3 illustrates the increase in surface area of a large particle (1 m in diameter) when it is broken into smaller uniform particles. The weight remains constant, while the number of particles and the collective surface area of the particles increase logarithmically. This increase in surface area becomes very dramatic as the resultant particles approach the colloidal size limit, that is, particles with a diameter of 1 μm or less (1 μm = 10^{-6} m).

Example

(a) Calculate the surface area of a spherical shaped soil particle that weighs 1 g. Assume that the particle has a density of 2.65 g/cm^3.
(b) Also calculate the collective surface area of particles made by breaking the 1 g particle into a million (10^6) spherical particles each weighing 0.000001 (10^{-6}) g.

Solution

(a) Surface area of 1 g sphere:

$$volume = (1 \text{ g})/(2.65 \text{ g/cm}^3) = 0.377 \text{ cm}^3$$

Since the volume of a sphere = $(4/3)\pi \, r^3$,

$$r^3 = (0.377 \times 3)/4\pi = 0.0901$$

$$radius = r = 0.448 \text{ cm} = 0.00448 \text{ m} = 4.48 \times 10^{-3} \text{ m}$$

$$surface \ area \ (SA) = 4\pi r^2$$
$$= 4 \times \pi \times (4.48 \times 10^{-3} \text{m})^2 = 0.000252 \text{ m}^2 = 2.52 \times 10^{-4} \text{m}^2$$

(b) Collective surface area of small spheres:

$$volume = \frac{10^{-6} g}{2.65 g/cm^3} = 3.77 \times 10^{-7} \text{ cm}^3$$

$$r^3 = \frac{3.77 \times 10^{-7} cm^3 \times 3}{4\pi} = 9.01 \times 10^{-8} \text{ cm}$$

$$r = 4.48 \times 10^{-3} \text{ cm} = 4.48 \times 10^{-5} \text{ m}$$

$$Surface \ area \ of \ each \ small \ sphere = 4\pi r^2$$
$$= 4 \times \pi \times (4.48 \times 10^{-5} \text{m})^2 = 2.52 \times 10^{-8} \text{m}^2/particle$$

Collective surface area SA = 2.52 x 10^{-8} m^2/particle × 10^6 particles = 2.52 × 10^{-2} m^2

Note: Breaking the larger 1 g particle into 1 million small particles increased the collective surface area 100-fold, from 2.52 × 10^{-4} to 2.52 × 10^{-2} m^2.

Many important chemical and physical properties of soil particles are functions of particle size and hence surface area. In practice, ranges of particle sizes exhibiting similar physical and chemical properties are grouped together into what are known as *size separates*. Particles in a specific size separate tend to have similar

TABLE 2.1 Classification of particle size ranges of soil separates

Source[a]	Soil separates			
	Gravel	Sand	Silt	Clay
USDA	>2 mm	2 mm–50 μm	50 μm–2 μm	<2μm
ISSS	>2 mm	2 mm–20 μm	20 μm–2 μm	<2μm
USPRA	>2 mm	2 mm–50 μm	50 μm–5 μm	<5μm
BSI, MIT, DIN	>2 mm	2 mm–60 μm	60 μm–2 μm	<2μm

[a]USDA, U.S. Department of Agriculture; ISSS, International Soil Science Society; USPRA, U.S. Public Roads Administration; BSI, British Standards Institute; MIT, Massachusetts Institute of Technology; DIN, German Standards.

properties, such as water holding and cation-exchange capacities. Several schemes (Table 2-1) exist for the classification of soil size separates.

This book will use the USDA classification of soil separates. In this system, gravel includes those particles with diameters greater than 2 mm, the sand-size separate contains particles with diameters ranging from 2 mm (2000 μm) to 50 μm, the silt-size separate contains particles with diameters in the range 50 to 2 μm, while the clay-size separate includes particles with diameters less than 2 μm. In general, sand is the largest class of particles recognized as soil material. Particles such as gravel or larger are considered to be rock material and are generally removed by sieving before performing particle-size analysis. The large particles affect soil mainly through mechanical impedance of tillage operations and reduced water erosion. Large particles also dilute important soil properties, since they add little to the chemical or physical properties of the soil.

2.1.2 Textural Classes

Soil texture refers to the content of sand, silt, and clay particles in the soil. Soils are placed into different textural classes based upon their percentages of sand, silt, and clay particles. Particles greater than 2 mm in diameter are removed from the soil and are excluded from textural determination. The presence of larger particles is recognized by the use of modifiers added to the textural class, such as gravelly, cobbly, stoney, cherty, slaty, or shaly, based on the size and composition of the larger particles. Soil organic matter and other potential cementing agents, such as calcium carbonate, are usually removed before determination of texture.

A soil's textural class is best expressed using a *textural triangle* (Figure 2-4). A textural triangle is constructed so that the bases represent 0% sand, silt, or clay and the opposite vertexes represent 100% sand, silt, or clay. The textural triangle defines 12 different textural classes. Each textural class defines a range in sand, silt, and clay contents that have similar physical and chemical properties as long as the clay mineralogy and humus content of the soils are not dramatically different. Hence all silt loam soils with similar humus contents and clay mineralogies will

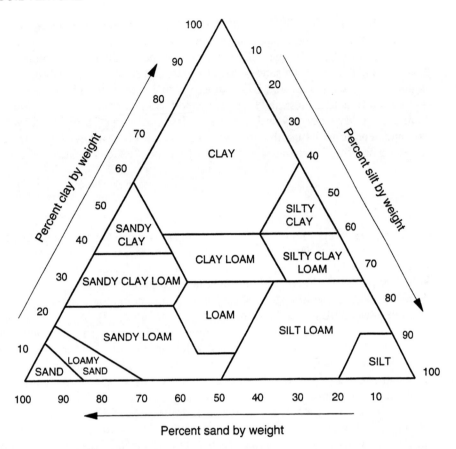

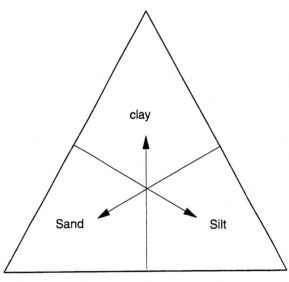

Figure 2-4 Textural triangle.

have similar water-holding and cation-exchange capacities (CECs). Silty clay loam soils from the same geographical region as the silt loam soils would have higher water-holding capacities and CECs because of their higher clay contents. Coarser-textured soils such as the sands, loamy sands, and sandy loams would have lower-water holding capacities and CECs due to their lower clay and silt contents.

Texture determines the "feel" of the soil. Soils that are high in clay tend to be hard and cloddy when dry and sticky and plastic when wet. Soils that have high silt contents tend to feel smooth and silky when wet. Soils that have high sand contents are loose and friable when dry and have little or no plasticity or stickiness when wet.

2.2 SOIL STRUCTURE

Soil structure refers to how the sand, silt, and clay particles are arranged into stable structural units. Structural units are known as *aggregates* or *peds*. Structure results when forces such as freezing–thawing, wetting and drying, root growth, or tillage activities push individual particles together. The stability of the resultant aggregates depends on the cementing agents, such as soil organic matter, secondary carbonates, and even water films that bind the aggregates together. Aggregates may be bound together to form larger units. A soil may contain large prismatic units that are relatively easily broken down into smaller, more coherent angular blocky aggregates. Natural aggregates or peds should be distinguished from clods, which are coherent masses of soil that have been produced by tillage or other operations on soils that are too moist.

Soil structure creates a framework of aggregates of different sizes and shapes, establishing the pore space in the soil (Figure 2-5). An ideal soil would consist of an arrangement of aggregates that creates not only an equal distribution of large and small pores, but also a continuous vertical arrangement of the large pores so as to facilitate the movement of air and water into and through the soil.

2.2.1 Classification of Soil Structure

Soils may be either structured or structureless. Structureless soils have no discernible peds or natural soil aggregates. Structureless material may be single grained or massive. *Single-grained material* results when individual soil particles show little or no tendency to adhere to other particles, an example being coarse-textured sandy soils. *Massive materials* result when the soil is composed of finer particles such as silts and especially clays. The natural cohesiveness of these size separates binds the individual particles together. The massive materials show little or no tendency to break apart under light pressure into smaller structural units.

Soil aggregates or peds are described by their shape (type), size (class), and strength (grade). Types or shapes of soil structure include granular, platy, blocky, subangular blocky, prismatic, and columnar (Figures. 2-6 and 2-7). Structural classes include very fine, fine, medium, coarse, and very coarse. The sizes that the

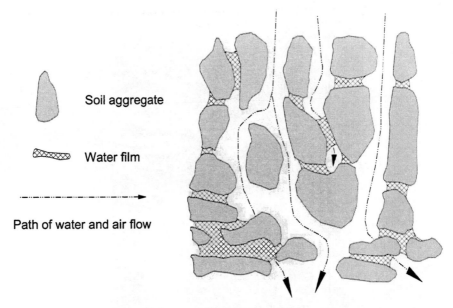

Figure 2-5 The soil framework, showing void spaces, aggregates, and water films.

various classes represent are dependent upon type. For example, fine granular structure is 1 to 2 mm in diameter, while fine columnar is 5 to 10 mm in diameter. Table 2-2 gives a summary of the various classes of the different types of soil

Figure 2-6 Examples of soil structure types.

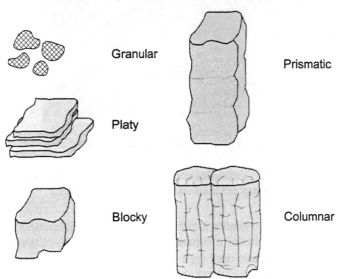

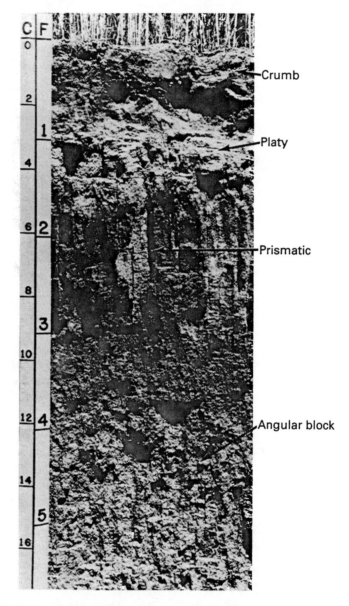

Figure 2-7 Examples of structure in the profile of a Fillmore soil (Nebraska). (Courtesy of USDA Soil Conservation Service.)

structure. Structure grades are classified by their distinctness, that is, how easy they are to distinguish in the soil profile and by their strength. Structure grades are more difficult to determine than type or class, especially in moist or wet soil. The three grades of structure are: weak, moderate, and strong.

TABLE 2-2 Classes and types of soil aggregates

	Type (shape and arrangement of peds)					
	Platelike	Prismlike		Blocklike		Spherodial
Class	Platy	Prismatic	Columnar	Angular Blocky	Subangular Blocky	Granular
Very fine or very thin	<1 mm	<10 mm	<10 mm	<5 mm	<5 mm	<1 mm
Fine or thin	1–2 mm	10–20 mm	10–20 mm	5–10 mm	5–10 mm	1–2 mm
Medium	2–5 mm	20–50 mm	20–50 mm	10–20 mm	10–20 mm	2–5 mm
Coarse or thick	5–10 mm	50–100 mm	50–100 mm	20–50 mm	20–50 mm	5–10 mm
Very coarse or very thick	>10 mm	>100 mm	>100 mm	>50 mm	>50 mm	>10 mm

Source: Soil Survey Staff, *Soil Taxonomy: A Basic System of Soil Classification for Making and Interpreting Soil Survey*, USDA Soil Conservation Service, Agriculture Handbook No. 436, 1975, p. 475, as compiled by R. W. Miller et al., *Soils: An Introduction to Soils and Plant Growth*, 6th ed., Prentice Hall, Englewood Cliffs, N.J., 1990.

Soil structure forms due to the action of forces that push soil particles together. These include wetting and drying, freezing and thawing, growth of plant roots, proper tillage, and the burrowing activities of soil animals such as moles, gophers, crayfish, and especially earthworms. Earthworms not only force soil particles together, but also pass large quantities of soil through their gut, resulting in an intimate mix of soil and organic materials into the worm casts. Forces that push soil particles together must be coupled with the presence of materials that bind the soil particles together to form soil structure. The binding materials include organic matter, clay particles, secondary or reprecipitated calcium and magnesium carbonates, and iron oxides.

Freezing and thawing and wetting and drying cycles coupled with the highest organic matter content of a given profile usually produce the granular structures of the surface horizons. Wetting and drying and the subsequent shrinking and swelling of certain clay minerals are the major forces responsible for subsoil structure formation. Subsoil structure tends to be composed of larger structural units than the surface structure. Subsoil structure units also tend to have the binding agents on ped surfaces (skins) rather than mixed throughout the ped. The translocation of clay and organic matter, as well as other materials out of surface horizons, and their subsequent deposition on subsoil ped surfaces is a major process in subsoil structure formation.

2.3 BULK AND PARTICLE DENSITIES

Soil structure provides *qualitative* information about the physical condition (tilth) of the soil. A soil with "good" granular structure in the surface horizons probably represents conditions conducive to good plant growth. That is, the soil is said to have *good tilth*. Tilth is the physical condition of the soil relative to the growth of higher plants. A soil with a well-structured subsoil probably has a "good" balance between macropores and micropores and hence a good balance between the water-holding capacity of the soil and soil aeration. *Quantitative* information about the soil's physical condition is provided by other parameters, such as mean particle density, bulk density, and porosity.

Density is the weight of an object divided by the object's volume. In soils we are concerned with the density of the soil solids, the density of the soil solution, and the density of the bulk soil. The density of the soil solution in most soils approximates the density of water, which is 1.0 g/cm^3 (62.4 lb/ft^3). In saline soils, soils affected by saltwater spills or by saltwater intrusion, the density of the soil solution may be slightly greater than 1.0 g/cm^3.

Mean particle or solid density.   The density of the soil solids is an average value for all solids in the soil. The mean particle or solid density (ρ_s) of most mineral soils is in the range of 2.6 to 2.7 g/cm^3. Soil solids tend to have densities close to quartz (2.6 g/cm^3) since most soil minerals are silicate-based minerals. The presence of iron and other heavy minerals, such as olivine, tends to increase the density of the soil solids, while the presence of organic matter (humus) tends to decrease the mean particle density of the soil. Mean particle density is defined by

$$\rho_s = \frac{M_s}{V_s} = \frac{ODwt}{V_s} \qquad [2\text{-}1]$$

where M_s is the weight of the soil solids and V_s is their volume. The weight (M_s) of the soil can be equated to the oven dry weight (ODwt) of the soil since the weight of the soil atmosphere is approximately zero. In general, the ρ_s of a soil can be considered a constant since it changes only upon the addition of humus or the weathering and loss of minerals, and these are very slow processes. The mean particle density may vary from horizon to horizon in the soil, but it usually is within the narrow range noted earlier for most mineral soils.

Dry bulk density.   Bulk density (ρ_b) is the weight of the soil divided by the total volume of the soil. Since bulk density is based on a water-free basis, the weight of the soil is equal to the soil's ODwt. The total volume (V_t) of the soil is equal to the volume of the solids (V_s) plus the volume of the pores (V_p).

$$\rho_b = \frac{M_s}{V_t} = \frac{ODwt}{V_s + V_p} \qquad [2\text{-}2]$$

Bulk density for an ideal soil that contains 50% pore space and 50% solid space would be equal to one-half the soil's particle density (Figure 2-8). Hence ρ_b for well-structured ideal soils ranges from 1.3 to 1.35 g/cm^3. Bulk density can range from 1.6 g/cm^3 or higher in sandy soils to values around 1 g/cm^3 in well-structured

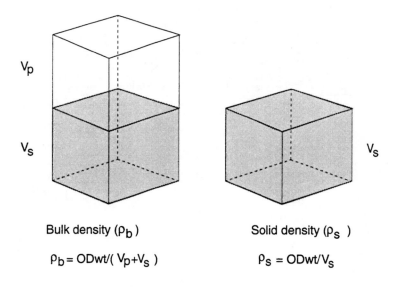

Bulk density (ρ_b)   Solid density (ρ_s)

$\rho_b = ODwt/(V_p + V_s)$ $\rho_s = ODwt/V_s$

Mass of solids = ODwt

Figure 2-8 Relationship between bulk density (ρ_b) and particle density (ρ_s).

medium-textured soils that have medium to high organic matter contents. Bulk densities may even be lower than 1 g/cm^3 in soils that have very high organic matter contents.

Bulk density, unlike particle density, is highly variable. The ρ_b of a soil varies from horizon to horizon in the soil, depending on the type and degree of aggregation of the horizon, texture, and organic matter content (Figure 2-9). Bulk density generally increases with depth, as soil structure changes from the granular structural units often found in surface horizons to the blocky and prismatic types found in lower horizons. Bulk density is also very sensitive to changes brought about by both proper and improper tillage operations. Proper tillage can increase aggregation, resulting in decreased ρ_b, whereas improper tillage operations, such as plowing when the soil is too wet, can result in destruction of soil pores and the subsequent increase in ρ_b. The fact that ρ_b is a function of the volume of the pores means that ρ_b is a valuable measure of the porosity or percentage pore space of a soil.

2.4 SOIL POROSITY

Air and water flowing into and through soils are functions of both the total amount of pore space in the soil, or soil porosity, and the distribution of the pore space between macropores and micropores, as well as the continuity of macropores through the soil.

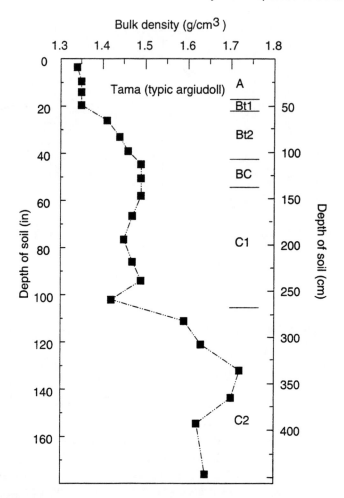

Figure 2-9 Variation of ρ_b within the soil profile. (Data from T. M. Goodard, The influence of leaching depth on the content and distribution of clay and phosphorus in selected loessial soils, M.S. thesis, University of Illinois, 1971.)

Percent pore space. Percentage pore space (% PS) of an ideal soil would be approximately 50%, with half the pore space large enough (macropores) to drain free of water due to the pull of gravity and the other half small enough (micropores) to retain water by capillarity against the pull of gravity. Water relations are discussed in later chapters. The % PS of mineral soils is normally in the range of 30 60%. Coarse-textured soils tend to have less pore space than medium- or fine-textured soils. Pore space in coarse-textured soils tends to be dominated by large pores, and as a result, these soils generally have good internal drainage and aeration but tend to be droughty. Fine- and medium-textured soils have greater porosities than do coarse textured-soils, but the pore space, especially in the clay soils, tends to be dominated by very small pores. This results in slow water movement and other water-related problems in soils with very high clay contents. The % PS in soils containing swelling-type clays will increase upon drying and decrease with wetting as the clay minerals shrink and swell.

Dry bulk density (ρ_b) can be related to the % PS of a soil.

$$\% \text{ PS} = \frac{V_p}{V_t} \times 100 \qquad\qquad [2\text{-}3]$$

$$\% \text{ PS} = \frac{V_t - V_s}{V_t} \times 100 \qquad\qquad [2\text{-}4]$$

$$\% \text{ PS} = 1 - \frac{V_s}{V_t} \times 100 \qquad\qquad [2\text{-}5]$$

Since $\rho_b = \text{ODwt}/V_t$ and $\rho_s = \text{ODwt}/V_s$, V_s and V_t can be substituted for in equation [2-5].

$$\% \text{ PS} = (\, 1 - \frac{\text{ODwt}/\rho_s}{\text{ODwt}/\rho_b}\,) \times 100 \qquad\qquad [2\text{-}6]$$

$$\% \text{ PS} = (\, 1 - \frac{\rho_b}{\rho_s}\,) \times 100 \qquad\qquad [2\text{-}7]$$

Equation [2-7] illustrates that the % PS in a soil is a function of the soil's particle and bulk densities. Since ρ_s can be considered to be constant for a given soil, % PS is only a function of the soil's bulk density. When compaction occurs, that is, when ρ_b increases, the term ρ_b/ρ_s increases and % PS decreases. Hence bulk density can be used as a quantitative measure of the total amount of pore space in a soil. Decreases in bulk density resulting from freezing and thawing or proper tillage represent increased pore space, while increases in bulk density due to compaction or improper tillage represent decreased pore space.

While bulk density can be used as a measure of the porosity of a soil, ρ_b is not a measure of the distribution of pore space between macropores and micropores. It is generally true that compaction, that is, increased ρ_b, destroys the larger, more fragile pores in the soil. Since water and air movement in the soil are dependent upon the amount and continuity of these macropores, compaction has its greatest effect on the flow of water into and through the soil and upon soil aeration.

2.5 SOIL COLOR

Soil color is an easily determined and informative soil property. Soil color directly affects the absorption of solar radiation and is one of the factors that determines soil temperature. Soil color is indirectly related to many other soil properties. Soil color can provide information about subsoil drainage and about the organic matter content of surface horizons. Soil color can also be used to help distinguish one soil horizon from another soil horizon.

Soil color is usually measured by comparison of the soil color with colors on a standard color chart. The collection of color charts generally used with soils is a modified version of the collection of charts appearing in the Munsell Book of Color and includes only that portion needed for soils. Figure 2-10 is an example from a soil color book.

The Munsell notation identifies color by use of three variables: hue, value, and chroma. *Hue* is the dominant spectral color, that is, whether the hue is a pure color such as yellow, red, green, or a mixture of pure colors, such as yellow-red. Mixtures are identified numerically according to the amount of yellow or red used to produce the mixture. 5YR is an equal mixture of red and yellow. As the number increases, the amount of the first color (yellow) increases and the second color (red) decreases.

Value and *chroma* are terms that refer to how the hue is modified by the addition of gray. Value is the degree of lightness or brightness of the hue reflected in the property of the gray color that is being added to the hue. A particular value (gray) is made by mixing a pure white pigment (10) with a pure black pigment (0). If equal amounts of white and black pigments are mixed, the value is equal to 5. Chroma is the amount of gray of a particular value that is mixed with the pure hue to obtain the actual soil color. A chroma of 1 would be made by adding 1 unit of pure hue to a certain amount of gray; a chroma of 5 would contain 5 units of pure hue to that amount of gray. The lower the chroma, the closer the color is to the pure gray of that value.

Interpretation of soil color.   Soil color is often due to the presence of organic matter and/or the oxidation status of iron compounds in the soil, although soil color can be a function of other minerals in the soil. For example, soils formed from basalt often are very dark colored even when they contain little or no organic matter. Soil organic matter (humus) imparts a brown to black color to the soil. Generally, the higher the organic matter content of the soil, the darker the soil (i.e., the lower the values). The oxidation status of the iron compounds in the soil, particularly in the lower horizons that contain little or no organic matter, also affect soil color. In better drained (i.e., aerated) soils, iron compounds occur in the ferric form [Fe(III)] and give soil a red or yellow color. In more poorly drained and hence poorly aerated soils, iron compounds are reduced and the neutral gray colors of ferrous [Fe(II)] minerals and the native colors of other minerals predominate.

Soil Organic Matter.   The organic matter content of soils can be estimated from a soil's Munsell color. Table 2-3 illustrates a general relationship between the amount of organic matter in Illinois prairie soils and the soil's Munsell notation. The most accurate estimates are obtained with medium- and fine-textured mineral soils. Soils with greater than 50% sand and less than 10% clay usually contain less organic matter than predicted. Since the performance of many agricultural chemicals is influenced by adsorption onto soil organic matter, estimation of soil organic

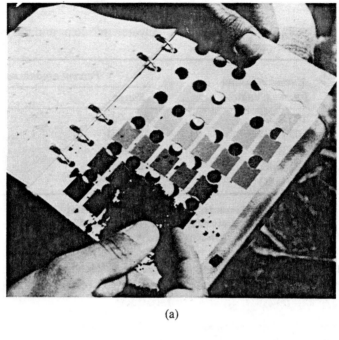

(a)

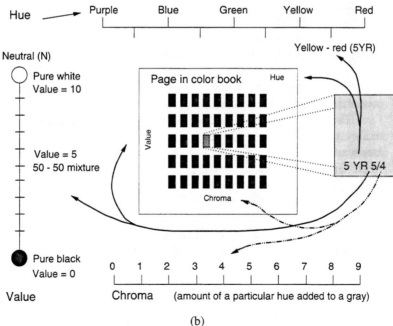

(b)

Figure 2-10 (a) Example of 10YR page of a Munsell Color Book. (Courtesy of USDA Soil Conservation Service.) (b) Sample page of a Munsell Color Book showing relationship of a color chip to hue, value, and chroma. Hue is the dominant spectral color that is modified by adding a particular gray (value) to produce the chroma of the chip.

TABLE 2-3   Relationship between soil color and the soil's organic matter content

| Munsell notation | Percent organic matter | |
for moist color	Range	Average
10YR 2/1	3.5–7.0	5.0
10YR 3/1	2.5–4.0	3.5
10YR 3/2	2.0–3.0	2.5
10YR 4/2	1.5–2.5	2.0
10YR 5/3	1.0–2.0	1.5

Source: J. D. Alexander, *Color Chart for Estimating Organic Matter in Mineral Soils in Illinois*, University of Illinois at Urbana–Champaign, College of Agriculture Cooperative Extension Service AG-1941.

matter from soil color can be helpful in selecting herbicides and determining application rates. Appropriate rates maximize weed control, while minimizing crop damage and potential environmental effects.

Soil Drainage Classes.   The drainage class and the aeration status of a soil can often be determined from the colors and color patterns in the subsoils. Red and brown colors generally indicate the presence of unhydrated ferric oxides such as Fe_2O_3(hematite), although partially hydrated ferric oxides and other minerals may also contribute red colors. The red colors may be inherited from the parent materials or developed by the oxidation of iron compounds during soil formation. In either case the red colors are stable only in soils that are well aerated, that is, are moderately-well to well-drained.

The yellow colors of soils are largely due to the presence of hydrated ferric oxides such as $Fe_2O_3 \cdot H_2O$(geothite). Soils with yellow colors tend to occupy landscape positions or occur in climates that are naturally wetter than associated red soils. Gray and neutral colors are caused by several substances, mainly quartz, kaolinite and other clay minerals, calcium and magnesium carbonates (limestones), and reduced iron (ferrous) minerals. The grayest colors, chromas of less than 1, occur in permanently water-saturated soil horizons; these soils may have bluish hues. The presence of reduced iron compounds means that the red and yellow colors of oxidized iron minerals do not mask the neutral colors of many soil minerals. The specific properties of soil natural drainage classes are presented in the sections on soil water.

2.6 TILLAGE PRACTICES AND PHYSICAL PROPERTIES

Tillage refers to mechanical operations that are designed to prepare the soil to receive fertilizers, herbicides, seeds, or other plant material, such as cuttings or seedlings; to prevent competition for water, nutrients, or sunlight by undesired plants (weeds) and to incorporate residues from previous crops into the soil. Tillage practices can improve the tilth of the soil when carried out properly or may have negative effects on soil tilth when done improperly or at the wrong soil water content.

The seedbed. Agricultural soils used for row crops consist of two zones requiring different tillage and management practices: (1) the seedbed or planting zone, which requires soil that is finely enough pulverized to provide proper soil–seed contact at planting, to facilitate germination, emergence, and seedling establishment and yet coarse enough to provide good contact with soil water and adequate aeration; and (2) the interrow zone, which requires less tillage and needs protection from wind and water erosion, as it remains bare for long periods of time before the crop canopy closes and provides some protection. The interrow zone provides water and nutrients to the crop as the crop matures and the roots grow out from the planting zone and enter the interrow soil.

Agricultural practices vary greatly in the way they manage the two zones. In older practices the zones were often treated in the same manner. The soils were

Figure 2-11 An Ohio farmer uses a ridge planter in his third year of no-till ridge planting. Tillage system differentiates between interrow region and planting zone. (Courtesy of USDA Soil Conservation Service.)

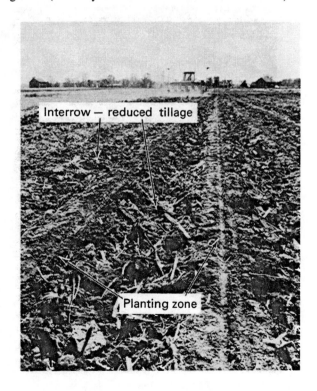

plowed with a moldboard plow and disked several times to create a finely pulverized seedbed consisting not only of the planting zone but also the interrow zones.
Modern practices use reduced tillage and often treat the planting and interrow zones
separately (Figure 2-11). Conservation and reduced tillage practices are discussed
in greater detail in the section on soil erosion.

Compaction and plant growth. Modern agriculture generally involves
the passage of equipment over the soil surface to prepare the seedbed, to cultivate
the crops to control weeds or aid seedling emergence, and to harvest the crop. Use
of this equipment creates the possibility of increased bulk densities, that is, increased soil compaction and destruction of soil structure (Figure 2-12). Some
compaction is unavoidable and most of this compaction is reversible. Most serious
compaction occurs when travel is random over the soil surface and is either excessive or at times when the soil has too high a water content.

Figure 2-12   Effects of compaction caused by plowing (plow sole pan) on
cotton root development. Cotton roots normally have deep tap roots that
penetrate more than 60 cm into the soil. These roots only penetrated 10 cm
and spread out above the plow pan. No roots were found below 10 cm depth.
(Courtesy of USDA Soil Conservation Service.)

TABLE 2-4   Effect of deep tillage on ρ_b values

Depth (cm)	Initial ρ_b (g/cm^3)	Depth of tillage	
		90-cm Final ρ_b (g/cm^3)	150-cm Final ρ_b (g/cm^3)
0–30	1.45	1.38	1.37
30–60	1.59	1.49	1.44
60–90	1.62	1.46	1.45
90–120	1.54	1.53	1.44
120–150	1.47	1.52	1.44

Source: P. W. Unger, Water-relations of a profile-modified slowly permeable soil, *Soil Sci. Soc. Amer. Proc.* 34:492–495 (1970).

Modern tillage practices minimize the number of field operations. These practices are intended to reduce compaction, to reduce operation costs, and in those cases where substantial crop residue is left on the soil surface, to reduce erosion. Tillage operations such as "zero till," "minimum tillage," "ridge till," "wheel track planting," and "plow plant" are designed to minimize the trips across the field and, in some cases, to distinguish between the planting and interrow zones.

The effect of reduced-tillage operations on compaction is not always as intended; recent research has shown that zero-till practices often have higher ρ_b values than those of conventional tillage. This is because the soil is not tilled to break up compacted zones mechanically, and if natural processes, such as freezing and thawing, are not fully effective, increased bulk densities persist.

Table 2-4 illustrates the effect of profile modification on soil bulk density values. The original profile had a B horizon that contained large amounts of clay and had high bulk density values. Tillage to depths of 90 and 150 cm diluted the clay by mixing and lowered the bulk density values of the lower horizons.

Crop rotations can also affect soil bulk density values. Figure 2-13 shows the effect of three crop rotations on soil bulk density values. The highest bulk density values occurred for continuous potatoes. The lowest bulk density values were found with the potato, wheat, and hay rotation. Similar decreases can be shown for other crop rotations. This is especially true when a deep-rooted perennial hay crop such as clover or alfalfa is included in the rotation.

Compaction can also be a serious problem with turf grasses in parks, playgrounds, golf course fairways and greens. Soil compaction results in (1) restricted root growth due to poor aeration and mechanical impedance, (2) decreased infiltration of water into the soil as well as reduced percolation of water through the soil (see Chapter 5), and (3) a general decline in turfgrass vigor and density. Soil compaction is usually corrected through turf cultivation, usually either by coring or slicing. Coring involves removing small soil cores, crushing the cores and then working the loose soil back into the turf. Slicing involves making deep

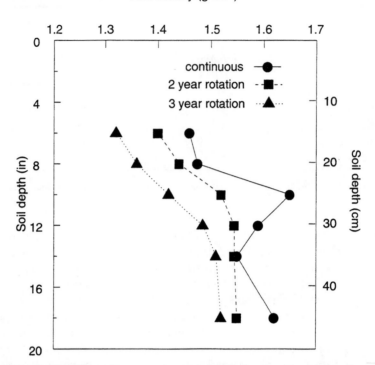

Figure 2-13 Effect of crop rotation on soil bulk density values. [Data from experiments by G. D. Brill and G. R. Blake as presented by J. A. Vomocil, Measurement of soil bulk density and penetrability: a review of methods, *Adv. Agron.* 9:159–175 (1957).]

vertical cuts in the soil. A 7- to 10-cm depth is usually recommended with both coring and slicing. Coring and slicing should be practiced as needed to correct soil compaction problems. Late spring is generally the preferred time for cool-season grasses, and spring or summer is suitable for warm-season grasses. During these periods the grasses are actively growing and the turf will recover quickly from the cultivation.

STUDY QUESTIONS

1. Why is texture an important soil property?
2. What other important soil properties are functions of the texture of the soil?
3. How does soil structure affect the tilth of the soil?
4. What is the significance of the term $(\rho_b/\rho_s \times 100)$?

5. What kind of information does soil color provide about a soil?

6. What are the benefits and drawbacks of tillage?

7. Calculate the dry weight (pounds) of a cubic foot of soil if $\rho_b = 1.42$ g/cm^3.

8. Calculate the dry weight (pounds) of an acre of soil 6 in. deep, assuming that $\rho_b = 1.27$ g/cm^3. What would a hectare of this same soil 15 cm deep weigh?

9. Calculate the % PS of three soils if their ρ_b values are 1.12, 1.62, and 1.85 g/cm^3, respectively. Assume that $\rho_s = 2.65$ g/cm^2 for all three soils.

10. Calculate the surface area of a cube of soil that measures 0.5 m on each side. What would the resultant surface area be if the cube were cut into cubes 1 μm (10^{-6} m) on each side with no loss of material?

11. Calculate the surface area of the largest sand, silt, and clay particles. Assume that the particles are spheres.

3

Soil Formation

Soil is the upper portion of the earth's crust. Soil is the zone of interaction, the interface, between the underlying geologic material and the earth's atmosphere. Because of its position, soil and its constituents are subjected to greater amounts of chemical and physical alteration and to greater amounts of biological activity than in the underlying material. Soil is often defined as the more chemically, physically, and biologically altered portion of the earth's crust.

Soil is a natural body. It is the result of the interaction of parent materials, climate, topography, organisms, and time. These variables are the five soil-forming factors. Two factors, climate and topography, and their interaction with organisms determine the environment of the soil. Organisms also determine the type, amount, and placement of organic materials that accumulate in the soil. In addition, organisms function to recycle nutrients from deep within the soil profile back to the soil surface, as well as mediate many important soil reactions.

Parent material is altered in response to the environment created by climate, topography, and organisms. Parent materials, particularly the minerals they contain, were formed under different conditions from those to which they are presently exposed in the soil. Parent materials were moved by geologic processes to their present locations or were formed in place by weathering or by the accumulation of organic matter. A major portion of soil science is concerned with the alteration of these materials and minerals in response to the ever-changing soil environment. The fifth factor, time, along with the ease of weathering of the parent material, determines the degree to which parent material has been altered by the other soil formation factors.

3.1 PARENT MATERIALS

3.1.1 Mineralogy

The earth's crust contains a variety of minerals. In unweathered igneous crustal materials, feldspars and quartz are the major minerals on a mass basis. In soils (weathered crustal material), quartz is the major mineral in the sand and silt separates. By contrast, the clay size separate of soils is dominated by such minerals as montmorillonite, kaolinite, and iron and aluminum oxides and hydroxides that may not have been present in the original crustal material. Primary minerals are minerals that were present in the original unweathered igneous crustal material, while secondary minerals are those that formed from weathering processes. Metamorphic and sedimentary rocks, such as shale, which form from eroded soil materials often contain secondary minerals. The secondary minerals found in these rocks were originally formed in soils that have since been partially or wholly destroyed by erosion. The sediments resulting from the erosion of the soils were incorporated into the sedimentary and subsequent metamorphic rocks completing the erosion cycle.

Examination of the minerals in the earth's crust and in soils demonstrates that nine elements dominate. Table 3-1 illustrates that not only is oxygen the most abundant element in the earth's crust, but it is also the only one of the nine major elements that occurs as an anion. Oxygen is therefore an important element that occurs in the structures of most minerals. Minerals can be classified into broad groups based on which cations are bonded to oxygen in their crystal structures:

silicon + oxygen = silicates
silicon + aluminum + hydrogen + oxygen = aluminosilicates
aluminum + oxygen and/or hydroxyls = metal oxides and hydroxides
iron + oxygen and/or hydroxyls = metal oxides and hydroxides
manganese + oxygen and/or hydroxyls = metal oxides and hydroxides

TABLE 3-1 Major elemental composition of crustal minerals

Element	Major ion	Percent mass basis
Oxygen	O^{2-}	60
Silicon	Si^{4+}	20
Aluminum	Al^{3+}	6
Hydrogen	H^+	3
Sodium	Na^+	3
Calcium	Ca^{2+}	2
Iron	Fe^{2+} and Fe^{3+}	2
Magnesium	Mg^{2+}	2
Potassium	K^+	1

cation + carbon + oxygen = carbonates
cation + sulfur + oxygen = sulfates

Analysis of the crystal structures of soil minerals reveals that there are several types of crystal building blocks that are repeated over and over in different minerals, as described below.

Tetrahedron.   A tetrahedron consists of four oxygens coordinated with one cation, normally silicon or aluminum. The tetrahedron derives its name from the four-sided figure that is created when the center of the oxygen atoms are connected with planes. The tetrahedron in the lower panel of Figure 3-1 is represented by a pyramid with four sides, consisting of a base plus three edge planes. The oxygen at the apex of the pyramid is called the *apical oxygen.* Figure 3-1 also shows two other ways of representing the tetrahedron structure.

Octahedron.   An octahedron is composed of six oxygens and/or hydroxyls coordinated with one cation (Figure 3-2). Aluminum, magnesium, or iron, as well as a large number of other metallic elements, are octahedrally coordinated with oxygen and/or hydroxyls. The octahedron derives its name from the eight-sided figure that is generated when the center of the oxygen atoms are connected with planar surfaces.

Cubes.   The cubic arrangement consists of eight oxygens coordinated with a single cation. Sodium, potassium, and calcium are examples of cations that form cubic structures (Figure 3-3).

Figure 3-1   Different representations of a tetrahedron structure.

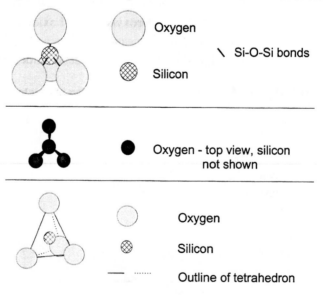

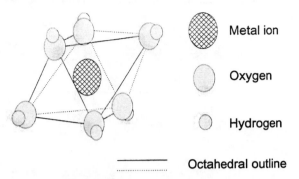

Figure 3-2 Structure of an octahedron.

Classification of silicate and aluminosilicate minerals. Many of the most important minerals found in parent materials or formed in soils by weathering processes belong to the silicate and aluminosilicate groups. It is important to understand how these primary and secondary minerals are related structurally to each other, as well as to understand the type of structural changes that take place during soil formation.

The ease of weathering of primary silicate minerals is a function of the number of Si–O–Si bonds. The Si–O–Si bond is a strong covalent bond and is not easily broken by weathering processes. The larger the number of Si–O–Si bonds per unit cell, the more resistant the mineral will be to dissolution and other weathering processes. A *unit cell* is the smallest repeating formula of a crystal. For example, NaCl is the smallest repeating formula of the sodium chloride crystal. The unit-cell formula is also known as the unit-layer formula when dealing with the layer silicates (phyllosilicates).

Neosilicates. Neosilicates are minerals composed of individual silica tetrahedrons (Figure 3-4). A silica tetrahedron (SiO_4) is composed of one Si^{4+} cation

Figure 3-3 Structure of a cube.

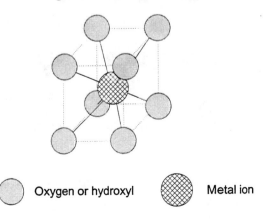

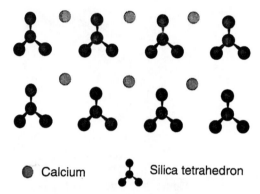

Figure 3-4   Structural arrangement of silicon tetrahedrons and metal ions in neosilicate minerals.

coordinated with four O^{2-} anions. This results in a structure that has four unsatisfied negative charges ($4 \times O^{2-} + 1 \times Si^{4+} = -8 + 4 = -4$). The neosilicate minerals are composed of individual silica tetrahedrons bound to cations such as Mg^{2+} or Ca^{2+} by ionic bonds. The individual tetrahedrons are bound together by their mutual attraction for the cations (Mg^{2+}, Ca^{2+}) that are dispersed at regular intervals between the tetrahedrons. There are no Si–O–Si bonds linking the individual tetrahedrons, consequently the neosilicates are unstable in the soil environment and are readily weathered. Examples are olivine and alite.

Figure 3-5   Structural arrangement of silicon tetrahedrons and metal ions in sorosilicate minerals.

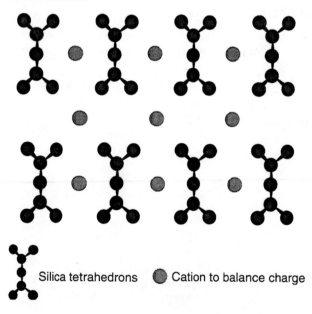

Sorosilicates. Sorosilicates are minerals composed of two silica tetrahedrons bound together by an oxygen that is shared with both tetrahedral structures (Figure 3-5). The basic structural unit is composed of two Si^{4+} cations and seven O^{2-} anions. The individual units have a negative charge of -6 ($7 \times O^{2-}$ + $2 \times Si^{4+} = -14 + 8 = -6$). This excess charge is satisfied by cations regularly dispersed between the structural units. Ionic attraction between these dispersed cations and the tetrahedral structures binds the minerals together. The degree of Si–O–Si oxygen bonding is higher than the neosilicates, and hence these minerals are not as easily weathered as the neosilicates, but they are still considered easily weathered. An example is epidote.

Cyclosilicates. These minerals are composed of cyclic patterns of silica tetrahedrons, such that each tetrahedron is bound to two other tetrahedrons by shared oxygens to form a ring structure (Figure 3-6). The units have a silicon/oxygen ratio of 1:3 and hence a net negative charge of -2 ($3 \times O^{2-} + Si^{4+} = -6 + 4 = -2$). This negative charge is satisfied by cations dispersed between the cyclic units. The cyclic units are bound together in the same manner as the neosilicates and sorosilicates, that is, by the mutual ionic attraction of adjacent cyclic units for the cations. Examples are beryl and tourmaline.

Inosilicates. Inosilicates are minerals consisting of single or double chains of silica tetrahedrons (Figure 3-7). The single-chain inosilicates also have silicon-to-oxygen ratios of 1:3, giving the units a negative charge of -2. In addition, from one-third to one-half of the tetrahedrons contain Al^{3+} instead of Si^{4+}. This reduces

Figure 3-6 Structural arrangement of silicon tetrahedrons and metal ions in cyclosilicate minerals.

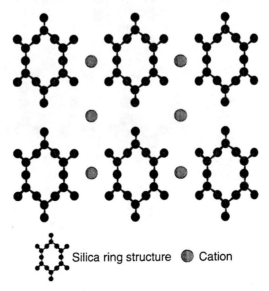

Silica ring structure ● Cation

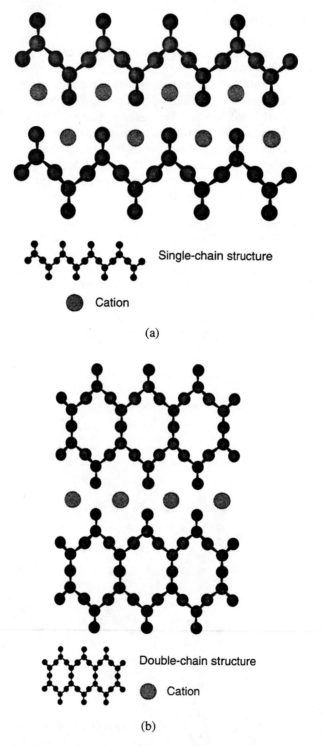

Single-chain structure

● Cation

(a)

Double-chain structure

● Cation

(b)

Figure 3-7 (a) Structural arrangement of single-chain inosilicates (b) Structural arrangement of double-chain inosilicates.

the amount of positive charge and results in additional negative charge on the mineral units. The substitution of like-sized cations of lower valence, such as Al^{3+} for Si^{4+}, in tetrahedrons or octahedrons in minerals is known as *isomorphous substitution*. An example of the single-chain inosilicates are the pyroxenes, of which augite is a specific example.

The double-chain inosilicates have silicon/oxygen ratios of 4:11 and would have a positive charge except for the isomorphous substitution of Al^{3+} for Si^{4+} in about one-half of the tetrahedrons. Both of these minerals consist of chains of silica and/or aluminum tetrahedrons bound to other chains of tetrahedrons by the ionic attraction for the cations that satisfy the negative charge of the chains. Close examination of the double-chain structure shows that it could also be considered to be a linear extension of the cyclosilicates. An example of double-chain inosilicates are the amphiboles, of which hornblende is a specific example.

Tectosilicates. Tectosilicates are minerals with a three-dimensional framework of tetrahedrons. Two major divisions of the tectosilicates are identified.

1. Quartz. The repeating unit is a silica tetrahedron in which each oxygen of the tetrahedron is linked to silicon atoms of adjacent tetrahedrons, that is, each oxygen is shared by two tetrahedral structures. The silicon/oxygen ratio is 1:2, giving rise to a neutral crystal (i.e., no dispersed cations required to maintain electroneutrality). Some research suggests that there may be limited substitution of Al^{3+} for Si^{4+} in the quartz crystal. The extreme resistance of quartz to weathering is the result of the large number of Si–O–Si bonds in the crystal and the absence of cations, held by ionic bonds, that can be hydrated and disrupt the crystal lattice.

2. Feldspars. These minerals also consist of a three-dimensional framework of tetrahedrons, except that half the tetrahedrons are occupied by Al^{3+} instead of Si^{4+}. This results in a negative charge on the crystal resulting in alkali (Na^+, K^+) and alkaline earth (Ca^{2+}, Mg^{2+}) cations dispersed in openings in the framework to maintain electroneutrality. The feldspars have a lower number of Si–O–Si bonds than quartz and, because of the lower valence of Al^{3+}, contain cations that are subject to hydration during weathering. Hence these minerals are less stable than quartz. Examples include microcline, zeolites, and albite.

Notice in Figure 3-8a that half of the tetrahedrons in the ring pattern have their apical oxygens oriented into the plane of the paper, away from the viewer, and half are oriented toward the viewer. The ring of tetrahedrons are bound to other rings to form a plane of tetrahedrons (Figure 3-8b). This plane of tetrahedrons is then bound to other planes of tetrahedrons to create a three-dimensional pattern. Figure 3-9 illustrates how a ring of tetrahedrons in one plane is bound through shared oxygens to a ring of tetrahedrons of another plane. Different representations of the ring patterns were used to emphasis the two different planes of tetrahedrons.

Phyllosilicates. These minerals consist of sheets of silica tetrahedrons bound through shared oxygens to sheets of aluminum or magnesium octahedrons to form

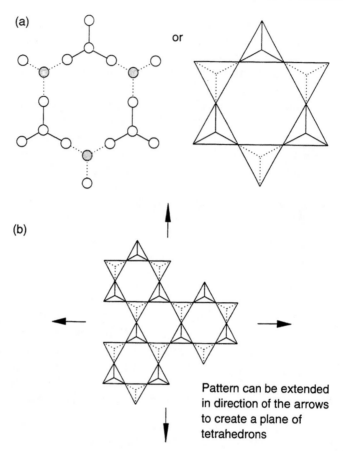

(a)

or

(b)

Pattern can be extended
in direction of the arrows
to create a plane of
tetrahedrons

Figure 3-8 Two ways of representing the tetrahedral patterns of
SiO_2(quartz) and feldspar.

layers. The silica tetrahedral sheet is a two-dimensional (x,y) expansion of the cyclic
pattern of the cyclosilicates. The minerals differ in the number and kind of
tetrahedral and octahedral sheets that are bound together to form a layer, and in the
amount of isomorphous substitution that has taken place in the tetrahedral and
octahedral positions. The negative charge due to isomorphous substitution is
balanced by a variety of cations that are either tightly held (fixed) or weakly held
(exchangeable) between the layers, that is, in the interlayers of the minerals. These
minerals have been called the layer silicates, the aluminosilicates, and the clay
minerals.

   Most secondary minerals, that is, minerals that were not part of the original
parent material but formed in the soil during the process of pedogenesis, are either
phyllosilicates or metal oxides and/or hydroxides. As soils develop there is a
gradual change from the primary minerals that characterized the parent material to
the phyllosilicates and metal oxides and/or hydroxides that distinguish the soil.

Top view

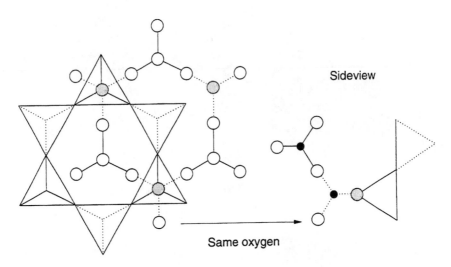

Sideview

Same oxygen

Figure 3-9 Arrangement of rings into the three-dimensional pattern of tectosilicates.

The phyllosilicates are composed of different arrangements of tetrahedral and octahedral sheets bound through shared oxygens into layers. These minerals consist of layers bound to similar or dissimilar layers by hydrogen bonding, van der Waals forces, or electrostatic attraction of adjacent layers for the interlayer cations.

The tetrahedral sheet is composed of individual tetrahedrons bound together by shared oxygens, that is, by oxygens that are common to two structures (i.e., Si–O–Si bonds) (Figure 3-10). Note the "hole" formed by the tetrahedral arrangement. This hole plays an important role in determining the nature of the clay's properties (i.e., cation fixation and swelling). The arrangement of the tetrahedrons in the tetrahedral sheet is not ideal, as shown in Figure 3-11a, but rather, the individual tetrahedrons are rotated in opposite directions to neighboring tetrahedrons, resulting in a distorted arrangement (Figure 3-11b). Diagrams of clay structures shown in the following sections are drawn with the ideal arrangement of tetrahedrons.

The octahedral sheet is composed of individual octahedrons bound together by oxygens that are common to two different octahedral structures (shared oxygens) (Figure 3-12). Octahedral sheets are classified as being either dioctahedral or trioctahedral. Dioctahedral sheets have one out of three of the individual octahedrons empty, that is, without a cation. Dioctahedral sheets are most common when all of the octahedral cations are trivalent, such as Al^{3+}. Trioctahedral sheets have all of the octahedral sites occupied by cations. This is most common when the octahedral cations are divalent cations, such as Mg^{2+}. Note that two Al^{3+} cations balance the same amount of negative charge as three Mg^{2+} cations.

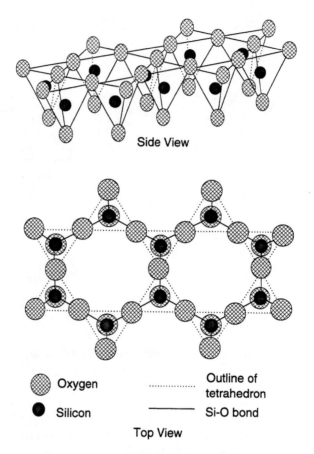

Side View

Top View

Figure 3-10 Top and side views of a tetrahedral sheet.

Figure 3-11 Ideal (a) and distorted (b) tetrahedral arrangements.

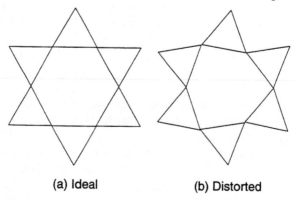

(a) Ideal   (b) Distorted

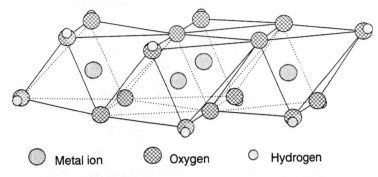

Metal ion   Oxygen   Hydrogen

Figure 3-12 Structural arrangement of an octahedral sheet.

The phyllosilicates (clay minerals) differ from the other silicon-containing minerals in that the silica tetrahedral sheets are bound through shared oxygens to octahedral sheets to form layers (Figure 3-13). The layers may be 1:1 type, 2:1 type, or mixed-layer types.

1. 1:1-Type phyllosilicates. 1:1 phyllosilicates or clay minerals consist of layers that have one tetrahedral sheet bound through shared oxygens to one oc-

Figure 3-13 Photograph of a phyllosilicate mineral (mica). Magnification is 730×. (Courtesy of Robert Darmody, Department of Agronomy, University of Illinois.)

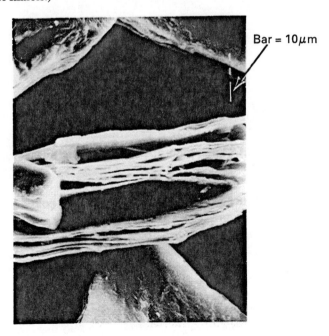

Bar = 10 μm

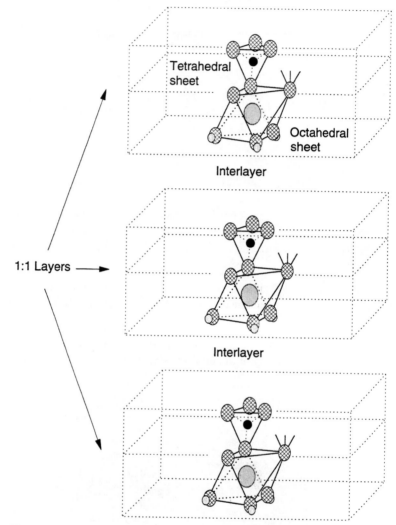

Figure 3-14 Tetrahedral and octahedral sheet arrangement for a 1:1 mineral structure.

tahedral sheet (Figure 3-14). The octahedral sheet is bound to the apical oxygens of the tetrahedral sheet, producing a layer such that one surface is a plane of oxygens from the tetrahedral sheet and the other surface is a plane of hydroxyls from the octahedral sheet. The mineral is composed of many 1:1 layers stacked on top of each other in such a manner that the tetrahedral sheet of one layer is adjacent to an octahedral sheet of the upper layer. An example is kaolinite.

2. 2:1-Type phyllosilicates. 2:1 minerals consist of layers that have two tetrahedral sheets bound through shared oxygens to one octahedral sheet (Figure

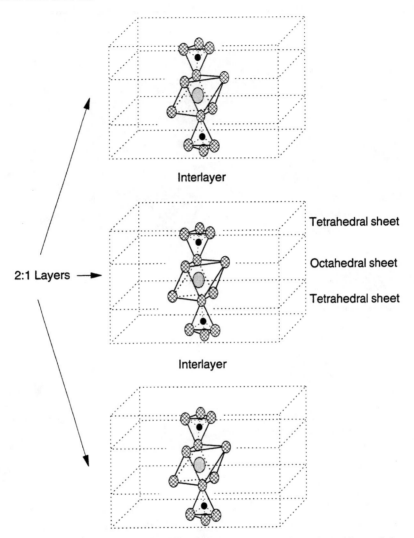

2:1 Layers ⟶

Interlayer

Tetrahedral sheet

Octahedral sheet

Tetrahedral sheet

Interlayer

Figure 3-15 Tetrahedral and octahedral sheet arrangement for a 2:1 mineral structure.

3-15). The octahedral sheet is bound to the apical oxygens of the two tetrahedral sheets in such a manner that the upper and lower surfaces of the 2:1 layer are planes of oxygens. The 2:1 minerals are composed of many 2:1 layers stacked on top of each other much like the pages of a book. 2:1-layer silicates differ from the 1:1-layer silicates in that the 2:1-layer silicates have a surface of oxygens on both the top and bottom of each layer, while 1:1-layer silicates have a surface of hydroxyls above and a surface of oxygens below the interlayer. This one major difference prevents hydrogen bonding between adjacent layers and determines many important proper-

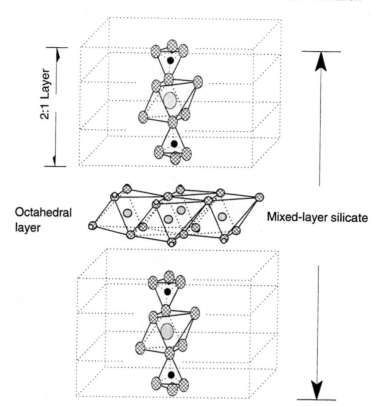

Figure 3-16   One possible tetrahedral and octahedral sheet arrangement for a mixed-layer structure.

ties of these clays, such as the potential for swelling and shrinking. Examples are muscovite, montmorillonite, illite, and vermiculite.

   3. Mixed-layer phyllosilicates. There are a variety of mixed-layer silicate minerals composed of different types of layers (Figure 3-16). The mixed-layer minerals may be regular in their stacking patterns or random. One portion of the crystal may exhibit the properties of swelling clays, while another portion of the crystal may be nonswelling. An example is chlorite.

   Sources of charge on the clay minerals.   The phyllosilicates are the dominant minerals in the clay size separate. The clay size separate is the particle-size fraction with effective diameters equal to or less than 2 μm. One of the most important consequences of small particle size is that surface area per gram of material increases exponentially with decreasing particle size. Hence the minerals that constitute the clay size separate have very high surface areas. Montmorillonite has been shown to have approximately 800 m^2/g of surface area. Coupled with the small particle size and the subsequent high surface area is the fact that phyllosilicates are negatively charged particles. This negative charge results in cations (counterions) being attracted to the clay's surface in order to maintain electroneutrality. Depending on the type of cation and the amount of negative charge, the counterions may be tightly held as part of the crystal lattice or weakly held as part of the pool of exchangeable plant-available cations in the soil.

Clay particles are negatively charged due to two separate mechanisms. These mechanisms result in what have been called the permanent and the pH-dependent charge. The permanent charge results during mineral formation from isomorphous substitution, while the pH-dependent charge results from the dissociation of weak acid groups in response to soil pH changes.

Isomorphous Substitution. Isomorphous substitution results when ions of like size, but different valences, substitute for each other in octahedral and tetrahedral positions during mineral formation. The substitution of Al^{3+} for Si^{4+} in tetrahedral positions or Mg^{2+} for Al^{3+} in octahedral positions are two prime examples. It should be stressed that isomorphous substitution occurs during mineral formation and hence does not change in response to changing soil conditions, unless these changes result in the formation of new minerals. Weathering processes tend to result in the formation of new minerals with lower and lower amounts of permanent charge due to isomorphous substitution.

The following formula is a unit-layer formula for an unsubstituted 2:1 clay mineral: $Al_4Si_8O_{20}(OH)_4$.

The amount of charge due to the cations is:	$4 \times Al^{3+}$	$= 12+$
	$8 \times Si^{4+}$	$= 32+$
The amount of charge due to the anions is:	$20 \times O^{2-}$	$= 40-$
	$4 \times OH^-$	$= \underline{4-}$
	net charge	$= 0$

In an unsubstituted mineral the positive charge balances the negative charge and a neutral mineral results. The mineral may still have some charge due to the pH-dependent mechanism, but there is no permanent charge due to isomorphous substitution.

Now consider a substituted 2:1 clay mineral in which 0.5 unit of Mg^{2+} has been substituted for 0.5 unit of Al^{3+} in the octahedral sheet of the "average" repeating unit:

$$[Al_{3.5}Mg_{0.5}]Si_8O_{20}(OH)_4$$

The amount of charge due to cations and anions is:

$3.5 \times Al^{3+}$	$=$	10.5	for aluminum
$0.5 \times Mg^{2+}$	$=$	1.0	for magnesium
$8.0 \times Si^{4+}$	$=$	32.0	for silicon
$20.0 \times O^{2-}$	$=$	-40.0	for oxygen
$4.0 \times OH^-$	$=$	$\underline{-4.0}$	for hydroxyls
		-0.5	net charge per repeating clay unit

The substitution of 0.5 unit of Mg^{2+} for 0.5 unit of Al^{3+} in the octahedral sheet results in a net charge on the clay of −0.5 per unit layer. This charge would be balanced by cations such as Ca^{2+} or K^+ held between adjacent layers, that is, in interlayer positions or on external surfaces.

Example: Calculation of unit-layer charge

(a) $Al_4[Si_7Al]O_{20}(OH)_4$

(b) $[Al_{3.7}Mg_{0.3}][Si_{7.2}Al_{0.8}]O_{20}(OH)_4$

Solution In unit-layer formulas the octahedral cations are written first, the tetrahedral cations second, and finally the anions. If there is more than one type of octahedral or tetrahedral cation, they are enclosed in brackets. The "type" of mineral, that is, whether it is a 1:1, 2:1, or mixed, can be determined from the ratio of octahedral to tetrahedral cations. If the ratio is 2:1 (dioctahedral) or 4:3 (trioctahedral), the mineral is a 2:1 mineral (see illite and vermiculite, Table 3-2). If the ratio is less than 4:3, it is a 1:1 mineral (see kaolinite, Table 3-2).

The amount of isomorphous substitution per unit-layer or unit-layer charge can be determined by summing the coefficients of the substituted cations. In formula (a), one Al^{3+} has been substituted for one Si^{4+} in the tetrahedral sheet giving rise to a unit-layer charge of −1. In formula (b), 0.3 Mg^{2+} has been substituted for 0.3 Al^{3+} in the octahedral sheet and 0.8 Al^{3+} for 0.8 Si^{4+} in the tetrahedral sheet giving rise to an unit-layer charge of −1.1 [−0.3 + (−0.8) = −1.1].

It is important to stress again that the unit-layer charge is due to isomorphous substitution that results during mineral formation. It does not vary with soil pH and is only changed upon weathering of the mineral. It is also important to note that substitution of Fe^{3+} for Al^{3+} in an octahedral sheet does not contribute to the unit-layer charge of a mineral, since both cations have the same valence.

pH-Dependent Charge. At the edge of the clay crystal the oxygens coordinated with silicon and/or aluminum are replaced with hydroxyl groups (OH^-) to balance the charge (i.e., one negative charge of oxygen is satisfied by hydrogen and one by the silicon). Hence the edges of the octahedral and tetrahedral sheets and points of surface irregularities are >SiOH and >AlOH groups, where > denotes that the groups are attached to the clay mineral. Both of these groups can potentially function as weak acid groups. When pH is raised (i.e., H^+ ions removed from the soil solution) the groups dissociate, producing a negative charge site on the mineral and the hydrogen ion is released into the soil solution. When pH is lowered (i.e., when the H^+ ion concentration is increased) the dissociated groups react with H^+ ions in the soil solution reforming the undissociated group, eliminating the negative charge of this site.

$$>AlOH \leftrightarrow >AlO^- + H^+$$

low pH   high pH   [3-1]

$$>SiOH \leftrightarrow >SiO^- + H^+$$

As pH increases the H^+ produced by dissociation of these groups is neutralized by the process that increased soil pH and the negatively charged site is balanced by some other cation (e.g., Na^+) from the soil solution.

$$>SiO^- \quad Na^+$$

In acid soils the groups are not dissociated: the H^+ is held very tightly and very specifically by the $>AlO^-$ and $>SiO^-$ sites and thus does not exchange with the cations in the soil solution.

Classification of phyllosilicates. The many varied properties of clay minerals, such as swelling, cation-exchange capacity, and K^+ and NH_4^+ fixation, can be explained by knowing the type of mineral (i.e., 1:1, 2:1 or mixed layer) and the amount of isomorphous substitution, that is, the unit-layer charge (X) of the mineral. The following classification system can be used to organize clay minerals into groups with similar properties. It should be emphasized that the boundaries between groups are often artificial, in that there may be greater similarities between end members of two adjacent groups than there are between opposite end members of the same group.

TABLE 3-2 Classification of the Phyllosilicates

Type	Unit-layer charge, X	Generalized formula[a]	Example
2:1	$X = 0$	$Al_4Si_8O_{20}(OH)_4$ $Mg_6Si_8O_{20}(OH)_4$	Pyrophyllite Talc
	$X = 0.5$ to 0.9	$[Al_{3.5}Mg_{0.5}]Si_8O_{20}(OH)_4$	Smectites: montmorillonite, beidellite, nontronite
	$X = 1.0$ to 1.5	$Al_4[Si_7Al]O_{20}(OH)_4$ $Mg_6[Si_7Al]O_{20}(OH)_4$	Illite Vermiculite
	$X \approx 2$	$Al_4[Si_6Al_2]O_{20}(OH)_4$	Micas: muscovite (dioctahedral) and biotite (trioctahedral)
1:1	$X \approx 0$	$Al_4Si_4O_{10}(OH)_8$	Kaolinite, Halloyisite

[a] Interlayer cations required to balance the unit-layer charge of the minerals are not shown in the generalized formulas.

Table 3-2 illustrates that given the formula of an unknown clay you can compare the amount of tetrahedral to octahedral cations to determine the type of mineral (2:1 versus 1:1). Once the type of mineral is known the amount of unit-layer charge can be used to further classify the mineral. For the 2:1 minerals, if X is approximately 0, it is classified as a talc or a pyrophyllite and the mineral in question will have similar properties to other talcs and pyrophyllites. If the unit-layer charge is in the range 0.5 to 0.9, the mineral is classified as a smectite and it will have properties similar to other smectites. If the unit-layer charge is in the range 1.0 to 1.5, the mineral will behave as either an illite or a vermiculite. In this case the properties will depend on the nature of the octahedral layer and on the dominant interlayer cation. The effect of the interlayer cation on the properties of illite and vermiculite is discussed in Chapter 8.

3.1.2 Classification and Distribution of Parent Materials

Classification of parent materials. Parent material is a passive soil formation factor. It is the material that has been altered by the actions of climate, topography, and organisms to produce the soil. Parent material tends to dictate the properties of young soils, but as soil development proceeds over time, climate and organisms usually become the dominant soil formation factors. The parent materials of soils may differ dramatically from one location to another (Figure 3-17). Parent materials are classified as residual or transported. Both the parent materials of organic soils and the formation of organic soils are discussed in other sections.

1. Residuum. Material that formed from rock that weathered in place. No transporting agent was involved. The composition of residuum can be highly variable, depending on the mineralogy of the rock material. Texture and mineralogy are highly variable, as are native soil fertility and other properties of soils formed from residuum.

2. Transported. Material that has been transported by some agent. The material is often sorted to some degree by the transporting process.

Lacustrine. Material that is moved by water and deposited in freshwater lakes. It is usually sorted into different-sized particles. Coarse and medium particles are found near the shore, while fine particles predominate toward the center of the lake. Soils formed from lacustrine materials tend to have high clay contents. In humid regions, the high clay contents and the depressional site positions often produce poorly drained soils.

Fluvial-Alluvial. Material that is deposited on floodplains by modern streams or at the base of slopes of hills and mountains when the velocity of water in streams slows due to decreasing slopes. *Fluvial* is usually used to designate material deposited on floodplains; while *alluvial* (fans) is usually used to describe the fan-shaped deposits at the base of hills and mountains. These materials are often

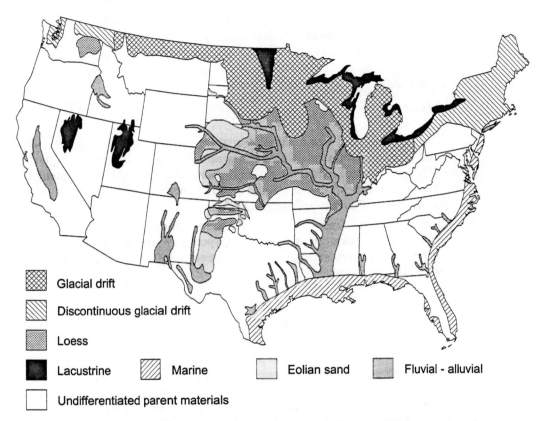

Glacial drift

Discontinuous glacial drift

Loess

Lacustrine   Marine   Eolian sand   Fluvial - alluvial

Undifferentiated parent materials

Figure 3-17 Major areas of transported parent materials in the United States (Adapted from C. B. Hunt, *Geology of Soils*, W. H. Freeman, New York, 1972, p. 126.)

sorted since each flood episode can deposit different-sized material at a given site resulting in stratified parent materials.

Colluvial. Material that is moved by gravity and deposited at the base of steep slopes. It is usually very coarse in nature. Examples include the talus slopes of the desert southwestern United States.

Marine. Material that is deposited in a marine environment. This tends to be the finer material that has been transported long distances by the streams, but also includes the coarse-textured materials deposited on old beaches.

Eolian

1. **Loess.** Silt-size material that is deposited by wind. The source of the major loess deposits were floodplains left barren after major rivers swollen with glacial meltwaters receded.

2. **Dune sand.** Dune sand is wind-deposited or wind-sorted material. It is usually of the same origin as loess, but dune sand, due to its larger particle size, was not transported as far from its source. Dune sand includes sand deposits, such as the sand hills of Nebraska, where wind action has removed the finer particles, concentrating the coarser sand.

Glacial drift

1. **Till.** Glacial till is a heterogeneous mixture of rock material that has been ground into an assortment of different-sized particles by the action of glacial ice. The texture and particular physical characteristics of tills are dependent on the original rock or other material moved by the glacier. Texture varies from coarse sand and gravel to fine clay.
2. **Outwash.** Fluvial material that was deposited by streams of glacial meltwaters. It is often found as lenses or layers on top of till in front of glacial moraines. Glacial moraines are ridgelike deposits of glacial till formed at the lateral and front edges of the ice sheets.

3.2 WEATHERING AND SOIL FORMATION PROCESSES

Weathering refers to the chemical and physical alteration of materials from which soil forms. Weathering occurs because rocks and minerals formed under very different regimes of temperature, moisture, pressure, and oxygen levels and are not stable when exposed to the conditions presently found in soil. Weathering also occurs because soils are continually leached by precipitation. Leaching removes large amounts of weathering products and allows the weathering processes to continue by shifting the chemical equilibria toward the products. Examination of highly developed (old) soils shows that large quantities of silicon, calcium, and potassium, as well as other materials, have been leached out of the soil into the groundwaters and eventually into the oceans.

Weathering processes can be classified as physical processes that alter the sizes and shapes of rocks and minerals and as chemical processes that alter their composition. Physical weathering is important because as the size of the particles is reduced, the surface area per mass of material is increased. Increasing the surface area increases the contact of the minerals with the soil solution and increases the rates of chemical processes. Increased surface area may even increase the solubility of certain mineral species.

3.2.1 Physical Weathering

Physical weathering results in reduction in particle sizes of the minerals. It is the result of processes such as freezing and thawing, wetting and drying, heating and cooling, and erosion. In freezing and thawing or wetting and drying, water is the active agent. When liquid water freezes it expands. If the water is in a crack in a rock the expansion creates a strain within the rock that eventually can disintegrate the rock into smaller fragments. Heating and cooling results in physical weathering because rocks are composed of a variety of minerals, each with its own coefficient of thermal expansion. Upon heating and cooling the minerals expand and contract to different degrees, creating strain within the rock. The eventual peeling or surface disintegration of the rock into smaller fragments due to heating and cooling and other physical processes is known as *exfoliation.*

Physical weathering can also occur when parent materials are moved and sorted by transporting agents such as wind, water, gravity and ice. The heterogeneous glacial tills of the upper midwest and northeast, which once were consolidated rocks of hills and mountains, as well as the graceful wind- and water-carved rocks of the arid west, are testimony to the power of these transporting agents.

The amount of work that can be done on a landscape by flowing water, wind, or ice is enormous. For example, the transporting power of a flowing stream varies with the sixth power of the current's velocity. "A current moving six inches a second will carry fine sand; one moving 12 inches a second will carry gravel; four feet a second, stones of about two pounds; eight feet a second, a stone of about 128 pounds; thirty feet a second, blocks of about 320 tons."[1] When suspended boulders collide with other rocks they are smashed into smaller fragments. The smaller fragments may be carried miles by the current, eventually being reduced to silt and sand particles and smooth rounded pebbles.

3.2.2 Chemical Weathering

Chemical weathering is the alteration of the original (primary) minerals of the rocks into new species of minerals (secondary) in the soil and into dissolved ions and molecules. The secondary minerals and soluble species may move down through the soil to be deposited in lower horizons or may be lost out of the soil. Most primary minerals were formed under higher temperatures and pressures than now exist in soils, and hence they are not stable in the present soil environment. Chemical weathering results in the unstable primary minerals being transformed to stable secondary minerals. Weathering also occurs because soils are constantly subjected to leaching by water containing weak inorganic and organic acids and other reactive chemicals. Some of the classes of chemical weathering are described below.

Hydration. Hydration is the chemical binding of water to cations and anions. Water is bound to the cation or anion as a water molecule and not split into H^+ and OH^- ions.

---

[1] J. S. Joffe, 1949, *The ABC of Soils*, Pedology Publications, New Brunswick, N.J., 1949, p. 25.

$$CaSO_4 + 2H_2O \leftrightarrow CaSO_4 \cdot 2H_2O \qquad\qquad [3\text{-}2]$$

Hydration results in an increase in the size of the cation or anion in the mineral and results in decreased stability of the mineral. Minerals such as the neo-, soro-, or cyclo-, or inosilicates, which are composed of silica tetrahedral structures interlaced with cations to balance the negative charge, are very sensitive to hydration. The ease of weathering of silicate minerals increases as the number of Si–O–Si bonds decreases and as the number of cations required to balance the negative charge of the tetrahedral units increases. Minerals such as quartz, a tectosilicate, are very resistant to weathering since they contain a high number of Si–O–Si bonds and few cations held by ionic bonds that are subject to hydration.

Hydrolysis. Hydrolysis is the splitting of water into hydrogen and hydroxyl ions and the reaction of these ions with minerals.

$$H_2O \leftrightarrow H^+ + OH^- \qquad\qquad [3\text{-}3]$$

$$KAlSi_3O_8 + H^+ \leftrightarrow HAlSi_3O_8 + K^+ \qquad\qquad [3\text{-}4]$$

The resulting species ($HAlSi_3O_8$) is not stable and eventually precipitates to form new secondary minerals such as allophane or halloysite [$Al_4Si_4O_{10}(OH)_8$]. The cations of weak bases and the anions of weak acids are very active in hydrolysis. Since these are important species involved in the control of soil pH, they are discussed in greater detail in Chapter 9. Hydrolysis is considered to be the most important of the chemical weathering processes, often resulting in the complete decomposition of minerals.[2]

Oxidation-Reduction. Many rocks and minerals contain species such as ferrous (Fe^{2+}) and ferric (Fe^{3+}) iron, and sulfide (S^{2-}) and sulfate (SO_4^{2-}) sulfur that are subject to oxidation and reduction in the soil profile. Reduced species (i.e., ferrous iron and sulfide) are subject to abiotic or biologically mediated oxidation in well-aerated soils. Oxidized species such as ferric iron and sulfate can serve as terminal electron acceptors for anaerobic organisms in anoxic environments and be reduced.

$$Fe^{3+} + e \leftrightarrow Fe^{2+} \qquad\qquad [3\text{-}5]$$

Oxidation or reduction changes the size and, most important, the valence of the species. This disrupts mineral lattices, facilitating the decomposition of the minerals.

Complexation. Many of the organic compounds produced by higher plants and by microorganisms are capable of complexing the soluble products produced by other chemical weathering reactions, especially the metallic ions. This results in

[2]S. W. Buol et al., *Soil Genesis and Classification*, Iowa State University Press, Ames, Iowa, 1980.

a shift in chemical equilibria toward the weathering products and increases the amount of weathering.

$$Fe^{3+} + \text{complexing agent} \leftrightarrow \text{complexed Fe} \qquad [3\text{-}6]$$

Complexed metal ions are less reactive and subject to much greater amounts of leaching and movement into lower horizons or out of the soil profile.

Dissolution. Water is an excellent solvent for the anions and cations that compose soil minerals. Water hydrates the ionic species, shielding the cations and anions, and reduces their electrical attraction for each other. Eventually, water results in the total separation of a mineral into cations and anions, that is, the dissolution of the mineral into soluble species.

$$CaSO_4 + 2H_2O \leftrightarrow CaSO_4 \cdot 2H_2O \leftrightarrow Ca^{2+} + SO_4^{2-} + 2H_2O \qquad [3\text{-}7]$$

Chemical weathering alters the original rock material into a variety of products. These include new secondary minerals such as smectites, kaolinites, and metal oxides and hydroxides, as well as soluble products such as Ca^{2+} ions. Weathering also concentrates the more resistant primary minerals, such as the tectosilicate, quartz (Figure 3-18). Weathering follows a general pattern that is dictated by soil acidification and the loss of silica from the soil profile.

Figure 3-18 Weathering sequence and products of primary minerals in soils.

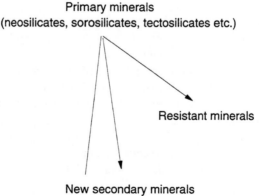

Primary minerals
(neosilicates, sorosilicates, tectosilicates etc.)

Resistant minerals

New secondary minerals
(phyllosilicates, metal oxides and hydroxides, etc.)

Soluble products
(cations, anions)

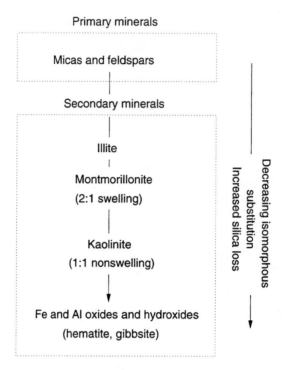

Figure 3-19   Weathering sequence of clay minerals in soils.

    The usual sequence of clay mineral formation and weathering is shown in Figure
3-19. In general, weathering results in a decrease in isomorphous substitution and a
shift from tightly held (fixed) interlayer cations to weakly held (exchangeable) cations.
Weathering also results in a general loss of silica, going from 2:1 to 1:1 clays and
eventually to the metal oxides/hydroxides. The amount of exchangeable cations bound
to the mineral, that is, the minerals CEC, first increases with the weathering of micas
to montmorillonites and then decreases with further weathering to kaolinites and
eventually the metal oxides/hydroxides. Figures 3-18 and 3-19 illustrate two major
principles affecting the ease and sequence of mineral weathering and formation in soils.
The first principle is related to the number of Si–O–Si bonds and the presence of cations
such as Ca^{2+} or Na^+ that are required to maintain electroneutrality when the number of
Si–O–Si bonds is low. As discussed previously, when the number of Si–O–Si bonds is
low, the mineral is more easily weathered. The second principle is related to what is
known as the *common ion effect*. When two or more minerals contain a common ion,
the most soluble of the minerals will dissolve and maintain a high equilibrium concentration of the common ion relative to the other minerals. The high concentration of
the common ion shifts the solubility reactions of the less soluble minerals toward the
solid phases and decreases their solubilities.

The common ion effect can be illustrated by considering the effect of $CaSO_4 \cdot 2H_2O$(gypsum) dissolution on the solubility of $CaCO_3$(calcite) and a Ca-feldspar.

$$CaSO_4 \cdot 2H_2O(\text{gypsum}) \leftrightarrow Ca^{2+} + SO_4^{2-} + 2H_2O \qquad [3\text{-}8]$$

$$CaCO_3(\text{calcite}) \leftrightarrow Ca^{2+} + CO_3^{2-} \qquad [3\text{-}9]$$

$$\text{Ca-feldspar} \leftrightarrow Ca^{2+} + \text{dissolved silicon} + \text{dissolved aluminum} \qquad [3\text{-}10]$$

The relatively high level of Ca^{2+} in equilibrium with $CaSO_4 \cdot 2H_2O$(gypsum) results in the dissolution reactions for $CaCO_3$(calcite) and calcium feldspar being shifted to the left because mass action principles decrease their solubilities. Once the $CaSO_4 \cdot 2H_2O$(gypsum) has weathered out of the soil, the next most soluble mineral $CaCO_3$(calcite) will control the level of Ca^{2+} in the soil solution. The calcium feldspar will not weather appreciably until both of the more soluble sources of the common ion (Ca^{2+}) are removed from the soil. It should be emphasized that aluminosilicate minerals share two very important common ions, dissolved silicon and aluminum. Table 3-3 shows the sequence of weathering of selected minerals in the fine-clay fraction. This sequence is due in part to the number of Si–O–Si bonds in the minerals and in part to the common ion effect.

TABLE 3-3 . Sequence of removal of minerals from the fine-clay fraction

1. Gypsum, halite	8. Vermiculite
2. Calcite	9. Montmorillonite
3. Hornblende, augite	10. Kaolinite
4. Biotite mica	11. Gibbsite
5. Feldspars; plagioclase	12. Hematite
6. Quartz	13. Anatase
7. Muscovite mica	

3.2.3 Factors Controlling Soil Formation

Different soils are formed by the same chemical and physical processes. Soils are different because they develop from different parent materials, but more important because of different intensities of the chemical and physical processes. The amounts of hydrolysis, hydration, oxidation-reduction, or dissolution that occur, as well as the differential movement of reaction products and deposition of organic materials, are all functions of the five soil formation factors.

Parent material. Parent material is a passive soil formation factor. Parent material is the raw starting material that is altered by chemical and physical processes to produce soils. Parent materials affect soil formation mainly through the amounts and types of minerals that they contain. Parent materials dominated by easily weathered minerals such as the neosilicates or sorosilicates are quickly altered to mature soils. Parent materials that contain large amounts of minerals that are very resistant to weathering, such as quartz sand (tectosilicate), delay normal soil development.

Parent materials may be composed of very narrow particle size ranges due to the sorting effect of transporting agents such as wind or water. Soils forming in sorted parent materials often reflect the texture of the parent material, particularly during the early stages of development. For example, soils forming on loess deposits tend to have silty surface textures (i.e., silt loam and silty clay loam), whereas soils forming from lacustrine deposits tend to have high clay contents.

A deficiency or excess of a plant nutrient in parent materials can also affect the development of soils. A deficiency of a nutrient or an excess of a nutrient or toxic metal in the parent material can dramatically affect the type and amount of vegetative growth. For example, the low level of calcium in soils forming from rhyolitic rocks can limit vegetative growth.

Composition of the parent materials determine the properties of "young" soils. As soil formation proceeds, climate and organisms become the dominant soil formation factors and determine the characteristics of the soil. Early pedologists noted that soils forming on different types of parent materials in similar climates eventually develop similar profiles, while soils forming on similar parent materials in different climates can have very different profile characteristics.

Climate.   Temperature and water control the intensity of soil weathering. Temperature affects the rates of chemical and biological reactions. The availability of water, along with temperature, determines the type and amount of plant, animal, and microbial activity, as well as the amount of water that is available for physical and chemical work on the parent material and the subsequent soil profile.

The amount of water that enters and moves to different depths and/or through the soil profile is a major factor in controlling the weathering of minerals and the formation of the soil profile. The movement of water through the soil (leaching) is the result of soil being an open system. Tremendous quantities of calcium, silicon, and aluminum, as well as many other elements, are lost from the soil profile due to leaching. On the other hand, water movement through the soil profile can also build soil horizons by redistributing soil particles such as clay from surface horizons to subsurface horizons.

Note: Eluviation is the term used to describe the loss of material from a horizon, while *illuviation* is used to describe the gain or deposition of material that has been lost (eluviated) out of other horizons.

Water that does not infiltrate the soil either ponds on the soil surface and is lost by evaporation, or runs off the soil surface and does varying amounts of work on the soil (Figure 3-20). The effects of runoff may be negative at one location and positive at another. That is, erosion may decrease the thickness of surface horizons at one site, but at another site lower in the landscape, the eroded material may be

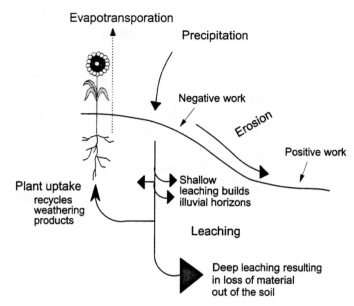

Figure 3-20   Role of water in soil formation processes.

deposited, increasing the thickness of surface horizons (Figure 3-21). Runoff and deposition can also modify the internal drainage of a site and affect the oxidation-reduction processes in the soil.

In general, precipitation in the United States increases from the east side of the Pacific mountain system to the east coast, and in the eastern United States from

Figure 3-21   Soil formation as affected by landscape position and erosion.

Surface horizon (thinned or thickened) by erosion

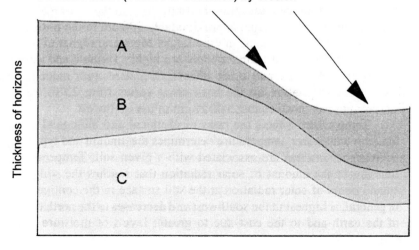

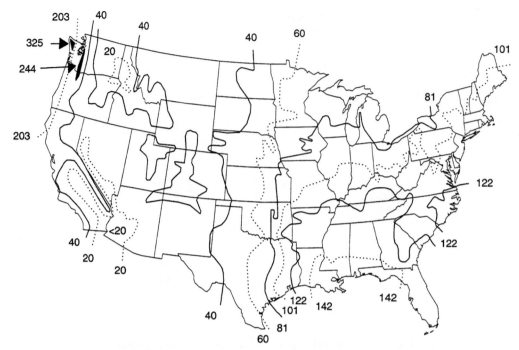

Figure 3-22   Annual precipitation amounts (cm/year) for the United States. (Adapted from *The National Atlas of The United States of America*, U.S. Department of Interior, Geological Survey, Washington, D.C., 1970, p. 97.)

north to south. Annual precipitation varies from less than 20 cm (8 in.) per year in the Great Plains states and in the Great Basin region of the Intermountain west to approximately 125 cm (50 in.) per year near the central eastern coastal areas (Figure 3-22). Precipitation increases from north to south in the eastern United States, with annual amounts ranging from approximately 100 cm in the northeast to more than 150 cm in Gulf coast regions. In the Rocky Mountain region of the western United States precipitation and temperatures are highly variable and are often related to changes in elevation and other effects associated with mountains such as rain shadows. Precipitation on the west coast varies from 20 to 40 cm in southern California to as much as much 300+ cm in the northwest.

    Temperature affects the rates of chemical and biological reactions. In combination with water, temperature determines the amount and type of vegetation and microorganisms that are associated with a given soil. Temperature is primarily a function of the amount of solar radiation that reaches the soil surface. The total annual hours of solar radiation at the soil surface in the contiguous United States, in general, is highest in the southwest and decreases to the north due to the curvature of the earth and to the east due to greater levels of moisture in the atmosphere (Figure 3-23).

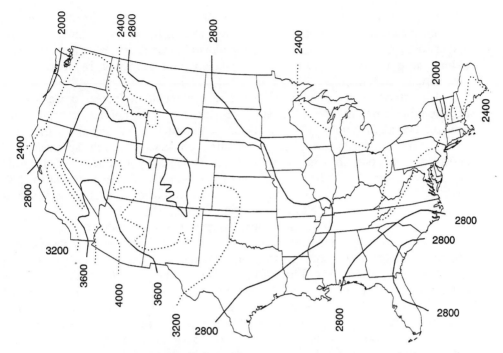

Figure 3-23 Total annual hours of sunlight at the soil surface for the United States. (Adapted from *The National Atlas of The United States of America*, U.S. Department of Interior, Geological Survey, Washington, D.C., 1970, p. 96.)

Temperature fluctuations are especially important mechanisms in the physical weathering of soils and in the formation of soil structure. Heating and cooling results in differential expansion of the minerals that compose rocks. This causes stress in the rocks and eventually leads to disintegration of outer layers (exfoliation). Freezing and thawing of soils, together with the effects of soil organisms, such as earthworms, results in the arrangement of soil particles into stable soil aggregates and produces soils with lower bulk densities. Freezing and thawing are also responsible for the disintegration of rock materials due to the expansion of water and the spreading of fractures upon freezing.

Topography. Topography is the configuration of the soil surface, including relief (slope) and the position of natural features. Topography modifies the macroclimate to produce the microclimate of the individual soils. Topography together with soil texture and aggregation affect the infiltration of precipitation and can result in runoff of the water from the soil if the precipitation rate exceeds the infiltration rate. Runoff results in less water to support soil development and plant growth. Runoff results in drier soils with thinner surface horizons and lower organic matter contents than soils forming in level areas that are not subject to runoff. The

thinner surface horizons are partially the result of less water to support soil formation and plant growth and partially the result of erosion. It should be emphasized
that erosion, especially accelerated erosion due to human activities, is a major threat
to the long-term productivity of the soil and also to the quality of the environments
downstream from the eroded soils.

Runoff of precipitation from one site can result in increased wetness or
occasional flooding at other sites. This occasional flooding not only changes plant
growth, but also dramatically affects the accumulation of organic matter and
modifies iron, nitrogen, manganese, and sulfur chemistry through the creation of
poorly aerated conditions in the soil profile. Topography, coupled with the soils
geographical location, also affects the amount of solar energy that reaches the soil
surface. Figures 3-24 and 3-25 illustrate some of the effects of topography in
modifying the macroclimate to produce the microclimate of the soil.

In the northern and southern hemispheres the solar radiation is less concentrated per unit of land area than at the equator. This results in the climate
becoming cooler with distance from the equator. The exposure (direction) and
aspect (angle) of the slope can increase or decrease this effect. A south-facing slope
in the northern hemisphere or a north-facing slope in the southern hemisphere will
present a more perpendicular land surface to the incoming solar radiation than a
level soil surface, resulting in warmer soils. North-facing slopes in the northern
hemisphere and south-facing slopes in the southern hemisphere receive less solar
radiation than the level soil and hence are cooler and usually more moist because

Figure 3-24 Effect of earth's curvature on the distribution of solar radiation at the soil surface.

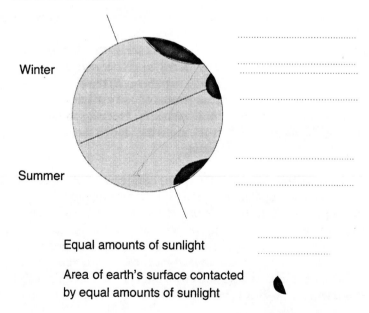

Winter

Summer

Equal amounts of sunlight

Area of earth's surface contacted
by equal amounts of sunlight

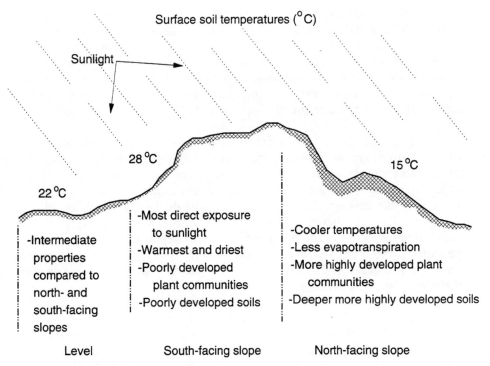

Figure 3-25 Effect of slope aspect (southern versus northern exposure) on soil temperatures.

of reduced evapotranspiration. Figure 3-25 illustrates the effect of exposure for a desert climate in the northern hemisphere.

The south-facing slope is much warmer (13°C) than the north-facing slope. The same trend is reflected in the air temperatures near the surface. As a result of warmer temperatures, the rates of evapotranspiration are greater for the south-facing slopes. Hence there is less water for chemical and biological processes in the soils on the southern slope. The differences are even greater for larger hills with development of different vegetation communities, markedly different moisture regimes, and subsequently, different soil types on the north- and south-facing slopes.

Organisms. Microorganisms (i.e., bacteria, fungi, actinomycete), animals (earthworms, moles, insects), and higher plants (trees, grasses, herbs) have a major influence on the formation of soils. Microorganisms are active in the oxidation of organic residues and the eventual production of humus. Microbes differ in the type of organic residues they can decompose and in the nature of the degradation products. Microbes also produce a variety of products that can complex metal ions, greatly modifying the chemistry of the soil solution and mobility of the metals.

Microbes mediate many of the oxidation-reduction reactions that take place in the soil. The oxidation of sulfur, ammonia, and nitrite, while they are all thermodynamically favored reactions in well-aerated soils, do not take place at significant rates without the mediation of soil microbes. Flooding of soils, while it isolates the soil from the atmosphere, does not drive reduction reactions. These reactions are driven by the soil microbes, primarily because cell metabolism dictates the need for a terminal electron acceptor in the absence of molecular oxygen.

Soil animals affect soil mainly through physical mixing. Earthworms ingest soil particles along with the detrital material on which they are feeding. The ingested soil particles are subjected to substantial amounts of chemical and physical alteration in the gut of these animals. Higher plants exert major effects on the nature of the soils that develop. Higher plants determine the amount, type, and placement of organic residues that are added to soils. Through nutrient uptake they are responsible for the recycling of nutrients from all depths in the soil profile back to the surface. The type of plant community that inhabits a given region is primarily a function of climate. The nature of the parent material and the degree of weathering of the soil also affects the type of plant community.

Trees and annuals, such as grasses, result in major differences in soils that develop from similar parent materials. Trees, which are perennials, tend to add organic materials primarily to the surface of the soils due to annual leaf or needle fall. Grasses, many of which are annuals or short-lived perennials, not only add organic matter to the soil surface upon death of the plant each fall, but they also add an almost equal weight of organic matter in terms of roots that are intimately mixed with the soil over several feet of depth. Grasses also tend to add organic materials that are richer in bases (cations such as K^+, Ca^{2+}, etc.) and hence tend to result in leachates that are less acidic than with tree species. Soils formed under grasses tend to have surface horizons that are thicker, higher in organic matter, and less acidic than soils formed under trees. In addition, soils formed under tree species tend to have more strongly developed and variable subsoils. Figure 3-26 shows the distribution of various vegetation communities in the United States. Comparison of this map with the distribution of soils in the United States (Figure 4-5) illustrates the close relationship between soil orders and vegetation communities.

Time.   Time is the fifth soil-forming factor. Generally, the longer a specific soil has been subjected to weathering processes, the more strongly developed the soil profile. This does not imply that soils with the same degree of profile development have the same chronological age. The ease or resistance of weathering of the minerals in the parent material, the amount of water available for soil development, and the temperature regime of the soil also interact with time to increase or decrease the amount of soil development that occurs in a certain time period.

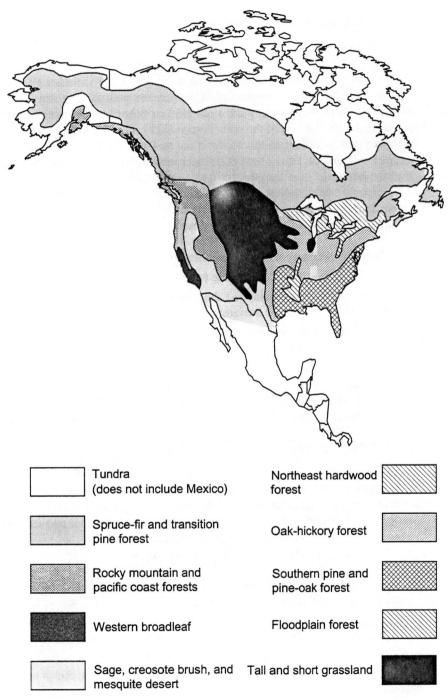

Figure 3-26 Distribution of major vegetation types in the United States and Canada. (Adapted from C. B. Hunt, *Geology of Soils,* W. H. Freeman, New York, 1972, p. 106.)

Legend:

- Tundra (does not include Mexico)
- Spruce-fir and transition pine forest
- Rocky mountain and pacific coast forests
- Western broadleaf
- Sage, creosote brush, and mesquite desert
- Northeast hardwood forest
- Oak-hickory forest
- Southern pine and pine-oak forest
- Floodplain forest
- Tall and short grassland

3.2.4 Soil Horizons

The soil profile is a vertical slice through the soil and includes all mineral zones formed or altered by soil formation processes, as well as the natural organic layers on the soil surface and the parent material or other layers beneath the solum that influence the formation of the soil (Figure 3-27). These zones (horizons) are formed approximately parallel to the soil surface and reflect differences in characteristics caused by the interactions between parent material, vegetation, climate, topography, and the length of time that the soil has been forming.

Specific horizons are distinguished from other horizons based on color, organic matter content, texture, and structural differences. Horizons are zones in which one or more of the following dominates:

1. Zones of organic matter accumulation.
2. Zones of loss (eluviation) due to movement of water through the soil profile.
3. Zones of gain (illuviation) due to the deposition of material that has been eluviated from other horizons.
4. Zones of transformations in place, such as found in weakly developed subsurface (Bw) horizons.

Soil horizons are designated as O, A, E, B, C, and R horizons. A given profile may contain all, some, or only one of these horizons. These horizons are known as *genetic* or *master horizons*.

Master horizons

Organic Horizons—O Horizons.   O horizons are horizons of organic matter accumulation, formed above or in the extreme upper portion of the mineral soil. O horizons are dominated by fresh or decomposed organic material such as leaves, twigs, moss, lichens, and humus. An O horizon is usually not present if the soil has been subjected to cultivation or other disturbance.

For water-saturated soils the O horizon will:

contain $\geq 18\%$ organic carbon if the mineral fraction is $\geq 60\%$ clay; or

contain $\geq 12\%$ organic carbon if the mineral fraction contains no clay; or

contain proportional amounts between 12 and 18% organic carbon if the clay content of the mineral fraction is between 0 and 60%.

For soils that are never saturated for more than a few days, the O horizon must by definition

contain $\geq 20\%$ organic carbon.

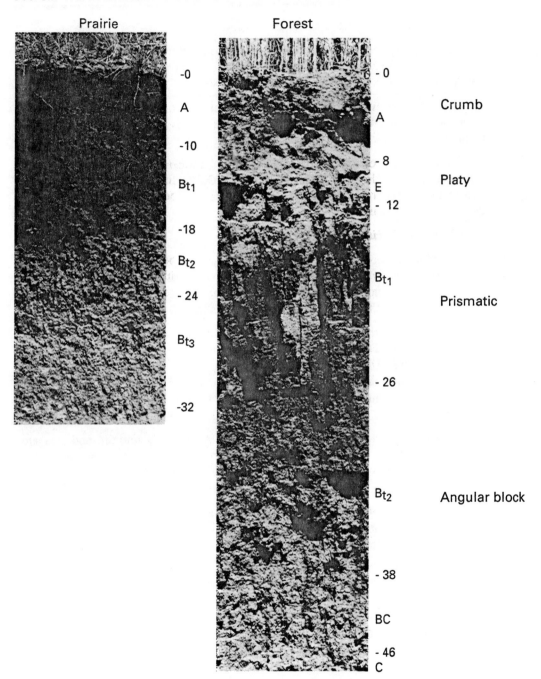

Figure 3-27 Master horizons for soil profiles formed under prairie and forest vegetation.

An O horizon can be further delineated as follows:

Oi. Slightly decomposed (fibrous) organic material that has plant parts readily recognizable. For example, this horizon might contain fresh or slightly decomposed oak leaves.

Oe. Organic material of intermediate decomposition (hemic). An examination of an Oe horizon might reveal, for example, the primary source of leaves as being from oak trees, but with considerable decomposition.

Oa. Highly decomposed organic material (sapric) which has no recognizable plant parts. An O horizon is designated as an Oa horizon if it contains adequate humus and the source of the organic materials cannot be identified, due to the high degree of decomposition.

Mineral Horizons

A. Mineral horizons that form at the surface or below an O horizon and are characterized by dark colors as a result of an accumulation of humified organic matter intimately mixed with the mineral matter. Some degree of eluviation may have occurred, but the accumulation of humus is the dominant process.

E. Mineral horizons in which the main feature is light colors caused by a loss (eluviation) of silicate clay, iron, or aluminum minerals, organic matter, or a combination of these. The result of this accelerated eluvial process is a residual concentration of sand and silt particles of resistant minerals such as quartz. The E horizon is lighter in color and lighter in texture than the horizons above or below it.

B. Mineral horizons that form below an A, E, or O horizon and that are dominated by obliteration of all or much of the original rock structure and/or by one or any combination of the following:

1. An illuvial concentration of silicate clay, iron or aluminum minerals, humus, carbonates, gypsum or silica, or a combination of these.
2. Evidence of removal of carbonates.
3. Residual concentration of resistant minerals such as sesquioxides.
4. Coatings of sesquioxides that make the color of these horizons conspicuously lower in value, higher in chroma, or redder in hue than overlying and underlying horizons without apparent illuviation of iron.
5. Alteration that forms silicate clay or liberates oxides or both.
6. Formation of granular, blocky, or prismatic structure.

B horizons in the midwest and on older geomorphic surfaces, such as in the southeastern United States, are commonly identified by an accumulation of clay and the total leaching of carbonates and other fairly soluble minerals out this zone. For soils formed in residuum, B horizons are also characterized by obliteration of all or much of the original rock structure. B horizons of very old highly weathered soils are often characterized by residual concentrations of sesquioxides and other resistant minerals, while B horizons of young soils may only show the formation of some weak structure and some changes in color.

C. Horizons are unconsolidated layers that are little affected by pedogenic (soil forming) processes and lack the properties of O, A, E, and B horizons. Most are mineral layers. The C horizon may be like or unlike the material from which the solum formed, that is, it may or may not be parent material. Some C horizons have accumulations of carbonates, silica, gypsum, or more soluble salts and are considered C horizons if they are *not* contiguous to an overlying genetic horizon; otherwise, these layers are considered to be B horizons.

R. Hard consolidated bedrock. Examples are granite, basalt, quartzite, and indurated limestone and sandstone. The bedrock may contain cracks or fractures, but these are small enough and far enough apart that few roots can penetrate.

Transition Horizons. There are two kinds of transition horizons:

1. Horizons dominated by the properties of one master horizon but with some properties of another master horizon.

   AB: A-horizon properties dominate, but some properties of both A and B.

   BA: B-horizon properties dominate, but of both A and B horizons. (Note that the master horizon symbol listed first is the dominate horizon.)

   Other examples: EB, BE, BC, etc.

2. Horizons in which distinct portions of the horizon have recognizable properties of a master horizon. In these types of horizons one section has the properties of one master horizon and another section has the properties of the other master horizon.

   E/B: pockets or areas of B surrounded by E material. E-horizon material makes up the greater volume.

   Other examples: B/E, B/C, etc.

There will always be gradients in soil properties at the boundaries of soil horizons. Where the change in properties from one horizon to the next is abrupt, no transition horizons are designated. Only when the transition zone exceeds a certain minimum thickness are these zones considered transition horizons and not part of the overlying or underlying horizons.

Subordinate Distinctions within Master Horizons and Layers. Lowercase letters are used as suffixes to designate specific kinds of master horizons and layers.

a   Highly decomposed organic matter (i.e., Oa)

b   Buried soil horizon (i.e., Ab)

c   Concretions or nodules of iron, aluminum, manganese, or titanium oxides or hydroxides (i.e., Ac)

e   Intermediately decomposed organic matter (i.e., Oe)

f   Frozen soil—permafrost in horizon (i.e., Cf)

g   Strong gleying—indicative of poor drainage and presence of reducing conditions (i.e., Bg)

h   Illuvial accumulation of organic matter (i.e., Bh)

i   Slightly decomposed organic matter (i.e., Oi)

k   Accumulation of carbonates, commonly calcium carbonates (i.e., Bk)

m   Strong cementation, roots can only penetrate these horizons through cracks (i.e., Bkm)

n   Accumulation of sodium Na^+ on colloids (i.e, Btn)

o   Residual accumulation of sesquioxides (i.e., Bo)

p   Plowing or other disturbance of the surface layer (i.e., Ap)

q   Accumulation of silica (i.e., Bq)

r   Weathered of soft bedrock, used only with C horizons (i.e., Cr)

s   Illuvial accumulation of sesquioxides such as iron (i.e., Bs)

t   Accumulation of clay (i.e., Bt)

v   Plinthite (i.e., Bv)

w   Color or structural B horizon (i.e., Bw)

x   Fragipan character (i.e., Bx)

y   Accumulation of gypsum (i.e., By)

z   Accumulation of salts (i.e. Bz)

Note 1: Master horizons may be subdivided into vertical subdivisions by using arabic numerals (i.e., A1, A2 or Bt1, Bt2). Master horizons can be subdivided with any change in color, structure, or texture that is not significant enough to define a different type of master horizon.

Note 2: Changes in parent materials within the solum are called lithologic discontinuities and are designated by using arabic numerals before the master and subordinate symbols (i.e., 2B, 2Bg2, 3C). The arabic number 1 for the first parent material is understood even though it is not written (Figure 3-28).

3.2.5 Development of the Soil Profile

The soil profile is formed by several general processes. These include the accumulation of organic matter, the formation of new (secondary) minerals due to the alteration of the primary minerals or precipitation of the decomposition products of

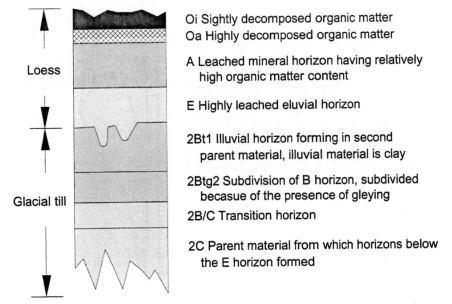

Figure 3-28 Hypothetical soil profile showing common subdivisions of soil horizons.

the primary minerals, the movement and deposition (translocation) of organic matter, minerals and decomposition products out of surface horizons into subsurface horizons, the loss of the decomposition products of primary minerals and soluble bases (i.e., Ca^{2+}, Mg^{2+}) due to leaching, and the organization of the soil material into structural units. The extent of these processes depends on the intensity of physical and chemical weathering, the type and amount of biological activity, and the amount of water available to do work on the soil profile. These are controlled by the soil formation factors.

Consider the sequence of events that results in the formation of a very mature forest profile from a calcareous loam glacial till. Initially (soil 1, Figure 3-29) the parent material will have a pH around 8 and will have high levels of Ca^{2+} due to the presence of $CaCO_3$(calcite) or other divalent carbonates. Clay content will be low and mineralogy will be dictated by the parent materials. Organic matter content and the associated nitrogen and sulfur levels will also be very low. In addition, the properties of the soil profile will be fairly homogeneous with depth since little translocation of weathering products will have occurred.

The next stages of soil development include increased accumulation of organic matter in and on the soil, as well as the dissolution of the more soluble minerals, specifically calcite. As calcite is removed from the upper portions of the profile, soil pH is decreased, resulting in increased formation and translocation of clays, humic materials, and other weathering products (soil 2, Figure 3-29).

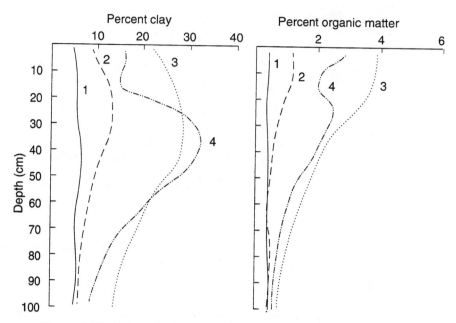

Figure 3-29 Effect of soil development on clay and organic matter distribution in the profile. Soil 1 is young weakly developed soil. Increasing numbers (2 through 4) represent increasing stages of soil development.

As weathering proceeds, organic matter deposition and clay formation and translocation continues and result in the mature soil profile (soil 3, Figure 3-29). At this stage neither soil pH nor soil fertility has decreased to the point where organic matter production begins to be reduced. The soil, because of a number of factors, is often the most productive at this stage of soil development. The higher organic matter levels result in higher N and S fertility, and at the same time, since the soil is only slightly acidic (pH 6 to 7), the soil has good fertility with respect to other nutrients, such as P and Ca.

The very mature profile is produced (soil 4, Figure 3-29) when clay formation and translocation have moved substantial amounts of clay and deposited it as clay films on ped (soil aggregates) surfaces in the B horizon. The weathering process will also have produced surface horizons showing large amounts of loss (eluviation), with the largest amounts of loss occurring just beneath the zone of organic matter accumulation. This zone of intense eluviation occurs at this point rather than at the surface because the surface mineral particles are coated or partially coated by humus and hence less subject to the extremes of weathering.

As weathering and leaching of weathering products out of the profile continue, the soil becomes more acidic, and clay type changes in response to reduction of pH and loss of silicon and basic cations, eventually producing the highly weathered acidic profile (Ultisol) typical of the southeastern United States.

S T U D Y   Q U E S T I O N S

1. Why are the neosilicate and sorosilicate minerals more easily weathered than the tectosilicate minerals?

2. Why are feldspar minerals more easily weathered than quartz even though they are both tectosilicates?

3. How are the tetrahedral and octahedral sheets of a 1:1 or a 2:1 layer bound together?

4. What are the sources of negative charge on the phyllosilicate (clay) minerals?

5. How does physical weathering differ from chemical weathering?

6. Identify the soil formation factors. Discuss the role of each factor in determining the type of soil that forms.

7. How do A, E, and B horizons form?

8. When is the C horizon not like the parent material of the A and B horizons?

9. What is implied about a soil by the terms "young," "mature," and "old" soils?

10. Discuss the concepts of a soil as a "natural resource" and a "natural product."

4

Classification of Soils

Soils are arranged into classification systems to facilitate the organization of knowledge, to improve understanding of relationships between individual soils, and to allow the transfer of information obtained from a specific soil to similar soils around the world.

4.1 USDA SYSTEM OF SOIL TAXONOMY

In 1965 after a series of approximations the USDA Soil Survey Staff[1] adopted a new system of soil taxonomy which was released in 1975 as the U.S. Comprehensive Soil Classification System. In this system, soils are classified according to properties of the soils in the landscape. These include soil temperature and moisture regimes as well as profile characteristics. The present USDA system differs from earlier systems in that earlier systems of classification were less objective or quantitative and were based largely on the soil formation processes that were responsible for soil formation. Identification of these processes was difficult for many soils. In the USDA system, soils are classified by the presence or absence of special types of pedogenic horizons (i.e., A, E, B, and C horizons) and by the presence or absence of other soil properties. These special types of pedogenic horizons are called *diagnostic horizons*. While soil formation processes are implied

---

[1]Soil Survey Staff, *Soil Taxonomy*, USDA Handbook No. 436, U.S. Government Printing Office, Washington, D.C., 1975.

Taxonomic level	Classification (example)
Order	*Mollisol*
Suborder	*Aquoll*
Great Group	*Haplaquoll*
Subgroup	*Typic Haplaquoll*
Family	*Fine-silty, Mixed Mesic Typic Haplaquoll*
Series	*Drummer*

Figure 4-1   Taxonomic levels in the USDA soil classification system.

in the USDA classification system they are not the major basis for distinguishing one soil from another soil as they were in older classification systems.

Soils are classified into six taxonomic levels (Figure 4-1). The highest level, *order*, broadly gathers individual soils into collections of soils that were formed by similar soil-forming processes as determined by the presence or absence of major diagnostic horizons. The lower levels are formed by considering the presence or absence of other diagnostic horizons, by differences in climate, drainage, texture, color, or other soil properties.

4.1.1 Diagnostic Horizons

Diagnostic surface horizons.   Diagnostic surface horizons are called *epipedons*, meaning the surface of the pedon or soil. Epipedons are special cases of A or O horizons. Epipedons are classified around the central concept of the thick, dark-colored (high organic matter), high-base-saturated, mineral A horizons of prairie soils known as mollic epipedons. *Note: Base saturation* refers to the presence of basic cations such as Ca^{2+}, Mg^{2+}, and K^+ on the cation-exchange complex, in contrast to acidic cations such as H^+ and Al^{3+} (see Chapters 8 and 9).

Mollic Epipedon.   Thick, dark-colored, high-base-saturated (>50%) mineral surface (A) horizons. Specifically, these horizons have sufficient soil structure that they are not both massive and hard or very hard when dry. Colors include chromas and values of 3.5 or darker when moist and values darker than 5.5 when dry.

Mollic epipedons must be more than 10 cm thick if lying directly over hard rock. If the soil contains a B horizon (i.e., an argillic, natric, cambic, or spodic diagnostic horizon) or a fragipan or duripan, the mollic epipedon must be at least one-third the thickness of the solum when the solum is less than 75 cm thick and at least 25 cm thick when the solum is thicker than 75 cm.

Although mollic epipedons contain high levels of organic matter (>1%) they must contain less organic matter than required to be O horizons (see Chapter 3). They must also contain less than 250 ppm of citrate extractable P_2O_5 or they become anthropic epipedons. Mollic epipedons must be moist 3 or more months of the year in 7 out of 10 years when the soil temperature is 5°C or higher at a depth of 50 cm.

Histic Epipedon.   Strongly developed organic (O) horizons of mineral soils. The horizons are from 20 to 60 cm thick. They are usually formed on soils that under natural conditions are saturated with water at least 30 days per year. The minimum organic carbon content necessary for a horizons to be classified as a histic epipedon ranges from 12 to 18% (20.7 to 31.0% organic matter), depending on the clay content of the soil.

Note: The organic matter content of a soil is usually estimated from the soil's organic carbon content. Organic carbon is measured by oxidizing the soil organic matter and measuring the amount of CO_2(gas) given off. The organic matter content is normally arrived at by multiplying the organic carbon content by 1.724. However, several studies have shown that the factor to convert the organic carbon content of a soil to soil organic matter is not constant but varies (1.7 to 2.5) with the source of organic matter.[2]

Umbric Epipedon.   Epipedons (A horizons) that meet all the requirements of mollic epipedons except that they are too low in base saturation (<50%), that is, too acidic, to be mollic epipedons.

Ochric Epipedon.   Surface (A or O) horizons that are too light in color, too low in organic matter, or too thin to be a mollic, umbric, anthropic, plaggen, or histic epipedon.

Anthropic Epipedon.   Surface horizons with the characteristics of mollic epipedons, but formed due to long-term human use of the soil. The long-term use has raised the citric-acid-extractable P_2O_5 levels above 250 ppm. The oyster-shell kitchen middens of the lower Potomac River in Maryland, which were formed when Indians deposited large quantities of oyster shells and other wastes on soils, are examples of anthropic epipedons. The calcium leached from the oyster shells has converted the underlying soils from low-base-status soils (Ultisols) to higher-base-status soils (Alfisols or Mollisols).[3]

Melanic Epipedon.   A thick black horizon with high concentrations of organic carbon (>6%), the organic carbon is usually associated with minerals such as allophane or aluminum–humus complexes. The intense black color of it is thought to be due to the decay of large amounts of root residues from graminaceous vegetation. These horizons can be distinguished from mollic epipedons in that these horizons have andic soil properties (see Andisols) throughout the horizon.

Plaggen Epipedon.   Man-made surface horizons, produced by long-term manuring, often thickened by the addition of soil. Artifacts are common and these soils are more prevalent in parts of western Europe.

[2]D. W. Nelson and L. E. Sommers, Total carbon, organic carbon, and organic matter, in *Methods of Soil Analysis: Part 2, Chemical and Microbiological Properties*, A. L. Page, R. H. Miller, and D. R. Keeney (eds.), American Society of Agronomy, Madison, Wis., 1982.

[3]D. S. Fanning and M. C. B. Fanning, *Soil Morphology, Genesis, and Classification*, John Wiley & Sons, New York, 1989, p. 66.

Note: The beginning soils student should use the concept of the mollic epipedon as a thick, dark-colored, high-base-saturated A horizon, while the ochric epipedon is too thin or too light in color to be mollic, and the umbric epipedon, while it is thick enough and dark enough to be mollic, is too acidic, that is, too low in base saturation.

Diagnostic subsurface horizons.   These are mineral horizons that form below the surface, although they may be exposed by erosion or other processes. They may be eluvial E horizons, illuvial B horizons that differ with respect to the type of illuvial material, residual B horizons of highly weathered soils, or they may be structural or color Bw horizons of "young" weakly weathered soils.

Agric.   Illuvial horizons formed under the plowed portion of soils. Agric horizons contain significant silt, clay, and humus, whose movement(s) have been induced by cultivation of the horizons above these zones.

Albic.   Strongly developed eluvial (E) horizons from which clay and free iron oxides have been removed. Generally lighter in color than the horizons above or below the albic horizons.

Argillic.   Illuvial horizons (zones of gain) in which the illuvial materials are silicate clay minerals (Bt). Since these horizons form due to the movement of clay minerals, they often have coatings of clay (clay skins) on ped faces and pore surfaces (Figure 4-2).

1. If any part of the eluvial horizon (zone of loss) has less than 15% clay, the argillic must contain at least 3 percentage points more clay.
2. If the eluvial horizon has between 15% and 40% clay, there must be 1.2 times more clay in the illuvial horizon.
3. If the eluvial horizon has more than 40% clay, the illuvial horizon must contain at least 8 percentage points more fine clay.

Argillic horizons must be at least one-tenth the thickness of the overlying horizons or have a collective thickness of more than 15 cm in sands and loamy sands with lamellas (bands of clay material).

Calcic.   Horizons of secondary calcium or magnesium carbonate accumulation. Calcic horizons are approximately equivalent to the combined horizons that are given the subscript k. Calcic horizons must be 15 cm or more thick and show evidence that carbonates have been moved and deposited.

Cambic.   Cambic horizons are illuvial horizons in which there has not been enough clay movement to qualify as argillic and/or where there is evidence of removal of carbonates. Also includes the structural or color Bw horizons.

Duripan.   Silica-cemented subsurface horizons, equivalent to horizons with

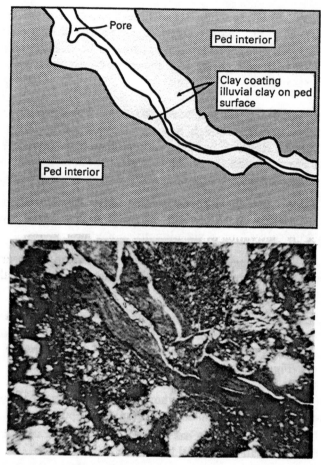

Figure 4-2 Clay coatings (skins) on ped surface in the Bt2 horizon of a Miami silt loam. (Courtesy of Dave Cremeenes and Robert Darmody, Department of Agronomy, University of Illinois.)

a qm subscript. Hardpans such as a duripan often must be mechanically disrupted with heavy tillage if the soil is to be used to grow crops. Such tillage has been shown to be economically feasible for the production of almonds in California.

Fragipan. Loamy (<35% clay) subsurface zones, often underlying B horizons. They are low in organic matter, have high bulk densities, and are hard to very hard and brittle when dry. The collective pedogenic horizons of a soil with a subscript x would constitute a fragipan. Fragipans are restricted to soils that form under forest vegetation. Fragipans have prismatic structure with bleached (gray) faces in polygonal patterns, when viewed in a horizontal cut, with brown to reddish-brown streaks of iron and manganese oxides on the vertical faces of the prisms.

Roots are excluded from the brittle portions of the fragipans but can penetrate between prisms and the bleached faces.

Gypsic.   Secondary enrichments of gypsum that are 15 cm or more thick and have least 5% more gypsum than the underlying C horizon. Equivalent to the horizons that have a y subscript.

Kandic.   Kandic horizons are special types of argillic horizons in which the illuvial materials are low-cation-exchange-capacity (kaolinitic) colloids.

Natric.   Natric horizons are special types of argillic horizons (Btn). In addition to the properties of argillic horizons, natric horizons have more than 15% exchangeable sodium or SAR (sodium adsorption ratio) values ≥ 13. The presence of high amounts of exchangeable sodium affects the pH and physical properties of these horizons (see Chapters 8 and 9).

Oxic.   Residual accumulations of hydrated oxides of iron and aluminum or both. Oxic horizons form because of the destruction and loss of less resistant minerals due to extremes of weathering acting over long periods of time. Equivalent to Bo horizons.

Petrocalcic.   Massive, carbonate cemented horizons, approximately equivalent to horizons with the subscript km. Petrocalcic horizons seem to be restricted to old land surfaces and hence are thought to be the product of ancient weathering, whereas the uncemented calcic horizons are the result of recent soil formation.

Petrogypsic.   Petrogypsic horizons are strongly cemented gypsic horizons. Roughly equivalent to horizons with a ym subscript. The gypsum contents of these horizons far exceed the minimum requirements of gypsic horizons and often exceed 60%.

Placic.   Thin dark-reddish pans cemented by iron, iron and manganese, or iron–organic matter materials.

Salic.   Horizons 15 cm or more thick that contain secondary enrichments of salts that are more soluble than gypsum. Approximately equivalent to horizons with a subscript z.

Sombric.   Subsurface horizons containing illuvial humus. The humus is not associated with aluminum as it is in spodic horizons, or with sodium. Often resembling buried A horizons, these horizons are restricted primarily to cool moist soils of high plateaus and mountains in tropical and subtropical regions.

Spodic.   Spodic horizons are illuvial horizons (Bh, Bs, Bhs) in which the illuvial materials are amorphous (noncrystalline) materials composed of humus and/or iron and/or aluminum hydrous oxides.

Sulfuric.   Acid horizons that are formed due to the oxidation of reduced sulfur materials. These horizons are composed of either mineral or organic matter that have pH values less than 3.5.

4.1.2 Soil Orders

Soils are classified into orders and lower levels of classification based on the presence or absence of diagnostic epipedons and subsurface horizons (Figures 4-3 and 4-4). The orders depict major differences in the type of epipedon coupled with major differences in the nature of the subsurface diagnostic horizons. Orders also distinguish between mineral and organic soils and, in one case, the determining factor for placing a soil in the proper order is the inhibition of soil development by the dryness of the climate.

Entisols. Soils that have no subsurface diagnostic horizons or a mollic or umbric epipedon. There is no illuvial B horizon development, although very sandy soils may have color (Bw) horizons. The general lack of diagnostic subsurface horizons indicates that these are developmentally young soils or soils forming on resistant parent materials.

Inceptisols. Soils with a cambic B horizon and/or an umbric, ochric, or a plaggen but not a mollic epipedon. These soils are either young soils with slightly more profile development than the Entisols or very old soils in which much of the horizon development has been destroyed by weathering.

Mollisols. Soils that have a mollic epipedon. If an argillic, kandic, or natric horizon is present, the soil must have at least 50% base saturation to a depth of 1.25 m below the epipedon. If there is no argillic, kandic, or natric horizon present, the soil must have at least 50% base saturation to a depth of 1.8 m or just above a lithic or paralithic contact, whichever is shallower. Also included are soils that have dark-colored argillic or natric horizons such that when mixed with the surface horizon, the mixed zone meets the requirements of a mollic epipedon.

Note: Lithic contact refers to the boundary between the regolith or soil and hard bedrock; if the underlying rock is soft, the boundary is referred to as a *paralithic contact.*

Alfisols. Soils that do not have a mollic epipedon but that have an argillic, kandic, or natric horizon with greater than 35% base saturation at a depth of 1.25 m below the upper boundary of the argillic, or kandic horizon. If a fragipan is present, it must underlie an argillic horizon or meet the requirements of an argillic horizon.

Spodosols. Soils that have a spodic horizon with an upper boundary within 2 m of the surface or that have a placic horizon (iron cemented pan) that meets all the requirements of spodic horizon except thickness and that rests on a fragipan or spodic horizon.

Ultisols. Soils without a mollic epipedon in a mesic (see definition later in

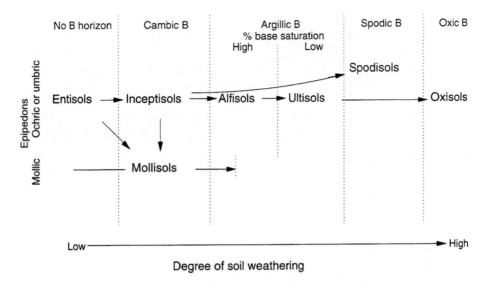

Figure 4-3   Weathering sequence of selected soil orders.

chapter) or warmer temperature regime that have an argillic or kandic horizon or a fragipan that meets all the requirements of an argillic. If the soil does not have a fragipan it must have less than 35% base saturation at a depth of 1.25 m below the upper boundary of the argillic or kandic horizon. If a fragipan is present, it must have less than 35% base saturation at a depth of 75 cm below the upper boundary of the fragipan or immediately above a lithic or paralithic contact, whichever is shallower.

Oxisols.   Soils that have either an oxic horizon or a kandic horizon that meets the weatherable mineral requirements of an oxic horizon and that do not have spodic horizons.

Histisols.   Soils with either more than half of the upper 80 cm composed of organic material or any thickness of organic material if resting on rock or fractured material with the cracks and interstices filled with organic matter.

Vertisols.   Soils that contain, after mixing of the surface 18 cm, 30% or more swelling type clay in all subhorizons down to 50 cm or more, have no lithic or paralithic contact within 50 cm, and have, at some time, cracks open at the surface. Cracks must be at least 1 cm wide at a depth of 50 cm. Vertisols form because surface soil adjacent to the wide, deep cracks falls into the cracks. Upon wetting the cracks close and the surface material becomes incorporated into the subsoil. The process repeats with each wetting–drying cycle causing the profile to *invert*, hence the name Vertisols. This self-mulching process prevents the formation of normal pedogenic horizons. These soils often have thick mollic epipedons but are included in this order rather than Mollisols because of their high content of swelling clay and self-mulching characteristics.

Cool - dry   Cool - wet

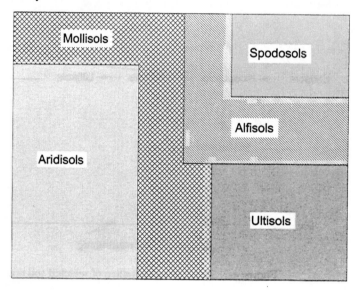

Hot - dry   Hot - wet

Figure 4-4   Climatic distribution of selected soil orders.

Aridisols.   Soils in an aridic moisture regime (see the definition later in the chapter); that is, they occur in climates that are dry enough to restrict normal soil development. They may have an ochric or anthropic epipedon. They may or may not have an argillic or natric horizon. If an argillic or natric horizon is present, either a salic, petrocalcic, calcic, gypsic, petrogypsic, or duripan must be present within 1 m of the surface.

Andisols.   Soils that normally formed from volcanic ash, but some may have formed from other pyroclastic rocks, sedimentary rocks, and basic extrusive igneous rocks. These soils contain less than 25% organic carbon and have andic soil properties throughout the subhorizons. *Andic soil properties* include: high levels of acid oxalate extractable aluminum and iron, and very high phosphorous retention capacities of the less than 2.0 mm fraction (both properties thought to be due to the presense of amorphous aluminosilicates such as allophane), and very low bulk densities (<0.90 g cm^{-3}).

As shown in Figure 4-3, most of the orders can be grouped according to the degree of soil development and type of epipedon. Soil orders with some exceptions may also be grouped according to the macroclimate of the soil. Figure 4-3 illustrates the increased soil development that occurs between the various soil orders. Entisols show only a little A-horizon development with no illuvial B-horizon development. Inceptisols also have weakly developed A horizons, but they have some illuvial B-horizon (cambic) development. Based on the degree of soil development, Entisols

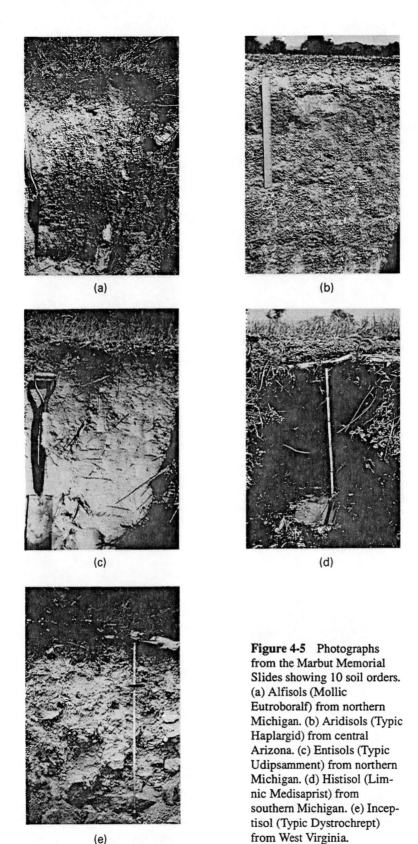

Figure 4-5 Photographs from the Marbut Memorial Slides showing 10 soil orders. (a) Alfisols (Mollic Eutroboralf) from northern Michigan. (b) Aridisols (Typic Haplargid) from central Arizona. (c) Entisols (Typic Udipsamment) from northern Michigan. (d) Histisol (Limnic Medisaprist) from southern Michigan. (e) Inceptisol (Typic Dystrochrept) from West Virginia.

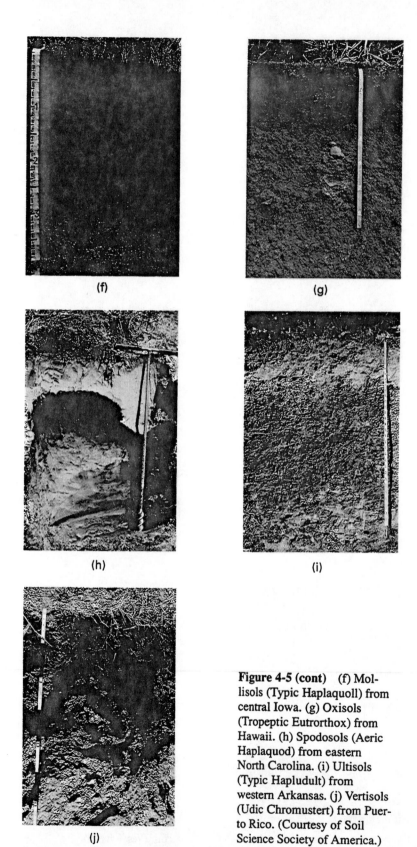

Figure 4-5 (cont) (f) Mollisols (Typic Haplaquoll) from central Iowa. (g) Oxisols (Tropeptic Eutrorthox) from Hawaii. (h) Spodosols (Aeric Haplaquod) from eastern North Carolina. (i) Ultisols (Typic Hapludult) from western Arkansas. (j) Vertisols (Udic Chromustert) from Puerto Rico. (Courtesy of Soil Science Society of America.)

can be considered to be "young" soils either due to their chronological age or due to the resistance of the parent material to weathering. Inceptisols are "immature" soils based on the degree of soil development. Inceptisols actually include several major groups of soils, soils that fit the concept of immature soils, soils in cold climates where temperature restricts normal soil development, and soils where the extremes of weathering have destroyed more highly developed soil horizons.

Alfisols are "mature soils" with A horizons that are either low in organic matter, thin or acidic (ochric or umbric epipedons), and have highly developed illuvial B (argillic, kandic, or natric) horizons. Ultisols are "old" soils very similar to Alfisols except that their profiles have been subjected to greater amounts of soil weathering, resulting in lower-percentage base saturations (<35%) in the lower portions of their B horizons. Spodisols are highly weathered soils in cool humid climates normally forming on coarse-textured parent materials under forest vegetation. Spodisols are "old" soils characterized by illuvial horizons in which the illuvial material is humic materials and/or iron and other metal hydrous oxides.

Figure 4-5 is a collection of color photographs showing a typical soil profile for 10 of the soil orders.

4.1.3 Suborders and Lower Levels of Classification

Suborders.   Suborders (Table 4-1) are subdivisions of orders based on differences in soil temperature and moisture regimes or by the presence of other major (unique) soil properties. Each particular order has a set of major soil properties that determines the appropriate suborder. Suborder names are formed from order names by taking a portion of the order name and adding a formative element (Table 4-1). For example, consider a Mollisol occurring in a humid climate that is not poorly or very poorly drained and that does not contain an albic horizon. This soil is classified as

$$\text{Mo}lli\text{sol} + h\text{um}id \; (ud\text{ic moisture regime}) = \text{Udoll}$$

Soil Temperature Regimes.   Soil temperature regimes are based on mean annual soil temperatures (MAST) at a depth of 50 cm, or at a lithic or paralithic contact, whichever is shallower, and on the differences between mean annual summer (MSST) and winter (MWST) soil temperatures. Mean winter and summer temperature differences (MSST − MWST) of the isotemperature regimes vary less than 5°C and are typical of tropical regions. For example, both the hyperthermic and isohyperthermic temperature regimes have mean annual soil temperatures ≥22°C, but the variation between summer and winter soil temperatures is less than 5°C for the isohyperthermic regime and equal to or greater than 5°C for the hyperthermic temperature regime. The following definitions of soil temperature and moisture regimes have been simplified from the detailed definitions given in the USDA soil taxonomy system[4]:

Pergelic. Mean annual soil temperature (MAST) is less than 0°C (32°F). Permafrost is present if the soil is moist, dry frost if excess moisture is not present.

Cryic. 0°C ≤MAST <8°C (47°F).

Frigid and Isofrigid. 0°C ≤MAST <8°C, but soil has warmer summer temperatures than the cryic temperature regime.

Mesic and Isomesic. 8°C ≤MAST <15°C (59°F).

Thermic and Isothermic. 15°C ≤MAST <22°C (72°F).

Hyperthermic and Isohyperthermic. MAST ≥22°C.

Note: The mesic temperature regime approximates the corn belt region, while the thermic temperature regime approximates the cotton belt region in the United States.

Soil Moisture Regimes.   Soil moisture regimes reflect the precipitation zone that the soil occurs in, but the regime may be modified by site position and other factors that affect internal soil drainage.

Aquic. Soils with an aquic moisture regime must include horizon(s) that contain high enough water contents to exclude air from the macropores for a sufficient time to create reducing conditions. When the aquic formative element is used to identify suborders, the whole soil must be affected. When the aquic formative element is used with subgroups, only the lower horizons need to be involved.

Aridic (Torric). Soils in arid climates that are dry in all portions of the soil profile more than half the time (cumulative) when soil temperatures at a depth of 50 cm are above 5°C and that are never moist in some or all portions for as long as 90 consecutive days when soil temperatures at 50 cm depth are above 8°C. Aridic may also include soils in semiarid climates if soil properties (e.g., high sand content) result in a droughty soil but does not include soils in pergelic or cryic temperature regimes.

Udic. Soils forming in humid climates such that in most years the soil is not dry in any part for as long as 90 days (cumulative). The udic moisture regime is common to soils that have a continental climate where precipitation is uniformly distributed throughout the year. Udic moisture regimes may also occur in less humid climates if the sum of stored soil water and rainfall exceeds evapotranspiration. The addition of the formative element "per" to udic (perudic) implies an extremely wet moisture regime.

[4]Soil Survey Staff, *Keys to Soil Taxonomy.* SMSS Technical Monograph No. 6, Ithaca, N.Y., 1987.

Ustic. Soils with moisture regimes that are intermediate between aridic and udic. These soils have limited soil water, but soil water is present at a time when conditions are suitable for crop growth. Soils are dry for 90 days (cumulative) in most years.

Xeric. Soils occurring in Mediterranean climates, that is, climates where winters are cool and moist and summers are hot and dry. In xeric moisture regimes soils are dry for 45 consecutive days following the summer solstice in 6 or more years out of 10.

The following suborders are for the orders Mollisols and Alfisols. Suborders for the other eight orders (Table 4-1) are formed in similar fashion (i.e., showing differences in major soil properties or temperature and moisture regimes). The following suborders and suborders in Table 4-1 are given in a priority order. This means that if a Mollisol has an albic horizon and an aquic moisture regime, it is classified as an Alboll since this property takes priority over the moisture regime.

TABLE 4-1   Suborders[a]

Alfisols	Aridisols	Entisols	Histisols
Aqualfs	Argids	Aquents	Folists
Boralfs	Orthids	Arents	Fibrists
Ustalfs		Psamments	Hemists
Xeralfs		Fluvents	Saprists
Udalfs		Orthents	

Inceptisols	Mollisols	Oxisols	Spodosols
Aquepts	Albolls	Aquox	Aquods
Andepts	Aquolls	Torrox	Ferrods
Plaggepts	Rendolls	Ustox	Humods
Tropepts	Xerolls	Perox	Orthods
Ochrepts	Borolls	Udox	
Umbrepts	Ustolls		
	Udolls		

Ultisols	Vertisols		
Aquults	Xererts		
Humults	Torrerts		
Udults	Uderts		
Ustults	Usterts		
Xerults			

[a] The suborders are listed within each order in a priority listing as they would key out in a taxonomic key. See Soil Survey Staff, *Keys to Soil Taxonomy*, SMSS Technical Monograph No. 6, Ithaca, N. Y., 1987.

Suborders of Order Mollisols   (Figure 4-6)

Albolls. Presence of an albic (strongly developed E) horizon.

Aquolls. Mollisols with aquic moisture regimes; normally, these are the poorly or very poorly drained Mollisols.

Rendolls. Mollisols that have no argillic or calcic horizon but have material including coarse fragments 7.5 cm in diameter that have 40% or more calcium carbonate equivalent in or immediately below any mollic epipedon.

Xerolls. Soils forming in Mediterranean climates (i.e., in xeric moisture regimes). These soils do not have cryic temperature regimes.

Borolls. Soils forming or that have formed in cold climates (i.e., frigid, cryic, or pergelic temperature regimes).

Ustolls. Soils forming in dry (ustic) moisture regimes.

Udolls. Soils forming in humid climates (i.e., udic moisture regimes).

Figure 4-6 gives the approximate geographical distribution of the suborders that are determined by soil temperature and moisture regimes. Within these climatic regions, if the Mollisol has the properties of an Alboll, this takes precedence over the other possible suborders. If the soil has the properties of a Rendoll, this takes precedence over the other suborders, with the exception of Albolls and Aquolls.

Figure 4-6   Generalized geographical distribution of mollisol suborders.

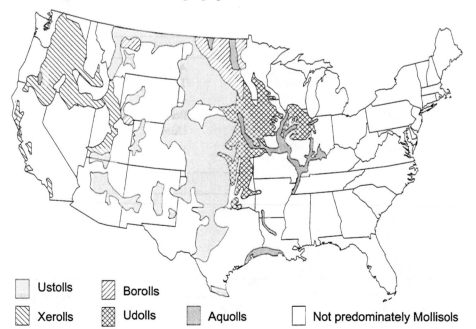

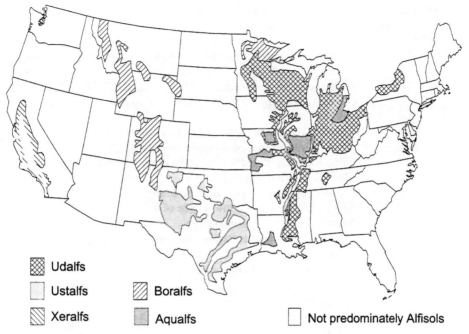

Figure 4-7   Generalized geographical distribution of alfisol suborders.

Suborders of Order Alfisols (Figure 4-7)

Aqualfs. Soils in aquic moisture regimes include soils that are somewhat poorly to very poorly drained.

Boralfs. Soils forming in cryic or frigid temperature regimes that do not have xeric moisture regimes.

Ustalfs. Soils forming in dry (ustic) climates.

Xeralfs. Soils in xeric moisture regimes.

Udalfs. Soils in udic moisture regimes, that is, humid climates.

If an Alfisol is somewhat poorly, poorly, or very poorly drained, the suborder is Aqualf (Tables 4-1 and 4-2). If the Alfisol has better internal drainage, its suborder is determined by the temperature or moisture regime of the soil. Notice that there are no suborders for Alfisols that have an E horizon or that contain coarse fragments of calcium carbonate. These are not distinct enough profile characteristics to justify separate suborders, since most Alfisols have E horizons and many, particularly in the western United States, contain coarse fragments of calcium carbonate. Also note that the somewhat poorly drained Alfisols are included in the suborder Aqualfs, whereas the somewhat poorly drained Mollisols are not separated from the climatic suborders. This is because the normal concept of Alfisols is one of soils occurring in better-drained landscape positions.

TABLE 4-2 Suborder formative elements not associated with soil moisture or temperature

Ar	Contains fragments of diagnostic horizons due to either anthropogenic or natural mixing of soil materials.
And	High content of vitric volcanic ash, cinders, or other pyroclastic materials. Cation-exchange complex is dominated by amorphous minerals.
Arg	Argillic horizon.
Ferr	Contains a high ratio of free iron oxides to carbon in all subhorizons.
Fibr	Three-fourths of surface organic materials are fibers derived from *Sphagnum* moss. Has a low degree of decomposition.
Flu	Floodplain of modern river.
Fol	Mainly leaf material, less than one-fourth of organic materials consist of *Sphagnum* fibers. Has a low degree of decomposition.
Hem	Intermediate decomposition stage.
Hum	High content of humus (organic matter) in the mineral horizon.
Ochr	Ochric epipedon.
Orth	Typical or common to the order.
Plagg	Plaggen epipedon.
Psamm	High content of sand.
Sapr	Most decomposed stage.
Umbr	Umbric epipedon.

Source: Soil Survey Staff, *Soil Taxonomy*. USDA Handbook No. 436, U.S. Government Printing Office, Washington, D.C., 1975.

Great groups and subgroups. Great groups and subgroups are formed by adding formative elements to the suborder names. The great group formative elements (Table 4-3) generally provide information about the nature of the diagnostic subsurface horizon, while the subgroup formative element provides information about drainage, lithic contact, mollic-like ochric epipedons, parent material, and even clay type. For example, the suborder Aquoll can be separated into two great groups based on whether the soil has a cambic or an argillic B horizon. If the soil has a cambic B, the formative element *hapla*, meaning "minimum horizon development," is added to the suborder name:

hapla + Aquoll = Haplaquoll

If the soil has an argillic B, then the formative element *argi* is added to the suborder name:

argi + Aquoll = Argiaquoll

Subgroup names are formed by adding one or more formative elements in front of the great group name. For example, a young Entisol-like Haplaquoll forming on a floodplain would be

fluv (fluvial) + entic (Entisol-like) + Haplaquoll = Fluventic Haplaquoll

TABLE 4-3   Formative elements that are used to form Greatgroup names from suborders[a]

Acr	Severe weathering	Med	Temperate climates
Agr	Agric horizon	Natr	Natric horizon
Alb	Albic horizon	Ochr	Ochric epipedon
And	Amorphous-volcanic materials	Pale	Old soil development
Bor	Cold temperature regime	Pell	Low chromas
Calc	Calcic horizon	Plac	Placic horizon
Camb	Cambic horizon	Plinth	Presence of plinthite
Chrom	High chroma	Psamm	Sand textures
Cry	Cryic temperature regime	Quartz	High quartz content
Dur	Durapan	Rhod	Dark red colors
Dystr	Low base saturation	Sal	Salic horizon
Eutr	High base saturation	Sider	Free iron oxides
Ferr	Iron	Sombr	Sombric horizon
Fluv	Fluvial site position	Sphagan	Sphaganum moss
Frag	Fragipan	Sulfi	Presence of sulfidic (reduced) materials close to surface
Gloss	Tonguing of albic into argillic horizon	Sulfo	Sulfuric horizon
Gyps	Gypsic horizon	Torr	Torric or aridic moisture regime
Hal	Soluble salts	Trop	Warm
Hapla	Minimum horizon development		
Hum	Humus	Ud	Udic moisture regime
Hydr	Water	Umbr	Umbric epipedon
Kandi	Kandic horizon	Ust	Ustic moisture regime
Luv	Illuvial deposit of humic materials	Verm	Presence of wormholes, wormcasts, or filled animal burrows
		Vitr	Glass (amorphous silica)
		Xer	Xeric mositure regime

Source: Soil Survey Staff, *Soil Taxonomy.* USDA Handbook No. 436, U.S. Government Printing Office, Washington, D.C., 1975.
[a]Combinations of formative elements may be used to form the greatgroup names (i.e., Nadur for natric horizon and durapan or fragloss for both tonguing and a fragipan).

Most users of soil information will have limited abilities to classify soils into their proper subgroups. However, it is possible with a minimal knowledge of the formative elements, along with a concept of how soils are placed into orders and lower levels of classification, to use the subgroup names to provide instant information about the soil, its properties, and its uses. For example:

Mollic Albaqualf

alf = Alfisol. This means that the soil has an argillic B horizon with greater than 35% base saturation. The soil does not have a mollic epipedon or a spodic or oxic B horizon.

aqu for the order Alfisol. This implies somewhat poor or lower internal drainage.

alba indicates a strongly developed E horizon.

mollic connotes that the ochric epipedon is higher in organic matter and hence darker in color than the typical ochric epipedon.

Fluventic Haplaquoll

oll = the order Mollisol. Thick, dark, high-organic-matter, high-base-saturated surface soil; mollic epipedon.

aqu refers to poor or very poor internal drainage.

hapla means minimum B horizon development. Not enough clay movement to qualify for argillic, hence a cambic subsurface (B) horizon.

fluv = soil forming in a floodplain of a modern river.

entic = Entisol-like, due to the recent origin of parent material.

Families and series.   The family level of classification provides information about the soil's temperature regime, mineralogy, and texture. The series name is the lowest category in soil taxonomy; in a sense it is the common name of the soil. For example, the Tama soil series belongs to the fine-silty, mixed, mesic family of the Typic Argiudolls subgroup. The fine-silty designation refers to the soil's texture, mixed to the mixture of minerals and mesic to the temperature regime.

The series name often refers to a town, village, or other geographic feature found in the region where the soil was first described (i.e., Tama, Muscatine, Preston, etc.). The series is the most commonly used category, but it provides the least information about soil characteristics.

4.2 LANDSCAPE RELATIONSHIPS

In the field, soils can be organized into groupings that are broader and more general than individual soil series. Soil associations (Figure 4-8) are broad groups of two or more related soil series that occur together in the same geographical area. Soils in an association tend to occur in a characteristic pattern in the landscape, and this pattern is often repeated across the countryside (Figure 4-9). The soils often have similar parent materials or at least parent materials that have some landscape relationship common to each other, such as fluvial material in the bottomland derived from glacial till or loess in the uplands. Soil series within an association may vary slightly with respect to degree of development. They may also differ in their internal soil drainage.

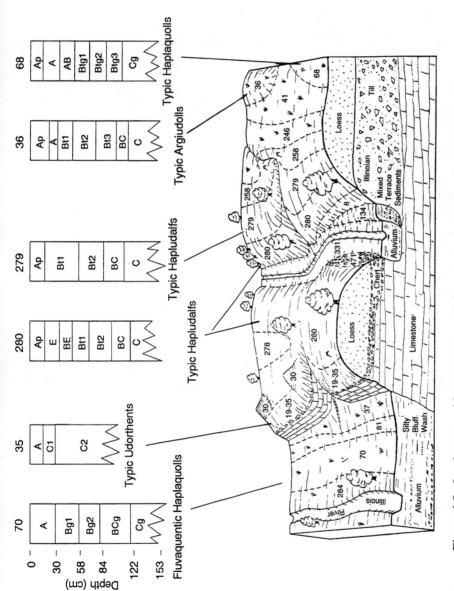

Figure 4-8 Landscape positions, parent materials, and sample profiles of soils in Jersey County, Illinois. Numbers refer to individual soil series and/or complexes. Compare to the soil map (Figure 4-11) for a portion of the same county. 284 (Tice silt loam), 70 (Beaucoup silty clay loam), 81 (Littleton silt loam), 37 (Worthen silt loam), 19-35 (Sylvan-Bold complex), 280 (Fayette silt loam), 278 (Stronghurst silt loam), 471 (Bodine), 331 (Haymond silt loam), 134 (Camben silt loam), 279 (Rozetta silt loam), 258 (Sicily silt loam), 246 (Boliva silt loam), 41 (Muscatine silt loam), 68 (Sable silty clay loam), 36 (Tama silt loam). (Soil survey: Jersey County, Illinois, University of Illinois Agricultural Experiment Station and USDA Soil Conservation Service.)

87

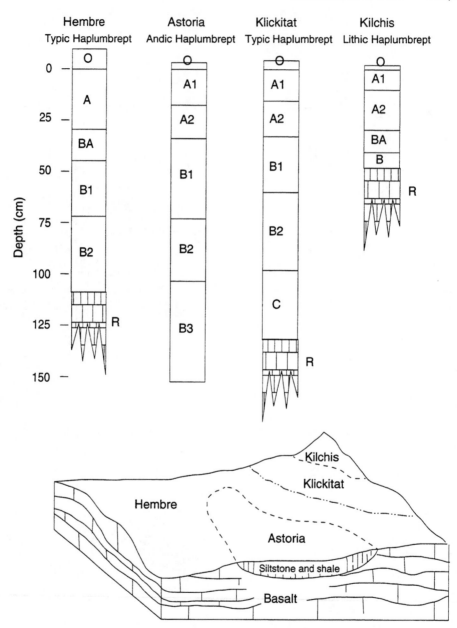

Figure 4-9 Landscape positions, parent materials, and sample profiles of soils, Yamhill area of Oregon. The soils are forming in colluvium and residuum of the Coast Range. Soils are strongly acid and used primarily for Douglas-fir production. (After the Soil Survey, Yamhill Area, Oregon, Oregon Agricultural Experiment Station and USDA Soil Conservation Service.)

Soil associations are generally grouped on the basis of the parent materials from which they have formed, their surface colors (i.e., organic matter content), major differences in soil development, and natural drainage classes. Soil associations are usually named for two or more of their most predominant soil series. The actual proportions of various soils included in an association change from place to place, depending mainly on slope and natural drainage of the landscape. A group of soils with similar parent materials that differ mainly in terms of their drainage class (well, moderately well, somewhat poorly, poorly, or very poorly drained) is known as a *catena*.

Both associations and narrower groupings have practical value in that they allow soil use and management considerations to be expressed on a larger geographical basis than the use of individual soil series would allow. Management of soil fertility or soil water is often dependent on the soils relationship within an association or catena. Soils occupying similar positions within different soil associations or catenas often require similar management. Information pertaining to the well-drained member of one catena is often applicable to the well-drained members of other catenas within an association.

4.3 THE SOIL SURVEY

Soils are distributed in sufficiently uniform patterns and exist as sufficiently recognizable entities to allow mapping and compilation of these maps, together with the properties and suggested uses of the soils into soil survey reports. Soil surveys are made cooperatively by federal, state, and sometimes local governments. The Soil Conservation Service is the federal agency that has the primary responsibility for soil survey activities. In some surveys, the Forest Service, Bureau of Indian Affairs, Bureau of Reclamation, or Bureau of Land Management may cooperate. Land-grant universities and colleges (usually, state experiment stations) bear the primary responsibility in representing the states. Other state agencies, such as the Department of Natural Resources or Conservation, and county or city governments may also cooperate. This joint effort is called the *National Cooperative Soil Survey*. The *Standard Soil Survey* is the major type of survey now being produced. The Standard Soil Survey is usually made on a county basis with the counties sharing in the cost of the survey. Less detailed surveys include reconnaissance and soil conservation surveys.

The surveys of soil resources are designed to meet the particular needs of each geographical area. Humid-region farmers have different needs than arid-region farmers and ranchers that depend on irrigation. Engineers, foresters, and conservationists all need different information about soils in a region. The Soil Survey is designed to provide a variety of information about the soils so that all potential uses of the soil can be explored with a high degree of rigor. The Soil Survey is often used to provide equitable tax assessment based on potential soil productivity.

Soil Survey reports are divided into two sections: one section contains descrip-

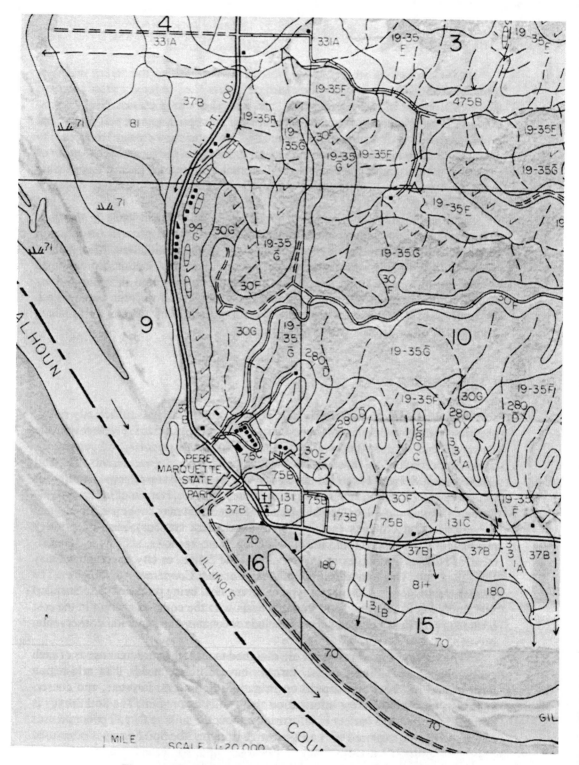

Figure 4-10 Soil map for a portion of Jersey County, Illinois. Compare with Figure 4-8. (Soil survey: Jersey County, Illinois, University of Illinois Agricultural Experiment Station and USDA Soil Conservation Service.)

tive material about the county, profile descriptions, soil classifications, tables giving information about engineering uses, soil productivity, wildlife and forest plantings, and other uses of individual soils. The other section contains detailed soils maps for the county (Figure 4-10). The maps are drawn using aerial photographs providing high accuracy and a high degree of detail. Soil survey reports are available at no cost from county and state Soil Conservation Service offices as well as from the land-grant university in the state. Many libraries have fairly complete collections of Soil Surveys, not only for their state but also for other states in the United States.

STUDY   QUESTIONS

1. Why do we classify soils?
2. What is the difference between pedogenic and diagnostic horizons?
3. Define the ochric, umbric, histic, and anthropic epipedons in terms of the mollic epipedon.
4. What are the illuvial materials in the argillic, kandic, natric, and spodic horizons?
5. What changes in soil development are implied in going from the order Entisols to Inceptisols to Alfisols to Ultisols?
6. In what order(s) would a soil be placed if development was hindered either by dryness or coldness of climate?
7. How do you determine in which suborder to place a soil classified as an Alfisol?
8. Provide as much information as you can about the following soils:
   (a) Fluventic Haplaquolls
   (b) Typic Natraqualfs
   (c) Typic Eutrochepts
   Attempt to draw a diagram of each soil illustrating the diagnostic horizons that would be present.
9. Compare the profile description and classification of Tama silt loam (36) and Sable silty clay loam (68) in Figure 4-8. Why is Sable classified as a Typic *Hapl*aquolls with a cambic B horizon, whereas Tama silt loam is classified as a Typic *Argi*udolls with an argillic B horizon, yet both soils have Bt horizons?
10. Why are Kilchis (Figure 4-9) and Bold (Figure 4-8) the shallowest soils of their respective soil associations?

5

Water in Soils

The content of water in soils affects many other important soil properties. These include plasticity, swelling and shrinking, consistency, ease of compaction, ability to support traffic, and aeration. Soil water is also a very important part of the hydrologic cycle. The hydrologic cycle (Figure 5-1) describes the movement of water from the ocean, to the atmosphere, to land, and back to the ocean or atmosphere. Soil serves to moderate the rate of flow of precipitation into streams, lakes, and other waterways. Precipitation that falls on a soil either runs off the soil directly into waterways, evaporates back into the atmosphere, or infiltrates into the soil. Vegetative cover slows the flow of water across the soil surface, allowing maximum infiltration while minimizing the amount of erosion. In humid regions, water that infiltrates into the soil either moves (leaches) slowly through the soil to the groundwater or is stored in the soil, eventually to be returned to the atmosphere by evapotranspiration. In arid regions there may not be sufficient water to leach through the soil. A portion of the water that is stored in the soil is available for use by higher plants and other organisms.

5.1 PROPERTIES OF WATER

An understanding of the behavior of water in soil and the important role of water in soil processes must be based on an appreciation of the structure of water and its unique properties. Water consists of two positively charged hydrogen ions covalently bonded to one negatively charged oxygen atom at an angle of approximately 105°. Oxygen being an electronegative element does not share the electrons equally

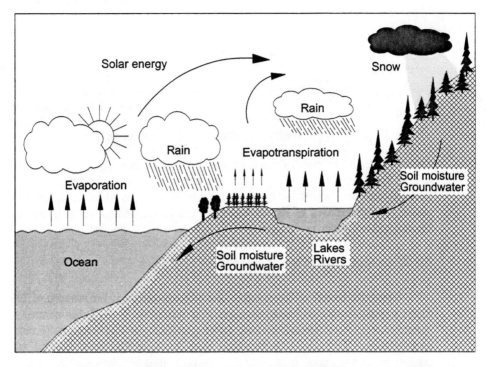

Figure 5-1 Hydrologic cycle. (After A. R. Bertard, Water conservation through improved practices, in *Plant Environment and Efficient Water Use*, American Society of Agronomy, Madison, Wis., 1967.)

with hydrogens in the covalent bonds, but rather, the electrons are pulled a little closer to the oxygen atom, giving the oxygen a slight residual negative charge and the hydrogen atoms a slight residual positive charge (Figure 5-2). This asymmetrical arrangement of the hydrogen atoms, coupled with the unequal sharing of electrons between the hydrogen and oxygen atoms, creates an electrical dipole that is responsible for many of the unique properties of water.

5.1.1 Flickering Cluster Model of Water Structure

Water molecules in liquid water are partially bonded together by hydrogen bonding between the oxygen of one water molecule and the hydrogen of a different water molecule. Hydrogen bonding is in part due to the attraction of the weak residual positive charge of a hydrogen of one water molecule for the weak residual negative charge of the oxygen of a different water molecule, but also includes some weak covalent bond characteristics. Hydrogen bonding results in water molecules in liquid water bonding together in clusters, while thermal energy causes the clusters to dissociate, creating a system of transitory or flickering clusters (Figure 5-3). The clusters have a crystalline icelike internal structure. These microcrystals or clusters

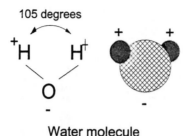

105 degrees

$^+$H   H$^+$

O

-

Water molecule

Figure 5-2 Two representations of the structure of an individual water molecule.

form and "melt" so rapidly and randomly that on a macroscopic scale water appears to be a homogeneous liquid. The clusters are visualized as short lived (10^{-10} to 10^{-11} second) and are continuously exchanging molecules with the adjacent "liquid" phase. The properties of liquid water are the properties of the clusters, as well as the individual water molecules.

5.1.2 Unique Properties of Water

Specific heat.   The specific heat of a substance is the amount of heat energy that 1 g must absorb or lose to change its temperature by 1°C. Substances with high specific heats must absorb or lose large quantities of heat energy to change their temperatures (Table 5-1). Substances with low specific heats exhibit marked increases or decreases in temperature, with correspondingly small gains or losses of heat energy.

The high specific heat of water has several important consequences in soil science. A wet soil must absorb or lose large quantities of heat to change its temperature, while a dry soil must absorb or lose smaller quantities of heat to

Figure 5-3   Flickering cluster model of water structure.

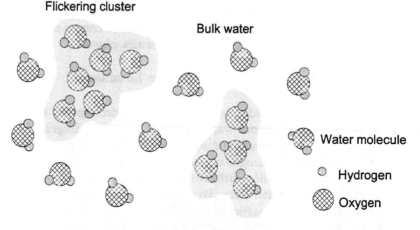

Flickering cluster

Bulk water

Water molecule

Hydrogen

Oxygen

Flickering cluster

TABLE 5-1   Specific heats of various substances

H_2O(liquid)	1.0 cal g^{-1} deg^{-1} at 15°C
Aluminum metal	0.215
Iron metal	0.16
Soil minerals	0.2 (average)

change its temperature. The high specific heat of water also serves to buffer organisms, especially seedlings, against rapid changes in temperature.

State changes of water.   When water changes from one physical state to another, large quantities of heat energy are consumed or released by the process.

$$H_2O(ice) \leftrightarrow H_2O(liquid), \quad 80 \text{ cal/g} \quad (at \ 0°C)$$

$$H_2O(liquid) \leftrightarrow H_2O(vapor), \ 540 \text{ cal/g} \quad (at \ 100°C)$$

$$H_2O(liquid) \leftrightarrow H_2O(vapor), \ 580 \text{ cal/g} \quad (at \ 25°C)$$

It requires 80 calories of heat energy to melt 1 g of ice in a frozen soil and produce liquid water at the same temperature (0°C) as the original ice. When 1 g of water evaporates from the soil surface at 25°C, it absorbs 580 calories of heat energy and cools the soil.

5.1.3 Water of Hydration

Water is a good solvent for ionic substances, primarily due to the strong attractive force between ions and water molecules. Metal ions such as Li^+, Na^+, K^+, Ca^{2+}, and Mg^{2+} react strongly with water molecules to form a layer of water referred to as the *primary hydration sphere* (Figure 5-4). The water in the primary hydration sphere is very strongly bonded to the ions. The water molecules in the primary hydration sphere form bonds with additional water molecules and these molecules form bonds with other water molecules. The strength of bonding decreases with each subsequent layer until it is equivalent to the bonding found between water molecules in the flickering clusters. At this point no additional water is added to the hydration shell of the ion. Water that is bonded to the primary sphere of water molecules is referred to as the *secondary hydration sphere*.

The strength of hydration of an ion is dependent on the size and valence of the ion. For example, Ca^{2+} a divalent cation is more strongly hydrated than the monovalent cation Na^+, while Mg^{2+} with an ionic radius of 0.82 nm is more strongly hydrated than Ca^{2+}, which has an ionic radius of 1.18 nm. In general, small, highly charged ions are the most highly hydrated. It should be noted that the H^+ exists in solution as a hydrated ion, the hydronium ion (H_3O^+). It is common practice when writing chemical equations to ignore the water of hydration of the hydrogen ion and other ions, but the strength and amount of hydration are often the reason that apparently similar ions behave differently.

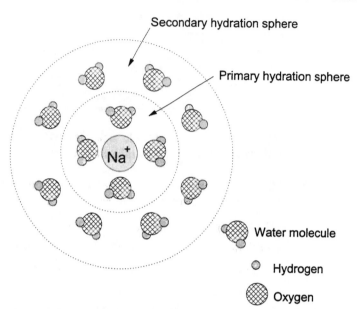

Figure 5-4   Primary and secondary hydration spheres.

5.2 ENERGETICS OF WATER IN SOIL

Water in the soil is bound to the soil solids. When the water content of the soil is low, the water is in close proximity to the soil solids and the water is bound by very high energies. Under these conditions a large amount of work must be performed to remove water from the soil. When the water content is high, some water is in close proximity to the soil solids and is tightly bound, while other water is at greater distances from the soil solids and is bound by lower energies. Under these conditions smaller amounts of work are needed to remove the less tightly bound water from the soil.

5.2.1 Matric Potential

Sites associated with the soil matrix (soil solids) that bond with water include functional groups that are associated with soil minerals and with organic matter, and ions that are partially or fully hydrated (Figure 5-5). The functional groups include hydroxylated surfaces of clay minerals, iron, manganese and aluminum hydroxides, and the polar phenolic and carboxyl groups associated with soil organic matter. The latter include hydrated cations (counterions) that are required to balance the negative charge of soil minerals. Water molecules that are in direct contact with the mineral surfaces or bound to counterions are tightly held, with subsequent water molecules forming layers of less tightly held water. Layers of water are added to the surfaces until the force of attraction between adjacent water molecules in the layers is no greater than between water molecules in the flickering clusters of the

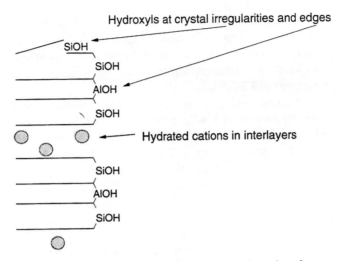

Figure 5-5 Sites of water adsorption on clay minerals.

nonbound "free" water. It should be noted that many mineral surfaces in soil are hydrophobic (reject water molecules) in nature. A surface composed of many Si–O–Si bonds will be hydrophobic, while the presence of >SiOH groups, >AlOH groups, or hydrated counterions will create a hydrophilic surface (attract water molecules).

Water-retention curves. The relation between the water content of a soil and the energy required to remove the water from the soil can be illustrated by the following experiment. A saturated soil (all pore space occupied by water) is placed on a porous plate and put into a pressure plate apparatus (Figure 5-6). When the

Figure 5-6 Pressure plate apparatus for soil-water–matric potential measurements.

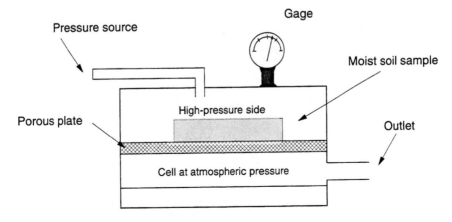

pressure in the upper chamber is increased, a portion of the water held in the soil is pushed through the porous plate into the lower chamber. If the pressure in the upper chamber is increased in steps and the soil is allowed to come into equilibrium with each new pressure, a relationship between the water content of the soil and the applied pressure can be obtained (Figure 5-7).

The pressure required to remove a certain amount of water from the soil is a measure of the energy with which the water is held in the soil. When the pressure is increased to a certain value, all water in the soil that was held with energies less than or equal to that pressure is forced through the porous plate into the lower chamber. Pressure is normally measured in bars, kilopascal (kPa) or joules per kilogram (J kg^{-1}). One bar is equal to 100 kPa or to 100 J kg^{-1} for water that has a density of 1000 kg m^{-3}. A bar is also equal to the suction (negative pressure) required to support a 1020-cm column of water. Figure 5-7 leads to the basic conclusion that as the water content of a soil decreases, the water remaining surrounds soil particles in thinner and thinner films and is held with higher and higher energies. Thus greater pressure (more energy) is required to remove the water from thin films than to remove water from thicker films. Water held tightly by soil particles is said to have a lower free energy than water that is held less tightly. The free energy of water can be lowered by various factors: for example, the presence of solutes in water, as discussed in the next section, or by attraction to the soil solids. The decrease of free energy due to the presence of solid surfaces in contact with water is denoted by the term *matric potential*. The matric potential of

Figure 5-7   Soil-water-retention curves for different textured soils.

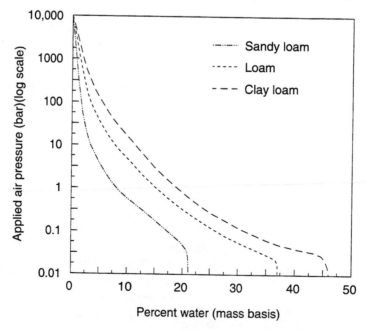

water in thick films not bound to soil solids is 0 bar. As water films become thinner and thinner, the matric potential becomes more and more negative, since a greater and greater amount of energy must be used to remove water from the soil. The matric potential of thin films may be as negative as –10,000 bar.

Capillarity. *Capillarity* refers to the phenomena by which water rises into soil pores from a water table, that is, from a zone where the matric potential is zero. When a soil is in contact with a water table, water will be attracted to the soil solids and will be chemically bonded to the soil matrix, and additional water will be bonded to the chemically bonded water. This results in water rising into the soil from the water table. Water will continue to rise until the gravitational pull on the mass of water balances the attraction of the soil solids for the water. Since the mass of water supported per centimeter of rise is less in small pores than in large pores, the height of rise of the water will be the greatest in small pores.

The attraction of water for solid surfaces (matric potential) results in the property of capillary rise. The maximum height of capillary rise can be related to the diameter of soil pores by the following formula:

$$h = \frac{2t}{\rho_w g r} \qquad [5\text{-}1]$$

where h = height of rise (meters)

t = surface tension of the soil solution $(0.073 \, \text{N m}^{-1})$

ρ_w = density of water $(1000 \, \text{kg m}^{-3})$

g = acceleration due to gravity $(9.8 \, \text{N kg}^{-1})$

r = radius of the soil pore (meters)

For example, calculate the height of capillary rise of water in a soil pore with a radius of 0.1 mm (0.0001 m).

$$h = \frac{(2)(0.073 \, \text{Nm}^{-1})}{(1000 \, \text{kg m}^{-3})(9.8 \, \text{N kg}^{-1})(0.0001 \, \text{m})}$$

$$= 0.15 \text{m}$$

The concept of matric potential and the related phenomenon, capillarity, lead to some interesting questions. For example, when a limited amount of water is applied to a soil, will the water enter into the large or small pores? When water flowing in a small pore comes in contact with a large empty pore, will the water flow out of the small pore into the large pore? Both of these questions will be discussed in the section concerned with water movement; for now try to answer the questions using the concepts of matric potential and capillarity.

5.2.2 Osmotic Potential

The presence of dissolved substances in soil water affects the free energy of water. The dissolved substances lower the free energy of water in much the same manner as the bonding of water to the soil matrix. The effect of dissolved substances on the free energy of water can be illustrated by the following experiment. When two solutions, one containing dissolved substances and the other pure water, are separated by a membrane (differentially permeable) that only allows the passage of water and not the passage of the dissolved substances, water will flow through the membrane until the water in the solutions have the same free energies. Figure 5-8a illustrates this phenomenon. In Figure 5-8b pressure is applied until the height of water in the two columns is equal. At this point the pressure applied to the column is equal to the free-energy differences of the two solutions, and the movement of water ceases. This pressure is known as the *osmotic pressure* of the solution. If greater pressure than is required to just stop the flow of water is applied, the flow of water will be reversed, that is, from the water plus solutes to water. This concept is the basis of reverse osmosis systems used to purify water.

The osmotic pressure of a solution is a colligative property. Colligative properties depend on the amount of solute present but not on its nature. In dilute

Figure 5-8 (a) The flow of water in response to an osmotic gradient. (b) Flow prevented by the application of pressure. The pressure that just corresponds to that required to stop the flow is known as the solution's osmotic pressure.

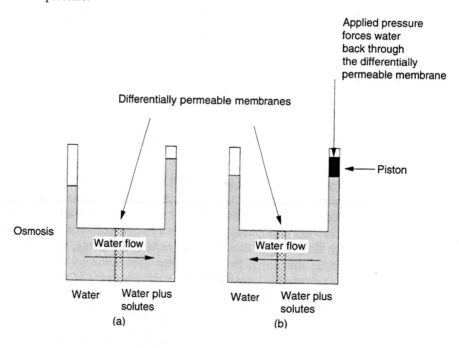

solutions, osmotic pressure is proportional to the concentration of solute particles and to temperature according to the following equation:

$$\Pi = MRT \qquad\qquad [5\text{-}2]$$

where Π = osmotic pressure (atmospheres; 1 bar = 1.01 atm)

M = total molar concentration of solute particles, ions, or molecules

R = gas constant (0.08207 liter atm/deg mol)

T = temperature (kelvin)

The lowering of the free energy of water due to the presence of dissolved solutes is denoted as osmotic potential (ϕ_o). The osmotic potential is equal in magnitude, but opposite in sign, to the osmotic pressure (Π) and represents the amount of work that would be required to remove water from a solution containing a given level of dissolved substances. For example, seawater at 25°C contains about 0.5 mole/liter of dissolved solute particles.

$$T = 25 + 273 = 298 \text{ K}$$

$$M = 0.5 \text{ mol/liter}$$

$$R = 0.082 \text{ liter atm/deg mol}$$

Hence the osmotic pressure of seawater is approximately

$$\Pi = (298 \text{ K})(0.5 \text{ mol/liter})(0.082 \text{ liter atm/deg mol})$$

$$= 12.27 \text{ atm} = 12.39 \text{ bar}$$

The osmotic potential (ϕ_o) will be equal to the osmotic pressure (Π), but is opposite in sign.

$$\phi_o = -\Pi = -12.39 \text{ bar}$$

5.2.3 Gravitational Potential

The *gravitational potential* is the amount of work that must be expended to move a given mass of water against gravity. The amount of work that must be expended depends on the water's position in the gravitational field. It is customary to select a reference point within or below the soil profile. The gravitational potential (ϕ_g) then becomes the amount of work required to move water from the reference point to the point of interest. Thus the gravitational potential at a point increases the higher that point is above the reference level. If the reference point is chosen at the soil surface, the gravitational potential is negative throughout the soil and becomes more negative the deeper one goes below the soil surface.

5.2.4 Total Soil-Water Potential

Soil water is subjected to a number of different potentials that determine the amount of work that must be expended to remove water from the soil. The total soil-water potential (ϕ_t) is the sum of all the individual potentials.

$$\phi_t = \text{matric potential + osmotic potential + gravitation potential + ...}$$

$$= \phi_m + \phi_o + \phi_g + ... \qquad\qquad\qquad\qquad [5\text{-}3]$$

When a differentially permeable membrane is not involved in the process, that is, when water and solutes move together, the osmotic potential can be ignored. If small differences in heights are involved, such as in water uptake by crops, the gravitational potential is often ignored. In soils with normal levels of dissolved salts, the total water potential is often taken as being equal to the matric potential.

Not all of the potentials act in same direction. The matric potential may result in water moving up the profile from a water table, and at the same time this movement would be against the gravitational potential.

5.3 CLASSIFICATION OF SOIL WATER

There is a continuous decrease in the energy with which water is bound to the soil solids as the thickness of the water films surrounding the soil particles increases. When water films are only one molecule thick, water has a matric potential of ≈−10,000 bar. As water is added to the soil, the water films become thicker and thicker and water in the outermost portion of the films is held less and less tightly to the soil solids. The physical classification of soil water separates this continuous range of matric potentials into regions of potentials that exhibit similar behavior and that correlate with observable phenomena.

When all the soil pores are filled with water, the soil is said to be *saturated*. Saturation corresponds to a matric potential of 0 bar. If the soil does not have restricted internal drainage, water will flow downward, from points where the gravitational potential is larger to points where it is smaller (a larger negative number). As water is removed from the soil, the soil hydraulic conductivity (discussed in the next section) decreases drastically, and drainage stops or at least becomes very slow. The water content at this point is called *field capacity* and matric potentials are in the range ≈ −1/3 bar. Ideally, field capacity could be defined as the point where the matric potential is in equilibrium with the gravitational potential, but drainage usually ceases before this point.

At field capacity water can still be lost from the soil by evaporation from the soil surface. Evaporation will continue until the energy with which water is held to the soil solids equals the energy that is driving the evaporation process. The matric potential at this point is usually in the range −20 to −40, but potentials can be more negative (i.e., −100 bar) when the soil is in equilibrium with a hot, dry atmosphere. When evaporation ceases, the soil is said to be air-dry or at the *hygroscopic coefficient*. A value of −31 bar is often used to represent soil at the hygroscopic coefficient even though the actual potential depends on the desiccating power of the atmosphere. Water can be further removed from the soil by placing the soil in a oven for 48 hours at 110°C. The soil is then said to be *oven-dry* and it contains only

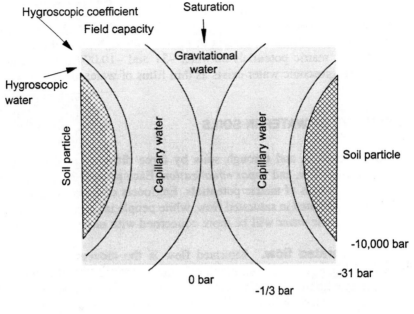

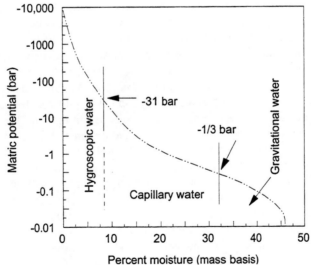

Figure 5-9   Physical classification of soil water.

strongly chemical bonded water, such as water of hydration. Matric potentials at this point are approximately −10,000 bar.

Water in the soil between saturation (0 bar) and field capacity (−⅓ bar), that is free to drain in response to the pull of gravity, is called *gravitational water* (see Figure 5-9). Water held in the soil between field capacity and the hygroscopic coefficient (−31 bar) is called *capillary water*. Capillary water is found in the micropores and as thick films around the soil particles. Capillary water is often considered to be the soil solution, even though gravitational water is important in soil formation and in movement of material through the soil profile. Water held in the soil at matric potentials between –31 and –10,000 bar is called *hygroscopic water*. Hygroscopic water exists as thin films of water around soil particles.

5.4 MOVEMENT OF WATER IN SOILS

Water moves in and through soils by three different processes: *saturated flow*, *unsaturated flow*, and *vapor equalization*. Each process is important and dominant in certain ranges of matric potentials. Engineers concerned with soil drainage will be most interested in saturated flow, while people interested in growth of plants and plant available water will be more concerned with unsaturated flow.

Saturated flow. Saturated flow is the movement of gravitational water. Saturated flow occurs during and shortly after the application of water to the soil surface by precipitation or irrigation. Saturated flow is governed by several factors. These include the rate of infiltration of water into the soil, the rate of flow of water through the soil, that is, the soil's hydraulic conductivity (permeability), as well as the amount and depth of the applied water. Infiltration, by definition, is water movement only into the surface layer of the soil. Infiltration results not only because of the movement of water into pores due to the hydraulic head of the water (gravitational potential), but also due to the attraction of water to the soil solids (matric potential).

Infiltration rates are primarily a function of soil texture and structure, but can also be affected by saturation of surface horizons due to the low hydraulic conductivities of lower horizons. In general, coarse-textured soils permit rapid infiltration because of the predominance of large pores, while the infiltration rates of finer-textured soils depend on the degree of aggregation of the surface soil. Processes such as compaction or surface crusting that reduce the degree of surface aggregation result in slower infiltration rates. In the field, infiltration rates generally decrease with time after a rainstorm begins. Figure 5-10 shows typical infiltration curves for three soils.

Decreases of infiltration rates with time after the initiation of rainfall are the result of (1) the breakdown of soil structure upon wetting, (2) movement of soil particles with the water that eventually become lodged and reduce the size of the pores, (3) the swelling of clay particles reducing pore space, and (4) pores are filled with water because the infiltration rate is greater than the permeability of lower horizons.

The permeability of the lower horizons is generally not constant, but varies from horizon to horizon depending on pore-size distribution, which is a function of the structure and texture of that horizon. Increases in clay content, such as found with strongly developed Bt horizons, usually reduce permeability. Abrupt changes in pore size, such as those encountered when the soil contains a sand lens, can decrease the permeability of "reasonably" dry soils, but would increase the saturated permeability of the soil. Figure 5-13c illustrates the effect of water moving

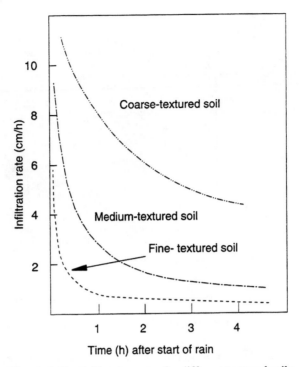

Figure 5-10   Infiltration rates for different textured soils.

in small pores in a medium-textured soil coming into contact with large pores in a coarse-textured sand lens. Does the water in the small pores flow spontaneously into the large pores? The answer is that it does not. Water in the small pores is held more tightly than the water would be in the larger pores, so water will not flow out of the medium- or fine-textured soil until the matric potential approaches zero as the soil-water content approaches saturation. Once water has moved into and saturated the sand lens, permeability should increase because water flow will not be against the matric potential gradient.

Darcy's law. Darcy's equation is the most commonly used expression to describe the vertical movement of water in response to the pull of gravity (Figure 5-11):

$$Q = \frac{K(dw)At}{ds} \qquad [5\text{-}4]$$

where Q = water quantity (cubic centimeters)

K = soil hydraulic conductivity (permeability)

A = soil area (square centimeters)

t = time (seconds)

dw, ds = height (centimeters) of water and soil above the depth where the matric potential equals zero

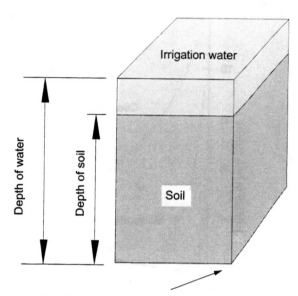

Depth where matric potential equals zero

Figure 5-11   Parameters for Darcy's equation.

Note that a soil's hydraulic conductivity is not constant but varies with texture, structure, and time of wetting. Some common values for hydraulic conductivity are given in Table 5-2.

Unsaturated flow.   Unsaturated flow is the movement of water in response to differences in matric potential. Unsaturated flow occurs from saturation, 0 bar, down to the point where water films become discontinuous, at about −31 bar, which is the hygroscopic coefficient. Unsaturated flow is responsible for lateral water flow during furrow or drip irrigation (Figure 5-13) and is the dominant method of water movement for capillary water.

TABLE 5-2   Hydraulic conductivities of some common materials

Material	K (cm/s)
Gravel	1.5×10^{-1} to 2.0×10^{-2}
Sand	1.2×10^{-1} to 2.0×10^{-3}
Loam	1.7×10^{-4} to 1.7×10^{-7}
Clay	2.5×10^{-9} to 1.0×10^{-9}

Source: R. J. Hanks and G. L. Ashcroft, *Applied Soil Physics*, Springer-Verlag, New York, 1980, pp. 62–65.

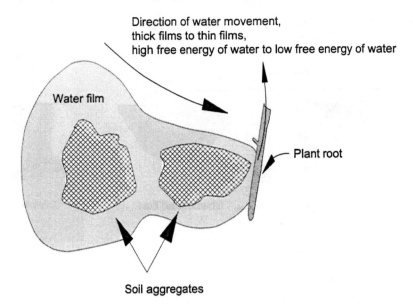

Figure 5-12   Unsaturated flow: adjustment of film thicknesses.

Figure 5-12 illustrates the phenomenon of unsaturated flow. Unsaturated flow is the movement of water from regions where water is weakly bound to the soil solids (thick water films) to regions where the water is strongly bound to soil solids (thin water films). Unsaturated flow requires that there be continuous films of water. Unsaturated flow can be in any direction dictated by the matric potential; water flow will be from small negative values (–1 bar) to larger negative values (–10 bar). Capillary rise, movement of water from drip tubes, and the lateral movement away from a saturated zone (Figure 5-13) are examples of unsaturated flow.

Vapor equalization.   Vapor equalization is the movement of water vapor from regions of high partial pressures of water vapor to regions of low partial pressures of water vapor. Vapor equalization functions over all soil-water contents, but it becomes the dominant method of water movement in very dry soils where noncontinuous water films make unsaturated flow slow or impossible. Vapor equalization does not result in the transport of large quantities of water. Vapor equalization generally results in movement from warm soil layers to cold soil layers and from wet soil layers to drier soil layers. A unique aspect of vapor equalization is that it does not result in transport of solutes as found with saturated and unsaturated flow.

5.5   METHODS OF MEASURING SOIL WATER

Soil water may be measured by a variety of techniques. Some measure the mass or volume of water in the soil by gravimetric techniques, other techniques measure the

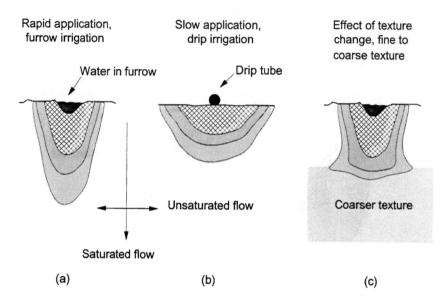

Rapid application,
furrow irrigation

Slow application,
drip irrigation

Effect of texture
change, fine to
coarse texture

Water in furrow

Drip tube

Unsaturated flow

Coarser texture

Saturated flow

(a)   (b)   (c)

Figure 5-13 Typical wetting patterns for saturated and unsaturated flow. (After R. L. Donahue et al., *An Introduction to Soils and Plant Growth*, Prentice Hall, Englewood Cliffs, NJ, 1983, p. 191.)

amount of soil water by correlation with electrical resistance or by production and detection of slowed neutrons, while still others directly measure the matric potential of soil water.

Gravimetric techniques.   The most common method of determining soil water is to sample and weigh the moist soil, place the moist soil in an oven for 48 hours at 110°C, and cool and weigh the dry soil. The percent water is then expressed as a percent on a dry weight basis.

$$W \% = (\frac{\text{wet weight of soil} - \text{oven–dry weight of soil}}{\text{oven–dry weight of soil}}) \times 100 \qquad [5\text{-}5]$$

$$W \% = \frac{\text{mass of water}}{\text{oven–dry weight of soil}} \times 100 \qquad [5\text{-}6]$$

$$= \frac{M_w}{\text{ODwt}} \times 100 \qquad [5\text{-}7]$$

where M_w = mass of water in the soil

  ODwt = oven–dry weight of the soil

  W % = percentage of water in the soil on a mass basis

Soil water may also be expressed on a volume basis, that is, the cm^3 of water per cm^3 of soil, which if the surface area of the soil is kept constant reduces to the depth of water (cm) per depth of soil (cm).

$$\theta \% = \frac{V_w}{V_t} \times 100 \qquad [5\text{-}8]$$

Where $\theta \%$ = percentage of water in the soil on a volume basis

   V_w = volume of water

   V_t = total volume of the soil

The mass percent and volume percent can be related in the following manner: Since

$$\rho_b = \frac{\text{ODwt}}{V_t} \qquad [2\text{-}2]$$

and

$$\rho_w = \frac{M_w}{V_w} \qquad [5\text{-}9]$$

it can be shown that

$$V_t = \frac{\text{ODwt}}{\rho_b} \qquad \text{and} \qquad V_w = \frac{M_w}{\rho_w}$$

Hence,

$$\theta \% = \frac{M_w/\rho_w}{\text{ODwt}/\rho_b} \times 100 \qquad [5\text{-}10]$$

or

$$\theta \% = \frac{M_w}{\text{ODwt}} \times 100 \times \rho_b/\rho_w \qquad [5\text{-}11]$$

which equals

$$\theta \% = W\% \times \rho_b/\rho_w \qquad [5\text{-}12]$$

where ρ_w is the density of water and the other terms are as defined previously. In most cases water content is expressed on a mass basis, but it can be very useful in some cases to express water content on a volume basis: for example, the volume of water contained in a specific depth or volume of soil.

Example

   How deep will a 0.5-in. rain wet a uniform soil that presently contains 28% water (W %)? Assume that water will enter into the soil and wet each layer of the soil from its present water content (28%) up to field capacity (42%), and then the excess (gravitational water) will leach and wet lower layers. Also assume that the soil has a bulk density (ρ_b) of 1.3 g/cm^3 and that there is no runoff.

Solution

(a) Storage capacity (W % basis) = field capacity − present water content
$$= 42\% - 28\% = 14\%$$
(b) Storage capacity (θ % basis) = 14% × ρ_b/ρ_w = 14% × 1.3/1.0 = 0.18%
This means that 1 cm^3 of soil will store 0.18 cm^3 of water, or that soil 1 cm deep will store 0.18 cm of water, or that soil 1 in. deep will store 0.18 in. of water.
(c) Depth of wetting:

0.5 in. of rain ÷ 0.18 in. of water stored per in. of soil = 2.78 in.

or

0.5 in. of rain × 2.54 in./cm ÷ 0.18 cm of water stored per cm of soil = 7.06 cm

Tensiometers.   Tensiometers are devices designed to measure directly the matric potential of soil water. The most common tensiometers (Figure 5-14) are plastic tubes filled with water. One end of the water column in the plastic tube is in contact with a vacuum gage, and the other end is in contact with a porous ceramic

Figure 5-14   Typical tensiometer used to measure matric potential of water in soils.

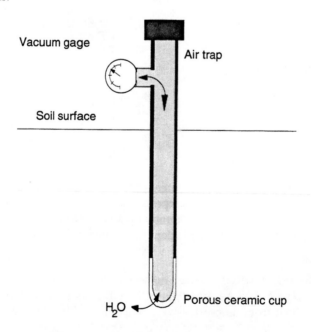

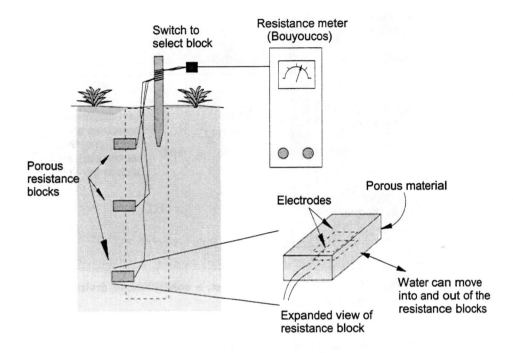

Figure 5-15 Typical arrangement of resistance blocks to measure water content of soils.

cup. When placed in the soil, water is pulled out of the porous cup as the soil dries, producing a reading on the vacuum gage.

Generally, the tensiometer is placed in the soil and left for long periods of time. When water is removed from the soil by evaporation or transpiration, the reading on the gage increases. When water is added to the soil by rain or irrigation, the reading decreases. Measurements made with tensiometers are generally restricted to the range 0 to −0.85 bar. This is due to the formation of air bubbles in the water column and the possible entry of air into the porous cup at lower (more negative) matric potentials.

Resistance blocks. A resistance block consists of two electrical leads embedded in a porous material such as gypsum, nylon, fiberglass, or Teflon (Figure 5-15). The blocks are placed in the soil and allowed to equilibrate with soil moisture. The electrical resistance between the two embedded electrodes is measured using a conductivity meter such as a Bouyoucos meter.

Measurements made using resistance blocks must be correlated with soil-water contents. Once the initial correlation is performed, the resistance blocks can be installed and left in the soil for long periods of time. Resistance blocks can be

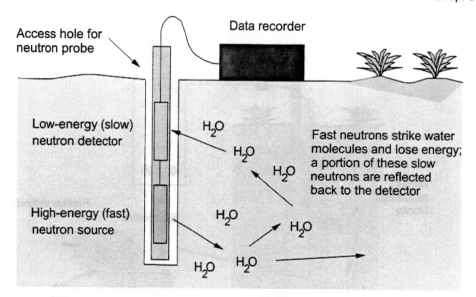

Figure 5-16 Neutron water meter, a nondestructive device used to measure soil water content.

connected to recording devices and can provide a continuous measurement of soil water content. Resistance blocks made of inert materials such as nylon, fiberglass, or Teflon can be affected by differences in soil salinity. Blocks made of gypsum ($CaSO_4 \cdot 2H_2O$) are less sensitive to salinity problems because they maintain a constant concentration of ions in the blocks due to the slight solubility of gypsum. The electrical leads that are embedded in the resistance blocks can be plates, screens, or wires in parallel or concentric arrangements.

Neutron water meter.   This device (Figure 5-16) consists of a source of fast (≈ 1600 km/s), high-energy neutrons [1 to 15 MeV (million electron volts)]. The fast neutrons are emitted into the soil and are slowed (≈ 2.7 km/s and 0.03 eV) by collisions with atomic nuclei. Slowing a fast neutron[1] requires about 18 collisions with hydrogen, 114 with carbon, 150 with oxygen, and, in general, for elements with a mass greater than hydrogen, it requires ($9N + 6$) collisions for nuclei with a mass of N. It is apparent that hydrogen is most effective in slowing fast neutrons. The water content of the soil determines how many fast electrons are slowed and reflected back to the detector and counted. Care should be taken when measuring water contents of soils with widely different organic matter contents since collisions with hydrogen and carbon atoms associated with soil organic matter could produce erroneous readings.

[1]D. Hillel, *Introduction to Soil Physics*, 2nd ed., Academic Press, New York, 1982, p. 62.

Normally, permanent access holes lined with aluminum or steel pipe are used with this method. These permanent access holes allow water measurements to be taken periodically over very long periods of time. The neutron meter readings must be correlated with soils of known water content. Once these correlations have been performed, neutron readings can easily be converted to volume water contents.

Care must be taken when using any instrument containing radioactive material. Usually, the probes are carried in shields containing a hydrogenous material, such as paraffin, that is highly absorptive of neutrons. If proper care is taken, the instrument can be used without undue risk. A major disadvantage to the use of neutron water meters is the relative high cost of the initial investment compared to tensiometers and resistance blocks. In addition, access tubes may interfere with field work when installed in cultivated fields.

Pressure plate apparatus.   This technique was discussed earlier (Figure 5-6) to illustrate the construction of a water-retention curve. The technique is primarily a laboratory technique and provides water contents and corresponding matric potentials over a wide range. The pressure plate apparatus is often used as a standard technique to calibrate tensiometers and neutron water meters.

5.6 SOIL-WATER AVAILABILITY TO PLANTS

Most crop, forest, turf, and ornamental plant species belong to a class of plants known as *mesophytes*. These plants grow in aerated soils and depend on water stored in the soil for their metabolic needs. Other classes of plants are: *hydrophytes*, which occur in water-saturated regimes; *phreatophytes*, which are capable of removing water from shallow water tables; and *xerophytes*, which have special water-conserving adaptations that allow them to inhabit deserts.

Mesophytes develop extensive root systems to supply the plant with water and nutrients. As do most plants, mesophytes also develop shoot and leaf patterns that expose large surface areas to the atmosphere and sun. The high surface areas are necessary to allow absorption of sunlight and carbon dioxide, which are both needed in the process of photosynthesis. As a consequence of the high surface areas of the leaves and the open stomates, which are required to facilitate gas exchange, most plants make very inefficient use of soil water. Of the water taken out of the soil by plants, less than 1% is used to meet metabolic needs. The remaining ≈99% is lost into the atmosphere in a process known as *transpiration*.

5.6.1 Soil–Plant–Atmosphere Continuum

Evaporation of water from the stomates and leaf surfaces (transpiration) provides the energy for water movement through the soil, into the plant root, and up the roots and shoots into the leaf tissues. This process is described as the soil–plant–atmosphere continuum (SPAC). Figure 5-17 illustrates the continuum of potentials that

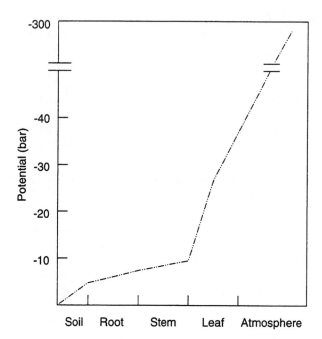

Figure 5-17 Water potentials through the soil–plant–atmosphere continuum (SPAC).

water is under as it moves from the soil to the root and then up through the plant to the atmosphere. This continuum of negatively increasing potentials is the driving force for water movement into and up through the plant to the atmosphere. Matric potentials in a moist soil may be close to $-\frac{1}{3}$ bar, increase through the root and stem, and reach values of around -20 bar in the mesophyll cells of the leaf. The greatest change in potential is associated with the phase change (evaporation) of liquid water in the leaf tissue to water vapor in the atmosphere. A dry hot atmosphere may have water potentials as large as -1000 bar or more.

As long as the movement of water through the soil and into the root can balance the loss of water through the stomatal openings to the atmosphere, the plant will remain turgid. If the loss of water becomes greater than the rate of water uptake, the plant loses turgidity (wilts) and the stomates close, preventing further water loss. This is a protective mechanism that prevents the desiccation and death of the plant, but it has the negative effect of preventing the gas exchange that is needed for photosynthesis. Prolonged periods of wilting often result in dramatic reductions of crop yields.

The long range transport of water in soil by unsaturated flow and/or vapor equalization is a slow process. Hence the movement of water to the root by these processes can only meet the atmospheric demands placed on the plant when the water is in close proximity to a root. If atmospheric demands become high or water films are discontinuous, the plant's needs cannot be met and the plant wilts.

Permanent wilting point. Does a matric potential exist at which soil water is bound so tightly that plants cannot remove water from the soil even when subjected to very low atmospheric demands? That is, is there a matric potential at which water becomes energetically unavailable to the plant? In classic experiments, sunflowers were grown in pots with the soil surface sealed with wax to prevent evaporation. With this design the only water loss from the pots would be due to transpiration from the sunflower leaves. The plants were grown in sunlight until the plants wilted. The plants were then placed in a dark, humid atmosphere and allowed to regain turgor. When the plants recovered, they were again placed in sunlight until wilted. The cycle was repeated until the plants were unable to remove water from the soil even in a dark, humid atmosphere. At this point the plants were permanently wilted. The matric potentials in these and similar experiments were generally close to –15 bar. This matric potential (–15 bar) is known as the *permanent wilting point* and soil water held in the soil between –1/3 and –15 bar is known as *plant-available water (PAW)*.

It should be pointed out that while plants vary in their permanent wilting points most mesophytes will be permanently wilted at approximately –15 bar tension. Some plants, especially xerophytes, can survive much lower soil matric potentials. It should also be emphasized that the soil seldom reaches the permanent wilting point since most plants wilt at much lower tensions. This nonpermanent wilting occurs any time the rate of water uptake by the plant cannot keep up with atmospheric demands. When soil water is managed, for example by irrigation, a 50% depletion figure is often used to schedule irrigations. For example, if a soil holds 8 cm of PAW in the root zone, the soil would be irrigated when 4 cm of the PAW had been used by the crop or lost by evapotranspiration.

The plant-available water-holding capacity of a soil is primarily a function of the soil's texture and organic matter content. Figure 5-18 illustrates the effect of texture, and, to a certain extent, clay mineralogy on the amount of plant-available water in the soil. The content of both –1/3- and –15-bar water is shown to increase as the percentage clay increases. Since the –1/3-bar water content increases faster than 15-bar water content, plant-available water (water at –1/3 bar - water at –15 bar) also increases as the texture becomes finer. Figure 5-18 illustrates that PAW increases until the clay content reaches about 25% for soils dominated by swelling clays and about 65% for soils dominated by nonswelling clays. Some of the greatest plant-available water-holding contents of soils are found for silt loam and silty clay loam textured soils of the midwest that contain 3 to 5% organic matter. These soils contain ≈0.22 to 0.24 cm of water per centimeter of soil. Table 5-3 gives some PAW values for different-textured soils.

Notice that gravitational water, that is, water between matric potentials of 0 and –1/3 bar, is not included in the definition of plant-available water. Energetically, this water is available to the plant, but since this water occupies the soil macropores at the expense of the soil atmosphere, it is generally not considered to be beneficial to the plant. If this water is present, soil aeration is usually adversely affected to the point that root growth is curtailed. This water has been called superfluous water in

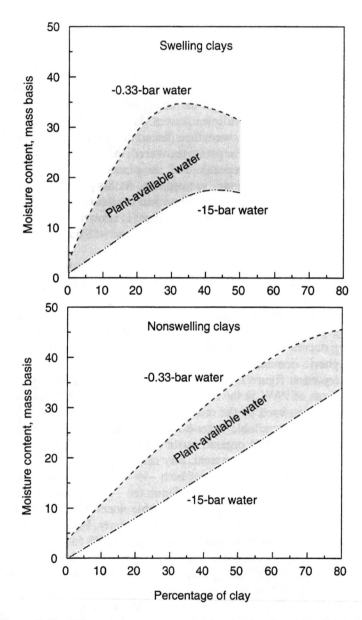

Figure 5-18   Effect of clay content and clay mineralogy on plant-available water. (After R. L. Donahue et al., *An Introduction to Soils and Plant Growth*, Prentice Hall, Englewood Cliffs, NJ, 1983, p. 174.)

TABLE 5-3 Effect of texture on plant-available water

cm water/cm soil	Textures
0.05	Sands, loamy sands
0.10	Loamy fine sands, coarse sandy loam, some gravelly textures
0.15	Sandy loams, sandy clay loams, fine sandy loam, silty clay, sandy clay, and clay
0.20	Loam, silt loam, silt, silty clay loam, clay loam

biological classifications of soil water. Figure 5-18 also shows that plant-available water does not include that portion of capillary water held at matric potentials more negative than −15 bar, nor does it include hygroscopic water.

5.7 MANAGEMENT OF SOIL WATER

Water is the most limiting factor in crop production. In arid and semiarid regions supplemental water must be supplied to compensate for the fact that potential evapotranspiration exceeds precipitation (Figure 5-19). In humid regions with dry summers, crop production is also frequently limited by insufficient precipitation during the growing season. But in most cases, stored soil water is sufficient to prevent a crop failure. In fact, crop production in humid regions is often hindered by an excess of soil water during planting and harvesting. Excess water creates soil problems related to leaching of plant nutrients and agricultural chemicals and related to erosion of the soil by the runoff of the portion of the excess precipitation that does not infiltrate the soil.

5.7.1 Efficiency of Water Use

Most crop species are extremely inefficient in their use of soil water. The inefficient use of soil water is due to the loss of water through open stomates. The stomates must be open to allow the exchange of O_2(gas) and CO_2(gas) that is required in photosynthesis. Some species of plants, including the Crassulacea and other succulent plants, have developed mechanisms that allow the needed gas exchange to occur during the night when the evapotranspiration demand of the atmosphere is low. Although these species use very little water, they are not highly productive. The high crop yields of modern agriculture require moist soils and open stomates and therefore use very large amounts of soil water.

The efficiency of water use can be expressed two ways: either in terms of the particular crop species (transpiration ratio) or in terms of the cropping system (consumptive use). The transpiration ratio is the kilograms water lost by the crop in transpiration divided by the kilograms of dry matter produced. Transpiration ratios are commonly in the range 300 to 700, with an extreme range 200 to 1000.

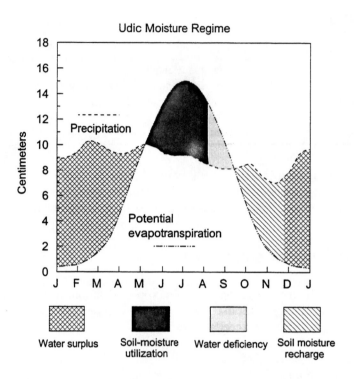

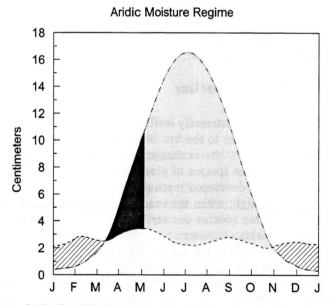

Figure 5-19 Precipitation and potential evapotranspiration for several moisture regimes.

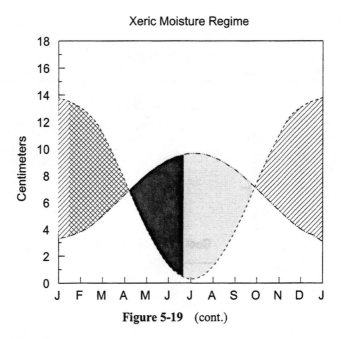

Figure 5-19   (cont.)

Transpiration ratios for crops grown in hot arid climates would approach the high end of the extreme range. Consumptive use is the depth (centimeters) of water lost by evaporation and transpiration or used by the plants' metabolisms during an entire growing season. Consumptive use is a better measure of the total water required to produce a crop, since it includes evaporation from the soil surface in addition to transpiration.

Water used by a crop is primarily a function of climatic and soil factors. If the soil is sufficiently supplied with water, water use will be a function of solar energy, atmospheric temperature, and wind speed. If the soil is not sufficiently supplied with water, the rate of water movement to the roots will be reduced and less water will be used by the plants than predicted from climatic data. Figure 5-20 illustrates that when the soil in the vicinity of the root zone is moist, evapotranspiration is established by the type of crop and the climatic conditions (initial flat portions of the curves). As the soil dries, evapotranspiration is controlled by the rate of water movement from moist soil zones to the root. Initial evapotranspiration would approximate evaporation from a free water surface (pan evaporation), but as the root zone becomes depleted of water, actual evapotranspiration decreases below pan evaporation.

Maximum water use efficiency is normally obtained with the highest water use. Optimum crop yields will be obtained when the soil is kept moist and well fertilized. Under these conditions water use will be at a maximum, but the ratio of kilograms of water used per kilogram of dry matter produced will approach a minimum. That is, the cropping system will show the most efficient use of water.

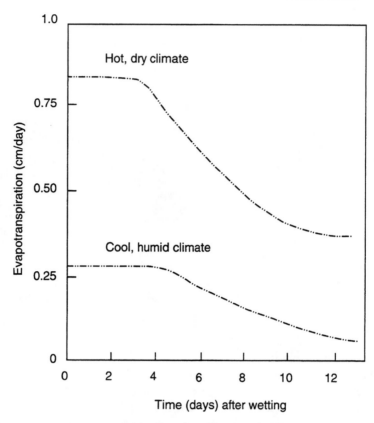

Figure 5-20 Evapotranspiration as affected by climatic conditions and
time of application. (After R. L. Donahue et al., *An Introduction to Soils
and Plant Growth*, Prentice Hall, Englewood Cliffs, NJ, 1983, p. 197.)

When the soil is dry or fertility is lacking, less water will be used but yields will be
proportionately lower, resulting in lower efficiency of water use.

Water use efficiency can be increased by controlling either water loss by
evaporation or water use by noncrop species (weeds). Evaporation from soil surfaces can be decreased through the use of tillage practices that leave substantial crop
residues on the surface and through the use of paper and plastic mulches (Figure
5-21). Mulches are generally used only on high-value specialty crops, such as
pineapples, strawberries, vegetable crops, or ornamentals.

Water use of noncrop species can be minimized either by cultivation or
through use of herbicides. The two major reasons for cultivation of standing crops
are, first, to control weeds and eliminate their water use, and second, to destroy soil
crusts that interfere with soil aeration and the infiltration of precipitation into the
soil. Care must be taken during cultivation to avoid damaging crop roots (root
pruning) in the interrow region. Root pruning is usually caused by cultivating too

Figure 5-21 Strawberry crop near Salinas, California, growing under plastic sheeting. Plastic sheeting provides water, weed, and erosion control. (Courtesty of USDA Soil Conservation Service.)

deep or too close to the base of the crop. For this reason cultivation of interrow regions should be shallow and usually early in the growing season before there is an abundance of crop roots in the interrow region. Once the crop canopy shades the interrow region, weed growth is inhibited and the need for further cultivation eliminated.

5.7.2 Irrigation

Irrigation is the practice of applying water to the soil to supplement precipitation. Irrigation can be applied either as furrow, flood, sprinkler, or drip irrigation (Figure 5-22). Irrigation becomes necessary when the potential evapotranspiration exceeds the monthly precipitation and the amount of stored plant-available water. Figure 5-19 shows the monthly amounts of precipitation and evapotranspiration for udic, xeric, and aridic moisture regimes. In xeric moisture regimes dryland production is common. The crop is planted just before the wet season begins and is harvested during the early part of the dry season. Production of other crops during the dry season generally requires irrigation. In aridic moisture regimes, production of most crops is not possible without irrigation. In these moisture regimes evapotranspiration exceeds precipitation for most of the growing season. In ustic moisture regimes, supplemental irrigation may be necessary to achieve optimum crop yields, since there may be extended periods when evapotranspiration exceeds precipitation. In udic moisture regimes, irrigation is generally not necessary except during a period of extended drought. In udic moisture regimes, stored soil water is generally capable

Figure 5-22 Examples of various irrigation techniques. (a) Sprinkler irrigation in citrus grove near Bradenton, Florida. (b) Center pivot system on newly planted corn field near Mabton, Washington. (c) Furrow irrigation using siphon tubes for uniform water application; there is a 2% slope in direction of irrigation. (d) Drip irrigation on young citrus trees. (Courtesy of USDA Soil Conservation Service.)

of supplying the crop through most periods of drought. In all of the moisture regimes, irrigation may be necessary for crop production on soils with limited plant-available water-holding capacities. These are generally coarse- textured sandy soils or soils with chemical or physical conditions that restrict root development.

Irrigation requires a source of good-quality water and careful management to prevent the buildup of soluble salts in the soil profile or the accumulation of sodium on the exchange complex. Both of these problems, as well as the definition of "good"-quality irrigation water, are discussed in Chapter 10.

5.7.3 Drainage

When soil conditions, such as heavy-textured subsoils, compacted zones, high-bulk-density horizons, or low-lying topographic positions, prevent the rapid removal of excess water from the macropores, surface and/or subsurface drainage can be used to remove the excess water (Figure 5-23). Drainage removes the "gravitational water" from the macropores and improves soil aeration. Most soils that have an aquic moisture regime require drainage to improve subsoil aeration and to allow timely tillage and harvest operations. In addition, irrigated soils often require drainage to allow for correct water management to prevent the formation of salt-affected soils.

Surface drains.   Surface drainage systems are designed to intercept and remove excess precipitation before infiltration into the soil profile. Surface drainage is most common with heavy-textured soils, whose low permeabilities prevent or make economically impractical the installation of subsurface drainage. Surface drainage is often achieved using landforming to create a level, slightly sloping soil surface that encourages surface runoff into a series of open ditches (Figure 5-24). Surface drainage system must be carefully designed to encourage the slow runoff

Figure 5-23   Poorly drained cornfield near Freeport, Illinois. These fine-textured soils require tile drainage to a suitable outlet. (Courtesy of USDA Soil Conservation Service.)

(a)

Original land surface
depressions subject to flooding (poorly drained soils)

Surface after landforming
Potholes filled with material from former ridges

(b)

Figure 5-24 Landforming to facilitate the removal excess precipitation
and/or prepare the land for irrigation. (a) Tractor-drawn land planner near
Winnsboro, Louisiana. (Courtesy of USDA Soil Conservation Service.)

of excess precipitation, yet not to cause excessive erosion. Surface drainage is often
used with fluvial- and lucustrine-derived soils, whose site positions and textures
make subsurface drainage impractical. Medium- and coarse-textured soils are
generally too susceptible to erosion to make surface drainage practicable.

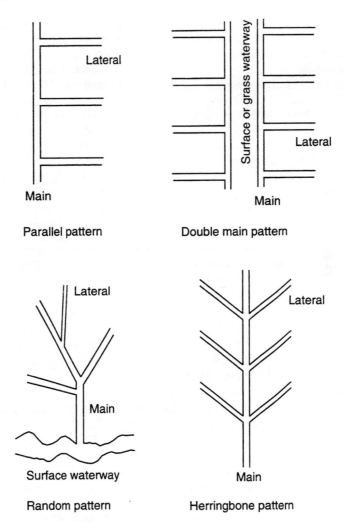

Figure 5-25 Patterns for subsurface drainage systems.

Subsurface drainage. For medium- and coarse-textured soils and for some heavier-textured soils with reasonable subsoil permeabilities, subsurface drainage provides the most effective means of removing excess water from the soil profile. A subsurface drainage system consists of subsurface laterals that collect the water, and submains or mains that receive the water from the laterals and eventually deliver the water to a surface ditch or waterway. It should be noted that a drainage system will function only as well as the outlet system. Figure 5-25 illustrates the basic patterns for subsurface drainage. The spacing and depths of the laterals depend on soil permeability, that is, how fast water flows through the soil, and also upon other soil factors, such as the depth of soil that can or must be drained (Table 5-4).

TABLE 5-4 Effect of permeability on tile spacing and depth of tiles

Rapidly permeable > 6 in./hr .	Mucks and peats 3 to 5 ft deep, 80 to 120 ft apart
Moderately rapidly permeable 2 to 6 in./hr	Mineral soils 3 to 4 ft deep, 100 to 150 ft apart
Moderately permeable 0.6 to 2 in./hr	3 ft deep, 80 to 100 ft apart; 4 ft deep, 100 to 120 ft apart
Moderately slowly permeable 0.2 to 0.6 in./hr	3 to 3.5 ft deep, 60 to 100 ft apart
Slowly permeable 0.06 to 0.02 in./hr	18 to 30 in. deep, < 70 ft apart
Very slowly permeable < 0.06 in./hr	*Note:* Current studies do not indicate clearly whether subsoil drainage is economical on these soils.

Source: Illinois Drainage Guide, Circular No. 1226, Cooperative Extension Service, College of Agriculture, University of Illinois, Urbana, Ill.

Drain tiles are usually constructed of clay, concrete, or plastic (Figure 5-26). Clay and concrete drain tiles are available in a variety of qualities to meet different field conditions. Specifications are established by the American Society of Testing and Materials (ASTM) for special conditions such as acid subsoils, soil with high sulfates, or tiles laid at greater than normal depths. In America plastic drain tiles are commonly constructed of high-density polyethylene, while polyvinyl chloride is commonly used in Europe. Plastic tubing is not affected by acid and chemicals normally found in the soil.

Figure 5-26 Soil drainage can be accomplished through use of surface or subsurface drainage. (a) Surface ditches encourage the removal of excess water and serve as outlets for subsurface drains. (b) Mechanized equipment laying subsurface plastic tiles. (Courtesy of USDA Soil Conservation Service and Agricultural Communications, University of Illinois.)

(a)   (b)

Benefits of land drainage. Benefits of land drainage include improved soil aeration, reduced frost heaving, faster warming of soils in spring, and improved timeliness of field operations.

Soil Aeration. One of the primary benefits of soil drainage is improved soil aeration resulting from the removal of gravitational water from the macropores. In poorly drained soils, root development is restricted by high water tables during the spring. In these soils, the water table often drops faster than the roots can grow. This creates a situation where root growth is inhibited by the high water table in the spring, and because of the restricted root system, the crop suffers from insufficient water during the summer months. Drainage also reduces the potential for the production of toxic substances by anaerobic organisms. Soil drainage also lessens other undesirable processes, such as the reduction of nitrate and its subsequent loss as a gas from the soil (see Chapter 13).

Temperature and Frost Heaving. Soil drainage lowers the specific heat of the soil. This is because the amount of heat energy required to warm a soil is primarily a function of the soil's water content (see Chapter 6). Soil drainage also minimizes the damage due to frost heaving. Frost heaving results when water expands upon freezing. When a poorly drained soil goes through several freeze–thaw cycles, considerable movement of soil material can result. From the viewpoint of soil structure, this is a desirable process, as soil particles are forced together to form aggregates and compacted zones are disrupted by soil movement. But from the viewpoint of objects built on the soil or plants growing in the soil, frost heaving can cause serious problems. Sidewalks and foundations are cracked and, in the case of plants such as alfalfa, sensitive tissues can be forced out of the soil and exposed to temperatures low enough to kill the plant.

Field Operations. Improved timeliness of field operations associated with planting and harvest is a major benefit of land drainage. In many locations of the country, excess soil water prevents farmers from planting during the optimum periods. This shortens the potential growing period, placing the crops at greater risk from drought during critical growth periods and from early frost in the fall. For example, corn planted in northern Illinois was found to yield 9406 kg/ha (150

TABLE 5-5 Effect of planting date on corn yields (bu/acre)[a]

	Northern Illinois	Central Illinois	Southern Illinois
Late April		156	102
Early May	151	162	105
Mid May	150		82
Early June	100	133	58

[a]Three-year averages.
Source: Illinois Agronomy Handbook, 1985–1986, University of Illinois, College of Agriculture Circular No. 1233.

bu/acre) when planted in mid-May, but only 6271 kg/ha (100 bu/acre) when planted in early June (Table 5-5).

STUDY QUESTIONS

1. What is the role of soil in the hydrologic cycle?

2. How does the structure of water affect its properties?

3. What sites associated with the soil matrix result in water being bound to soil solids?

4. What is a soil water-retention curve? How is pressure, such as measured in a pressure plate apparatus, related to matric potential? (Compare Figures 5-7 and 5-9.)

5. Why cannot ocean water be used to grow crops?

6. Define *field capacity, permanent wilting coefficient, hygroscopic coefficient, gravitational water, capillary water, PAW,* and *hygroscopic water.*

7. Describe saturated and unsaturated flow.

8. What forces are responsible for water movement into and through the plant and eventually into the atmosphere?

9. Why does a plant wilt? When a plant wilts, does it mean that all PAW has been removed from the soil?

10. List the benefits and disadvantages of tensiometers, neutron meters, and resistance blocks for measuring soil water. Judge which would be best for use in an irrigated field.

11. Calculate how many inches of water are needed to wet a uniform soil to a depth of 3 ft if the soil presently has a matric potential of -10 bar and has a water-potential relationship as shown in Figure 5-9. Assume that the soil has a bulk density of 1.4 g/cm^3 and that water has a density of 1.0 g/cm^3.

12. Calculate how many pounds of water are present in an acre of soil 4 ft deep if the soils contains 23% water (W %) and has a bulk density of 1.35 g/cm^3. How many kilograms of water would be in a hectare of this soil 2 m deep?

6

Soil Aeration and Temperature

6.1 SOIL AERATION

The roots of most higher plants, microorganisms, and soil animals, such as earthworms, depend on the oxygen in the soil atmosphere to meet their metabolic needs. Few plants have the capability to translocate oxygen, from aboveground stems or leaves, at rates sufficient to maintain actively growing root systems. Many microorganisms have the capability to oxidize organic matter in poorly aerated soils, but the end products, the efficiency of their metabolisms, as well as the oxidized compounds or ions that serve as terminal electron acceptors, are quite different than in well-aerated soils. A terminal electron acceptor is an oxidized species, such as O_2(gas), that receives electrons during the oxidation of organic matter by an organism (Figure 6-1). Electrons are a waste product of an organism's metabolism, and these electrons are combined with an oxidized species for removal from the organism. The oxidized species, serving as the terminal electron acceptor, is reduced upon combining with the electrons.

A well-aerated soil is defined as a soil in which neither the amount nor composition of the soil atmosphere is limiting to the growth of higher plants. Root, animal, and microbial respiration consumes O_2(gas) and produces CO_2(gas), hence the soil atmosphere generally contains lower partial pressures (concentrations) of O_2(gas) and higher partial pressures of CO_2(gas) than the atmosphere above the soil. Root growth can be limited by either low levels of O_2(gas) or by high levels of CO_2(gas). The atmosphere above the soil contains $\approx 21\%$ O_2(gas) and $\approx 0.03\%$ CO_2(gas). The surface horizons of a soil contain similar amounts of gases as the

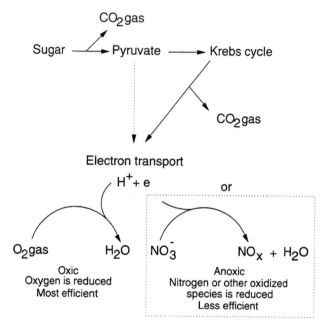

Figure 6-1 Role of oxygen and other oxidized species as terminal electron acceptors in oxic and anoxic conditions.

aboveground atmosphere, but horizons deeper in the soil profile usually contain lower amounts of O_2(gas) and 10 to 100 times more CO_2(gas).

A well-aerated soil is a soil with sufficient pore space, that is, empty of water, to provide for a rapid exchange of gases between the soil horizons and the aboveground atmosphere. The pores that are empty of water at field capacity, the macropores, are the pores that expedite the rapid exchange of gases. A well-aerated soil must first have a sufficient volume of macropores, and second, the macropores must provide a continuous network for gas exchange between lower soil horizons and the soil surface. There is a trade-off between sufficient macropore space to facilitate good soil aeration and sufficient micropores space to store plant-available water (PAW). Sandy soils tend to have a predominance of macropores and are well aerated, but at the expense of PAW. Finer-textured soils tend to have a predominance of micropores, so they contain larger amounts of PAW, but often at the expense of soil aeration. An ideal soil would contain 50% pore space equally divided between micro- and macropores and would have a continuous network of macropores from lower horizons to the soil surface.

Soil conditions that give rise to poor soil aeration are usually associated with poor soil drainage or the destruction of soil structure. Poorly drained soils are poorly aerated soils since the macropores are filled with "free" water. One of the major benefits to land drainage is the improvement of soil aeration. Compaction increases bulk densities and decreases the amount of pore space in a soil. Compaction also tends to reduce the macropore space to a greater extent than the micropore space. In addition, it only requires a thin compacted zone to reduce the aeration of all horizons below the compacted zone. Other soil conditions, such as high contents of swelling-clay minerals or high-bulk-density parent materials, can affect soil aeration negatively.

Mechanisms of soil aeration.   Gas exchange occurs by two mechanisms in soils, mass flow (convection) and diffusion. *Mass flow* describes the process where the soil atmosphere moves as a unit in response to temperature or pressure differences. That is, all the gases in the soil atmosphere move at the same rate, in the same direction. *Diffusion* is the movement of individual species of gas in response to a concentration gradient. Oxygen may be moving in one direction and CO_2(gas) in an opposite direction.

Mass Flow.   Mass flow occurs when a temperature or pressure difference is created between the soil atmosphere and the aboveground atmosphere. A temperature difference is created twice each day (diurnal) because of the faster rate at which the soil warms in the morning and the faster rate that the atmosphere directly above the soil cools in the evening. The soil warms faster in morning than the aboveground atmosphere because of greater absorption of solar energy by the soil. The aboveground atmosphere cools faster in the evening because of its lower specific heat. A lower specific heat means that the atmosphere above the soil needs to lose less energy than the soil to achieve a certain amount of cooling, and hence it cools more rapidly than the moist soil. During the remaining portions of the day and night, the soil and the atmosphere just above the soil will be basically at the same temperature, and therefore there is no driving force for mass flow. Figure 6-2 illustrates the diurnal temperature fluctuations of the soil and atmosphere directly above the soil in the absence of wind (convective energy).

Mass flow can also occur when wind blows across the soil surface, creating a pressure difference between the aboveground atmosphere and the soil atmosphere. Airflow in this case will be from high pressures to low pressures. Water uptake by plants can also result in mass flow. As water is removed from the soil, void space is created, and this results in a pressure difference between the soil atmosphere and the aboveground atmosphere and results in mass flow into the soil. Mass flow is important primarily in surface horizons, shallow soils, or soils with a predominance of macropores.

Diffusion.   Diffusion is the movement of gases in response to concentration gradients. Since concentrations of gases are usually expressed in terms of partial pressures, diffusion is the movement of gases from high partial pressures to low partial pressures. Normally, O_2(gas) will be diffusing from the aboveground atmosphere into the soil and from surface horizons into lower horizons, and CO_2(gas) will be diffusing from lower horizons toward surface horizons and then into the aboveground atmosphere. Diffusion is the result of the random kinetic movement of the gas molecules. Diffusion is a slow process. For example, the oxygen diffusion

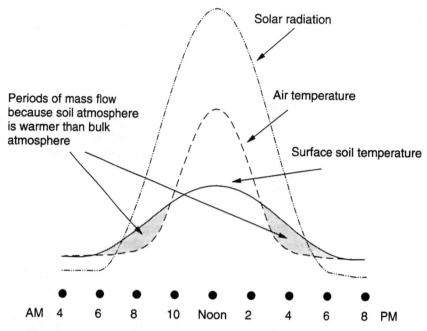

Figure 6-2 Diurnal temperature fluctuations of the soil surface and the atmosphere close to the soil surface.

rate (ODR) of a soil has been shown[1] to vary from 10 to 80×10^{-8} g O_2(gas) cm^{-1} min^{-1}. The same studies showed that root initiation was reduced or stopped when ODR was in the range 18 to 23×10^{-8} g O_2(gas) cm^{-1} min^{-1}.

Effects of soil aeration. In well-aerated soils, soil minerals, organic matter, roots, animals, and microorganisms are in contact with approximately the same partial pressure of O_2(gas) as found in the aboveground atmosphere. Because of these high partial pressures of O_2(gas), reduced species such as ferrous or sulfide minerals or organic matter are not stable, but subject to oxidation. Many, if not most, of these oxidations are mediated by soil organisms. In these oxidation reactions, the organisms obtain energy to drive their metabolisms from the oxidation of the reduced species and use O_2(gas) as a terminal electron acceptor. The decomposition of organic residues and humic substances (see Chapter 11), the oxidation of pyrite, as well as the oxidation of ammonium ions (NH_4^+) to nitrate (NO_3^-), are examples of these processes (see Chapter 13).

In poorly aerated soils, the partial pressures of O_2(gas) are much lower. In these environments, the oxidation of ferrous and sulfide minerals, ammonia, and most other inorganic reduced species ceases, but the oxidation of organic materials

[1]L. H. Stolzy and J. Letey, Characterizing soil oxygen conditions with a platinum microelectrode, *Adv. Agron.* 16:249–279 (1964).

can continue. Anaerobic organisms continue to use organic materials as a source of energy, but the end products of their metabolisms, as well as the terminal electron acceptors that they use, are different. Methanol, ethanol, acetic acid, and methane are examples of anaerobic waste products, while nitrate (NO_3^-), ferric ions (Fe^{3+}), and sulfate ions (SO_4^{2-}) are examples of oxidized species that can serve as terminal electron acceptors. Most reactions in soils involving the reduction of an inorganic species are coupled to the anaerobic oxidation of organic materials. The species that are reduced, generally, are serving as terminal electron acceptors for the anaerobic organism's metabolism.

The capacity for an oxidized species to serve as a terminal electron acceptor and be reduced is a function of both the oxidation-reduction (Eh) status of the soil and pH. The Eh of a soil is a measure of the potential for a species, such as nitrogen or sulfur, to be reduced or oxidized. Sulfide (S^{2-}) can be spontaneously oxidized to sulfate (SO_4^{2-}), with the release of energy, only when Eh and pH are above certain values. If soil Eh and pH values are above the sulfide-sulfate line (see the triangle and square data points in Figure 6-3), then sulfide can be oxidized to sulfate with the liberation of energy to drive the microorganism's metabolism. If soil Eh and pH values fall below the sulfide-sulfate line, the circle data point, sulfate can serve as a terminal electron acceptor for anaerobic organisms and can be reduced to sulfide. Figure 6-3 also contains the Eh-pH stability line for nitrite (NO_2^-) and nitrate (NO_3^-). When soil Eh and pH values fall above the nitrite-nitrate stability line

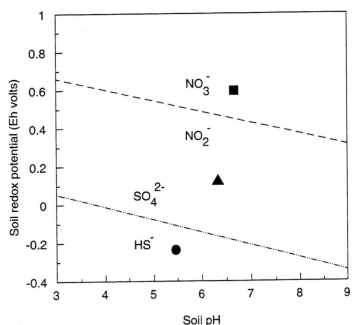

Figure 6-3   Eh–pH diagram for sulfide–sulfate and nitrate–nitrite.

(square data point), nitrite can be oxidized to nitrate (nitrification), and when values fall below the line (triangle and circle data points) nitrate can serve as a terminal electron acceptor and be reduced (denitrification; see Chapter 13). Note that nitrate can serve as a terminal electron acceptor at much higher Eh values than sulfate. Note also that there are Eh and pH values, triangle data points, when NO_3^- can be reduced and sulfide can be oxidized.

In poorly aerated soils, the soils are effectively isolated from $O_2(gas)$ in the aboveground atmosphere by excess soil water, soil compaction, or other processes. In isolated systems such as these, organic matter oxidation is observed first to occur coupled to the reduction of $O_2(gas)$. As $O_2(gas)$ is consumed, soil Eh values are reduced and organic matter oxidation becomes coupled to the reduction of nitrate (NO_3^-). When the nitrate is consumed, soil Eh values decrease to the point that organic matter oxidation uses ferric iron (Fe^{3+}) as a terminal electron acceptor. When the ferric iron has been consumed, Eh values decrease and organic matter oxidation is associated with the reduction of sulfate (SO_4^{2-}). Along with the shift in species that serve as terminal electron acceptors is a change in waste products. Carbon dioxide gas is the end product of aerobic respiration in well-aerated soils. As $O_2(gas)$ is consumed and Eh values drop, methanol, ethanol, and short-chain organic acids become end products. Methane (CH_4) becomes the end product of anaerobic respiration at about the same Eh levels as sulfate reduction.

It is important to emphasize that reduced materials can be oxidized in well-aerated soils since they are exposed to a strong oxidizing agent, $O_2(gas)$. Oxidized species are reduced in poorly aerated soils only when their reduction is coupled with the oxidation of organic matter by anaerobic organisms. If an oxidized species occurs in a poorly aerated zone that is below the zone of biological activity, reduction will generally not occur. This could be either deep within the soil, in an aquifer, or deeper within the regolith. The lack of reduction in the absence of organisms is because there are no reducing agents, comparable to the oxidizing agent $O_2(gas)$ of well-aerated soils, found in poorly aerated anoxic (without oxygen) zones.

6.2 SOIL TEMPERATURE

Temperature refers to the kinetic energy of molecules and ions. As the soil absorbs heat energy the kinetic energy of the molecules and ions increases, and hence its temperature. Two bodies have the same temperature when heat flow between the two bodies is equal in both directions; that is, they are at thermal equilibrium. Soil temperature affects the rates and directions of soil physical and chemical processes. Physical processes such as freezing and thawing, heating and cooling, evaporation, the flow of water vapor, and soil aeration are all examples of processes affected by soil temperature, temperature fluctuations, and temperature gradients. Soil temperature also has pronounced effects on biological processes. Biological processes such as seed germination, root growth, water and nutrient absorption, as well as the myriad of microbial processes, all exhibit minimum, optimum, and maximum temperatures in soils.

For example, germination of alfalfa seed requires a minimum soil temperature of 5°C and has an optimum temperature of 18°C; alfalfa growth has an optimum range from 10 to 25°C. Nitrification, the biological oxidation of ammonium ions (NH_4^+) to nitrate (NO_3^-), becomes very slow at a temperature of 10°C. Hence recommendations for fall-applied nitrogen, which is usually applied in some form that produces ammonium ions, generally suggest waiting for soil temperatures to fall below 10°C before application. This prevents the conversion of a cationic form of nitrogen to an anionic form of nitrogen and minimizes leaching losses (see Chapter 13).

6.2.1 Factors Controlling Soil Temperature

Soil temperature is a function of radiant, thermal, and latent energy exchange processes and soil parameters, such as the specific heat capacity and thermal conductivity of the soil. Soil can receive heat energy as solar energy from the sun, as convective energy from hot winds, or as heat energy that moves by conduction from deeper within the regolith.

Modes of energy transfer. Energy can be lost or gained by four mechanisms.

Radiation All bodies can lose or gain energy by absorption or emission of radiant energy. The major source of heat energy that a soil receives is in the form of solar radiation. The soil can also lose energy by radiation. All bodies emit radiation; the total amount lost and the wavelength emitted are functions of their temperature. Soil temperature ranges from a few degrees below 0°C to ≈57°C. In this temperature range, soil emits infrared, long-wave radiation.

Convection. Convection involves the movement of heat energy by a heat-carrying mass such as wind or water. Convective energy can be very important in arid regions, where warm, dry desert winds blow from the desert across irrigated fields and greatly increase the amount of evapotranspiration. In addition, the climates of many coastal regions are moderated by convective energy from warm- and cold-water currents and winds. Vapor equilization can also transport heat energy from warm layers to cool layers within a soil.

Conduction. Conduction is the flow of energy within a soil or other body due to internal molecular motion. The kinetic energy of a molecule is a function of its temperature. High temperatures result in molecules having higher vibrational and translational energies. When these "hot" molecules collide with "cooler" molecules, energy is transferred, increasing the motion of the "cooler" molecules and decreasing the motion of the "hot" molecules. This results in the flow, conduction, of heat from warm regions of the soil to cool regions of the soil.

Latent Heat Transfer.   Latent heat transfer is the loss or gain in heat energy resulting from changes of state. The state or phase changes associated with water are by far the most important. For example, when ice melts at 0°C, the solid is converted to liquid also at 0°C. The melting of ice consumes 80 cal g^{-1}, while the freezing of water would liberate the same amount of heat per gram. The evaporation of liquid water to produce water vapor without any change in temperature consumes 580 cal g^{-1} at 25°C. When water evaporates from the soil surface, 580 cal g^{-1} of heat energy is lost into the atmosphere. Conversely, when a dew forms, that is, when water condenses on the soil surface, a similar amount of heat energy is released that can be absorbed by the soil.

The role of solar radiation.

Solar radiation and heat energy conducted from lower portions of the regolith are the two potential sources of heat energy in soils. Solar radiation is the most important source of heat energy for the upper portions of the soil profile. The amount of solar radiation that can potentially reach the earth's surface, if there is no loss in the atmosphere, is a function of the soil's position relative to the north and south poles and the equator. Figure 3-23 illustrates that soils near the equator receive the greatest amounts of energy per unit of soil surface, while soils near the poles receive lesser amounts of solar energy per unit of soil surface.

Slope aspect (Figure 3-24) can greatly modify the relationship of the soil surface to the incoming solar radiation. A south-facing slope in the northern hemisphere or a north-facing slope in the southern hemisphere receive greater amounts of solar radiation than level soils in the same region. Hence the potential amount of solar radiation that could reach a soil's surface is a function of both of the soil's location and its aspect.

The actual amount of solar radiation that reaches the soil surface can be very different from the potential amount. Figure 3-22 shows the mean hours of sunlight for various regions in the United States. The differences from one region to the other are due to losses in the atmosphere. Solar energy can be lost due to reflection from the clouds, diffuse scattering, and absorption. The solar energy that reaches the soil surface is basically the energy that is not reflected, scattered, or absorbed, plus a portion of the scattered light that is called *sky radiation*. The intensity of solar radiation that reaches the soil surface varies from ≈550 g cal cm^{-2} (langleys) in the arid southwest to ≈250 g cal cm^{-2} in northern New York, Vermont, and Washington. The arid southwest receives ≈4000 hours of sunlight per year, while humid northern regions receive < 2000 hours annually.

Solar energy that reaches the soil surface can be absorbed by the soil or reflected back into the atmosphere. The amount absorbed is dependent primarily on soil color, but also on the angle with which sunlight strikes the soil particles. Dark-colored soils absorb greater amounts of solar energy than do light-colored soils. Their color is evidence of the absorption. Light-colored objects appear light colored because they reflect much of the light that strikes them. Dark-colored objects appear dark because they absorb most of the light and little light is reflected to the viewer's eye. Solar energy that is absorbed by the soil can be lost due to latent heat transfer, convection, or radiation. Latent heat transfer is due primarily to the evaporation of water from the soil surface, while the convective losses are due to warming of the soil air and its mass flow into the aboveground atmosphere. Radiation losses occur because the soil, like any other body, emits radiation that is a function of its temperature. The soil emits radiation in the infrared region of the spectra. The amount of solar radiation that actually heats the soil varies from region to region, but a value of ≈10% is an average figure for the central United States.

Specific heat of soils. The actual temperature change that occurs in a soil upon absorption of heat energy is primarily a function of the specific heat of the soil. The specific heat of a substance is the amount of heat energy per gram of material that is required to change the temperature of the substance 1°C. Table 6-1 lists the specific heat of several substances. The actual specific heat of a soil is dependent on the soil's moisture content. A soil that contains 25% water on an ODwt basis could have a specific heat of 0.36 cal g^{-1} deg^{-1}, while a soil that contains 50% water has a specific heat of 0.47 cal g^{-1} deg^{-1}. The specific heat of a moist soil can be calculated from the mass of dry soil and water and their respective specific heats.

In the spring, a soil that contains 50% water would require twice as much heat energy as would a dry soil to warm the soil to a given temperature. One of the benefits of land drainage is that soils warm faster in the spring. This allows earlier planting and a longer growing season for crops.

Heat movement in soils. The temperature of a soil varies with time of day, time of year, and position within the soil profile. Figure 6-4 shows the temperature variations that occur within a 24-hour period for a hypothetical soil. In the morning, sunlight warms the soil, with the surface reaching a maximum temperature around noon. Once past the maximum temperature, the surface soil losses heat energy, back into the atmosphere by radiation and convection, and down into lower horizons by conduction. Since the surface soil is losing heat energy faster than it gains heat from solar energy, the surface soil cools. The minimum temperature is reached ≈12 hours after the maximum temperature. From this point until sunrise the surface soil is cooler than lower horizons and heat flow is reversed, that is, from lower horizons back toward the surface and the surface soil is warmed. Both the soil

TABLE 6-1 Specific heat of soil and other substances

Substance	Specific heat (cal g^{-1} deg^{-1})
Iron metal	0.16
Aluminum metal	0.215
Dry soil	≈0.2
Water	1.0
Soil + 25% water	0.36
Soil + 50% water	0.47

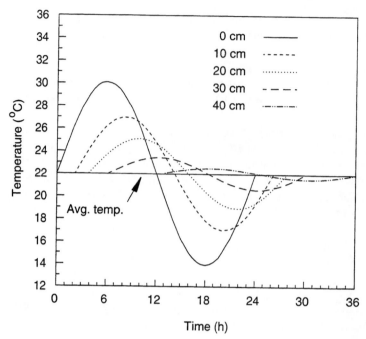

Figure 6-4   Hourly temperature fluctuations (daily temperature wave) of a hypothetical soil. (After D. Hillel, *Introduction to Soil Physics*, Academic Press, New York, 1982, p. 169.)

surface and the lower horizons go through a 24-hour cycle (daily temperature wave) of heating and cooling, but the maximum and minimum temperatures are offset by the length of time it takes heat energy to flow to different depths in the soil. In Figure 6-4 the temperature lag is about 12 hours at the 40-cm depth. Notice that the differences between the maximum and minimum temperatures decrease with depth. This means that at ≈60 to 80 cm (2 to 3 ft) of depth there are no hourly temperature fluctuations; that is, the daily temperature wave terminates at these depths.

Soil temperature is also a function of time of year. Figure 6-5 shows the fluctuation of soil temperatures for a soil located at College Station, Texas[2], as a function of time of year and depth. In January, the temperature of the surface soil is at a minimum and the surface is colder than the lower soil horizons. Heat flow in January is from lower horizons toward surface horizons. As the days lengthen, the surface soil becomes warmer and reaches a maximum in about mid-July. In the summer months the surface soil is warmer than the lower horizons and heat flow is from the surface toward lower horizons.

The soil surface reaches a maximum temperature in late summer, while the lower horizons, depending on their depth, lag behind the surface. The maximum temperature at the 8-ft depth occurred 74 days after the maximum at 2 in. of depth[3].

[2]B. J. Fluker, Soil temperature, *Soil Science*. 86:35–46 (1958).
[3]Ibid.

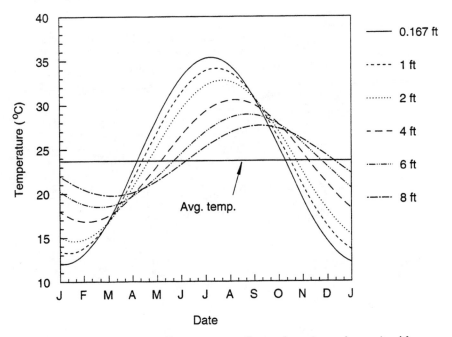

Figure 6-5 Annual soil temperature fluctuations (annual wave) with depth. [Based on data from B. J. Flucker, Soil temperature, *Soil Sci.* 86:35–46 (1968).]

Soil temperature becomes increasingly constant at greater depths (Figure 6-6). For this example, temperature fluctuations were predicted to be less than 0.5°C at a depth of 23 ft. As a general rule, temperature fluctuations from day to day, month to month (i.e., the annual temperature wave) stop at ≈19 times the depth of where the daily temperature wave terminates[4]. It is interesting to note that 19 is the square root of 365, the number of days in a year, and also illustrates the relationship between daily and annual temperature waves.

S T U D Y   Q U E S T I O N S

1. What constitutes a well-aerated soil?
2. What conditions result in a poorly aerated soil?
3. What are the consequences of poor soil aeration?
4. Discuss the mechanisms that are responsible for gas exchange between the soil and the aboveground atmosphere.
5. Why is nitrate reduced in the B horizons of a poorly drained soil but not in

---

[4]N. J. Rosenbert, *Microclimate: The Biological Environment*, Wiley-Interscience, New York, 1974, p. 68.

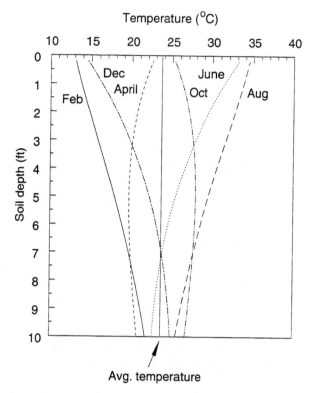

Figure 6-6 Monthly soil temperature fluctuations with depth. [Based on data from B. J. Flucker, Soil temperature, *Soil Sci.* 86:35-46 (1968).]

the saturated underlying C horizons of the regolith that are even more isolated from the aboveground atmosphere?

6. Discuss the general conditions that result in the oxidation or reduction of a substance such as sulfur (i.e., from sulfide to sulfate or sulfate to sulfide).

7. Describe the mechanisms by which heat energy can be lost or gained by a soil.

8. What factors determine the temperature of the soil?

9. Why does the surface soil begin to warm before sunrise? (See Figure 6-4.)

10. Discuss the annual and daily temperature waves.

11. What is specific heat capacity? Calculate the specific heat capacity of a soil that contains 30% water (W %). Use values from Table 6-1.

12. Define *heat, temperature,* and *thermometer.*

13. Provide a mechanism showing how cool rain or irrigation water might alter subsoil temperatures. What soil factors may influence the degree of cooling or depth of cooling?

14. Why is sprinkler irrigation sometimes used in an attempt to protect an orchard from a frost?

7

Soil Erosion

Soil erosion is an extensive and chronic problem in the United States and the world. Erosion results in the loss of the surface horizons, the topsoil, which are often the key to successful crop production. Erosion of this valuable resource results in loss of nutrients, decreased tilth, lower levels of soil organic matter, and decreased crop production. The effect of erosion on productivity depends primarily on the quality of subsoil that is exposed by the erosion. In general, the physical, chemical, and biological properties of soils that affect crop production are less favorable with increasing soil depth. Most topsoil loss is caused by water erosion, but wind erosion can be significant, particularly in arid and semiarid regions. If agriculture is to meet future needs for food and fiber, the productive topsoil of the world's agricultural lands must be maintained and protected from erosion.

Erosion not only reduces crop production, or in extreme cases renders the land useless for crop production, but also contributes to water and air pollution. Erosion represents a real cost to the public taxpayer and private industry in cleaning up the environment. It has been estimated that less than 1% of the eroded soil reaches the oceans. Most of the eroded material accumulates in ditches, fields, streams, rivers, reservoirs, harbors, lakes, and ponds. Eroded soil not only fills river channels and reservoirs, but it also carries nutrients and agricultural chemicals with it into the waterways. Much of agriculture's contribution to environmental pollution can be controlled by minimizing erosion.

While some erosion is inevitable, as long as rain falls and wind blows, consideration should be given to the fact that the natural process of soil formation is extremely slow. Soil formation is on the order of 1 cm of topsoil formed in 40 to

80 years or longer. It is far more prudent to reduce soil erosion to tolerable levels than to attempt to replace topsoil.

7.1 WATER EROSION

Water erosion is by far the most common and extensive form of soil erosion. A National Resource Inventories report[1] in 1977 found that 39 million hectares (97 million acres) of cropland experienced rates of soil erosion that exceeded 11 mt/ha (5 tons per acre) per year from the two most common types of water erosion. Thirty-nine million hectares of land represents the equivalent of the total land area of Iowa, North Carolina, and Ohio. Water erosion is not limited to any particular geographic region of the United States, but the extent of erosion is far greater in approximately the eastern one-half of the United States, where rainfall is the highest.

7.1.1 Mechanics of Water Erosion

Soil erosion, by the action of water, involves two basic phases. The first phase is the *detachment* of soil particles from the natural state of aggregation. This is a loosening action, which is followed by the second phase of erosion, *transportation,* which is the actual movement of soil particles from one physical position to another.

The extent of detachment of soil particles is dictated by the strength of aggregation of the soil and the force or work of water striking or passing over the surface does on the soil. Falling raindrops exhibit considerable kinetic energy, which is expended upon impact (Figure 7-1). The amount of kinetic energy an individual raindrop exerts on the soil depends on several factors, including raindrop size, acceleration by wind, and interception prior to striking the soil by vegetation or ground cover. Falling raindrops are also very effective in reducing or destroying the surface structure of the soil. The beating action of raindrops can result in the formation of soil crusts, which hinder emerging plants and impede the infiltration of future rains. Frost action, that is, freezing of water, as well as the action of flowing water, can also detach soil particles, making them vulnerable to transportation. In general, well-aggregated fine-textured soils are less subject than coarse-textured soils to the detachment forces of raindrops or other actions of water. Strength of aggregation is primarily a function of the soil's organic matter content. Organic matter functions as a binding agent between mineral particles. Compare the number of raindrops required to disrupt aggregates in Table 11-2. This difference is due primarily to the higher organic matter content of the earthworm casts.

Transportation or actual movement of the soil particles occurs when water runs off the soil surface. Runoff results when water is applied to the soil faster than the water can infiltrate. Runoff can be the result of rainfall, melting snow, applica-

[1]National Resource Inventories report.

Figure 7-1 Effect of a raindrop splash on soil erosion. Soil is suspended by the splash and may be transported as much as a meter from the point of impact. (Courtesy of USDA Soil Conservation Service.)

tion of irrigation water, or stream overflow. Runoff from rainfall accounts for the largest amount of soil erosion in most areas. Raindrop splash, especially in heavy rains, is a very effective transport mechanism. Research has shown that splashed particles can reach heights of two-thirds of a meter or more and that splashing can move a particle 1½ m or more on level surfaces. On sloping surfaces both splash and runoff can result in rapid movement of large quantities of soil.

Soil erosion, due to water, can result in sheet erosion, rill erosion, gully erosion, or streambank erosion. *Sheet erosion* (Figure 7-2a) is the removal of somewhat uniformly thin layers of soil from the soil surface. The effects of sheet erosion on a landscape are difficult to determine, as they are not easily seen. Lighter-colored hillslopes in production fields are evidence that sheet erosion has removed most or all of the topsoil. As water erosion becomes more severe, small erosion channels or gullies form. If these channels are small enough to be filled easily by standard tillage equipment, the erosion is termed *rill erosion* (Figure 7-2b). Rill erosion is easier to determine than sheet erosion because the rills are visible to the naked eye. Rill erosion commonly occurs when moderate to heavy thunderstorms follow very soon after planting or after the land has been tilled for planting. Channeling the water, such as in rills and gullies, significantly increases the velocity of flow, increasing the total amount of soil as well as the size of soil particles that can be transported effectively. *Gullies* (Figure 7-2c and d) are defined as erosion channels that are too large to be removed easily with standard tillage equipment. Gullies are considered to be active when the cut sides are not supporting vegetation, and inactive when the cut sides have been stabilized by vegetation. If conditions causing the gullies to form initially are not corrected, small gullies (less than 1 m in depth) will often continue to erode, resulting in moderate or large gullies (greater than 5 m in depth). Gullies tend to form when runoff is substantial in combination with steeper-sloped soils. Banks of perennial streams are sites of

Figure 7-2   (a) Severe sheet erosion, note the lighter-colored knolls where the lighter-colored subsoil has been exposed by sheet erosion. (b) Severe rill erosion; rills are small gulleys that can be crossed by tillage equipment. (c and d) Gulley erosion as shown in (c) is best corrected by installing a grass waterway; see Figure 7-5d. (e) Streambank erosion on the Napa River, California. (Courtesy of USDA Soil Conservation Service.)

streambank erosion (Figure 7-2e). Streambank erosion is dependent on conditions of high flow and usually follows periods of high rainfall. From a national perspective, streambank erosion influences a very small percentage of land, but it can result in devastation of some of the most productive agricultural areas, bottomland soils. Streambank erosion can also cause legal concerns with changing property boundaries and public concern when damage to bridges or water-diverting structures occurs.

7.1.2 Predicting and Controlling Water Erosion

The amount of erosion at a given location is dependent on a combination of physical variables and management practices. Erosion can be classified as either natural (geologic) or accelerated. Natural erosion carves landscapes and is part of the geologic cycle of soils, sediments, sedimentary rocks, metamorphic rocks, regolith, and once again, soils. Natural erosion can range from very slow to rapid and, at times, is even cataclysmic in nature. Accelerated erosion is erosion caused by the activities of humans and is generally in excess of natural erosion. Accelerated water erosion caused by sheet or rill erosion can be estimated by the *universal soil-loss equation* (USLE). This equation, developed by the U.S. Department of Agriculture Soil Conservation Service and state experiment stations, is designed to predict the long-time average soil losses due to runoff. The USLE predicts erosion from five factors: rainfall (R), soil type (K), length and steepness of slope (LS), crop management (C), and conservation practices (P). The equation, which predicts soil loss (A) in metric tons per hectare per year is given as follows:

$$A = R \times K \times LS \times C \times P \qquad\qquad [7\text{-}1]$$

The amount of erosion predicted from the USLE is then compared to the tolerable soil loss of a particular soil. If the predicted soil loss (A) is greater than the predetermined tolerable soil loss, management practices should be adjusted to reduce soil loss.

Tolerable soil losses.   Some loss of soil by water erosion is inevitable as long as rain falls and runs off the landscape. A goal of zero soil loss is, therefore, neither achievable nor practical. Rather, soil scientists have defined tolerable soil loss as the maximum amount of loss that can be tolerated without effecting the long-term productivity of the soil. Tolerance (T) values vary from soil to soil, depending on a number of factors, such as soil depth, physical and chemical properties of the substratum, ratio of surface and subsoil fertility, organic matter content, climatic factors, previous erosion, and the use of runoff control practices. The T values for a few selected soils are shown in Table 7-1. Tolerance values for most soils in the United States vary from approximately 11 mt/ha for soils best able to tolerate loss, to less than 5 mt/ha for soils that already have severe restrictions and cannot afford significant additional soil loss without a decline in productivity. Eleven mt/ha/yr can initiate a mental picture of huge amounts of soil cascading down a landscape. Keep in mind that 1 ha of an average soil to a depth of 15 cm represents approximately 2,000,000 kg or 2000 metric tons of soil. [*Note:* 1 metric ton (mt) is equal to 1000 kg and is also equal to 1.101 U.S. tons (2000 lb).] Thus 11 mt of soil spread over a hectare would represent a soil loss of approximately *0.08 cm in depth.* At this rate of loss, it would take over 180 years to experience a loss of 15 cm of soil, which approximates the amount of time required to form that amount of soil under ideal conditions. The rate of formation of soils, throughout the world, has been estimated to vary from 0.001 to 0.8 cm/yr, and estimated rates of 0.01 to 0.02 cm/yr are not uncommon. Rates of soil formation are much greater in humid climates and where subsoils and C- horizon materials (substratum) can be improved by the addition of fertilizers and incorporation of organic matter. Although 11 mt/ha/yr is not inconsequential, especially as it may influence water quality downstream or silting in of reservoirs and flood control structures, it must be kept in perspective with rates of soil formation and with the amount of soil loss that will affect soil productivity.

Rainfall and runoff factor (R). The rainfall factor is calculated for a given geographic region and is based on long-term records from the National Weather Service. The erosive potential for a rainfall event is based on both the total amount of rainfall and the intensity with which it is expected to fall. The R factor of the USLE is based on energy times intensity, or the EI value of individual rainstorms. The EI value of an individual storm is the total storm energy (E) times the maximum 30-min intensity I_{30}, for that storm. In practice, this means that two rainstorms with the same total amount of rainfall could have considerably different erosion potentials. A long gentle rain may result in very little or no measurable erosion, while a torrential cloudburst with the same amount of rain may result in considerable soil movement. The calculation of the R value is also complicated by the fact that in some areas, such as the Pacific Northwest, as much as 90% of the erosion from steep wheatlands is associated with runoff from snowmelt and surface thawing, rather than rain. In these cases, erosion from snowmelt and thaw, as well as the effects of light rains on frozen soils, are included in the computation of R by adding a subfactor to the rainfall for the area. The rainfall factor used in the USLE represents an average of all weather records over more than a decade and is a modified EI value for conditions such as snowmelt. These values vary greatly depending on the geographic location within the United States. The lines on the map (see Figure 7-3) are referred to as *isoerodent lines,* meaning that they connect points of equal rainfall erosivity (R values). Values in the western states are generally less than 100, while the value for southern Florida is 450.

Soil erodibility factor (K). The erodibility factor is a term that utilizes the physical properties of the soil to predict "potential erosion" but should not be confused with actual erosion measured by the combination of all five factors in the USLE. The K factor accounts for the relative rates of runoff and infiltration. A very sandy soil would be expected to have very high infiltration and thus a low K value. Soil conditions such as high organic matter, very fine granular structure, and rapid

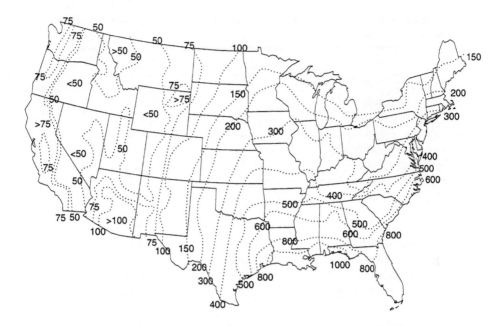

Figure 7-3 Rainfall erosion index (R) values in units of 10^7 J/ha for the United States. (After Troeh, F. R., Hobbs J. A., and Donahue, R. L. *Soil Water Conservation for Productivity and Environmental Protection,* Prentice Hall, Englewood Cliffs, New Jersey, 1980.)

permeability would be expected to increase infiltration and to reduce runoff, resulting in lower K values. K values have been determined experimentally for a wide range of soils and reflect the rate of soil loss per erosion index unit on a plot of land, with up-and-down slope tillage, that is, 22.1 m (72.6 ft) long, with a uniform lengthwise slope of 9% in continuous fallow. Under these conditions, LS, C, and P are each equal to 1 and substituting in the USLE equation, K would be soil loss per erosion index unit, or A/R. K factors are always less than 1, thus reducing the value for the amount of erosion predicted only from R (shown in Figure 7-3) to a lower number. For example, if LS, C, and P were equal to 1 and R was 200, a soil with a K value of 0.5 would lose 100 kg/ha/yr, while a soil with a K value of 0.1 would lose 20 kg/ha/yr. K values for selected United States soils are shown in Table 7-1 and for general soil conditions in Table 7-2. Table 7-2 illustrates that soils with higher infiltration rates would be expected to have lower K values. Other soil properties, such as structural strength, content of organic matter, and sand, also affect the K value of a specific soil. The Soil Conservation Service has assigned K values to most soils based on known soil properties and observed erosion.

K factors can be predicted using a nomograph[2] which has been developed for

[2]W. H. Wischmeier, C. B. Johnson, and B. V. Cross, A soil erodibility nomograph for farmland and construction sites, *J. Soil Water Conserv.* 26:189–193 (1971).

TABLE 7-1 Soil-erodibility values (K) and tolerable soil-loss values (T) for specific soils

Soil	Location	K (mt/J)	T (mt/ha-yr)
Albia gravelly loam	N.J.	0.04	—
Freehold loamy sand	N.J.	0.11	9
Tifton loamy sand	Ga.	0.13	9
Plainfield sand	Ill.	0.20	11
Zaneis fine sandy loam	Okla.	0.29	9
Cecil sandy loam	Ga.	0.30	7
Boswell fine sandy loam	Tex.	0.33	11
Cecil clay loam	Ga.	0.34	9
Ipava silt loam	Ill.	0.36	11
Cecil sandy loam	S.C.	0.37	7
Honeoye silt loam	N.Y.	0.37	7
Mexico silt loam	Mo.	0.37	7
Austin clay	Tex.	0.38	4
Hagerstown silty clay loam	Pa.	0.41	9
Mansic clay loam	Kans.	0.42	9
Ida silt loam	Iowa	0.43	11
Marshall silt loam	Iowa	0.43	11
Cecil sandy clay loam	Ga.	0.47	7
Fayette silt loam	Wis.	0.50	11
Shelby loam	Mo.	0.54	11
Keene silt loam	Ohio	0.63	9
Dunkirk silt loam	N.Y.	0.91	7

farmland and construction sites (Figure 7-4). Five soil parameters are used to predict soil erodibility using the nomograph: percentage silt plus very fine sand (0.05 to 0.1 mm), percentage sand greater than 0.1 mm, organic matter content, structure, and permeability. For 100 K factors estimated from the nomograph, 68 were within 0.02 of the measured value and 95 were within 0.04 of the measured value. The nomograph appears to be equally accurate for surface and subsurface materials.

TABLE 7-2 Soil erodibility for general soil classes with varying infiltration rates[a]

Soil class	Permeability	K value
Loose sands	Rapid	0.22
Sandy loam soils	Moderate to rapid	0.26
Light colored forest soils	Moderate	0.48
Light colored forest soils	Slow	0.56

[a]General soil classes illustrating the relationship between soil infiltration rates and soil erodibility. Detailed K values are available for specific soils in most states.

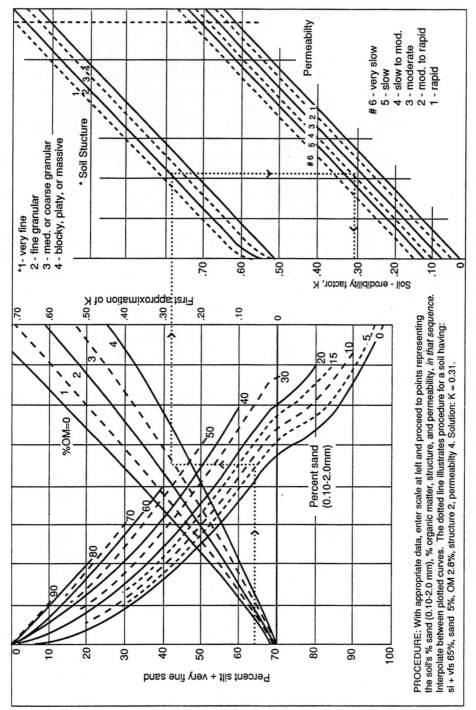

PROCEDURE: With appropriate data, enter scale at left and proceed to points representing the soil's % sand (0.10-2.0 mm), % organic matter, structure, and permeability, *in that sequence.* Interpolate between plotted curves. The dotted line illustrates procedure for a soil having: si + vfs 65%, sand 5%, OM 2.8%, structure 2, permeabilty 4. Solution: K = 0.31.

Figure 7-4 Nomograph for estimating soil erodibility from physical properties.

Length and steepness factor (LS).   The *LS* factor in the USLE is also based on soil loss per unit area relative to a field 22.1 m in length with a uniform 9% slope. This means that a field with these parameters will have a *LS* factor of 1.00 in the equation. Where soil loss is expected to be greater, the *LS* factor will be larger than 1.00, while for those situations where loss would be expected to be lower, the *LS* factor will be less than 1.00. Calculated *LS* values for specific combinations of slope length and steepness are given in Table 7-3.Values for *LS* other than those given in Table 7-3 can be determined by the following formula:

$$LS = (\frac{\text{field slope length}}{22.1})^m (0.065 + 0.045S + 0.0065S^2) \qquad [7\text{-}2]$$

where *S* is the slope length in meters and *m* is 0.5 if the percent slope is 5 or more, 0.4 on slopes of 3.1 to 4.9%, 0.3 on slopes of 1 to 3%, and 0.2 on uniform gradients of less than 1%.

The data in Table 7-3 show, for example, that a 2% slope 25 m long would reduce the predicted soil loss value by a factor of approximately 5, whereas a 16% slope 100 m long would increase the predicted soil loss value by approximately a factor of 5. In general, the longer the slope, the greater the expected loss. However, the increase in soil loss with increased slope length is not linear but increases in an quasi-exponential fashion (see equation [7-2]). For a 6% slope 60 m long the percentage of soil lost in the first 20 m would be 19% of the total, while the percentage of soil lost in the next two 20-m segments would be, respectively, 35

TABLE 7-3   Values for length and steepness of slope (*LS*) for the USLE[a]

Slope	Slope length (m)								
(%)	15	25	50	75	100	150	200	250	300
0.5	0.08	0.09	0.10	0.11	0.12	0.13	0.14	0.14	0.15
1	0.10	0.12	0.15	0.17	0.18	0.21	0.23	0.24	0.25
2	0.16	0.19	0.23	0.26	0.29	0.32	0.35	0.37	0.40
3	0.23	0.27	0.33	0.37	0.41	0.46	0.50	0.54	0.57
4	0.30	0.37	0.48	0.57	0.64	0.75	0.84	0.92	0.99
5	0.37	0.48	0.68	0.84	0.96	1.18	1.36	1.52	1.67
6	0.47	0.60	0.86	1.05	1.21	1.48	1.71	1.91	2.10
8	0.69	0.89	1.26	1.55	1.79	2.19	2.53	2.83	3.10
10	0.96	1.24	1.75	2.15	2.48	3.04	3.50	3.92	4.29
12	1.27	1.64	2.32	2.84	3.28	4.02	4.64	5.18	5.68
14	1.62	2.09	2.96	3.63	4.19	5.13	5.92	6.62	7.25
16	2.02	2.60	3.68	4.52	5.21	6.38	7.37	8.24	9.02
18	2.46	3.17	4.48	5.50	6.34	7.77	8.97	10.03	10.98
20	2.94	3.79	5.36	6.58	7.58	9.29	10.72	11.99	13.13

[a]Based on the equation $LS = (\text{length}/22.1)^m(0.065 + 0.045s + 0.0065s^2)$, where $m = 0.2$ for $s < 1\%$, $m = 0.3$ for $s = 1$ to 3%, $m = 0.4$ for $s = 3.1$ to 4.9%, and $m = 0.05$ for $s \geq 5\%$.

and 46%. Therefore, it becomes obvious that practices which reduce effective slope length, such as the installation of terraces, will reduce expected soil loss. Increases in slope steepness also increase the *LS* factor, but not in a linear fashion. As shown in Table 7-3, for a slope of 15 m in length, increasing the slope steepness from 3 to 6% approximately doubles the *LS* factor, while an increase in steepness from 6 to 12% results in an increased *LS* factor of nearly three times. The shape of the slope is also important. Convex slopes lose slightly more soil than concave slopes of the same length and average gradient. *LS* values can be adjusted for this slope irregularity.

Cover and management factor (*C*).   The *C* factor in the USLE attempts to account for the percentage of surface covered by crops and/or residue during all months of the year, but especially during those months when rainfall exceeds infiltration. For example, prepared seedbeds in the spring can be tilled very smooth or can be left rough prior to planting. Crop residues can be completely removed, left on the surface, incorporated to various degrees, or untouched prior to planting. Crop residues or crop canopies greatly reduce the energy expended by falling raindrops or flowing water on the soil surface. One very important contribution to decreased erosion that deserves some discussion and results in increased *C* factors is conservation tillage.

Conservation or reduced-tillage systems are those that leave crop residues on the surface and/or result in larger soil clods than conventional tillage (Figure 7-5). For example, conventional tillage in the corn belt often includes moldboard plowing, disking, planting, and cultivating. Conservation or reduced tillage may include a chisel plow, disk, sweep, blade, or rod weeder as the primary tillage tool. Stubble mulch tillage is one form of reduced tillage where undercutting tools such as sweeps, blade machines, or rod weeders are used to till the soil but leave the stubble mulch on top. These implements cut or stir the soil with minimal disturbance on the surface. Also, crops are sometimes planted using *reduced-tillage systems*, such as plow-plant, wheel-track-planting, or till-plant. In these systems, land is prepared for seeding with one or two passes across the field. As the names imply, *plow-plant* and *wheel-track-planting* are accomplished by planting immediately after plowing, while the soil is still moist and without additional tillage. *Till-plant* refers to a once-over operation where tillage is performed ahead of the planter by sweeps and/or rotary hoe sections, but both are pulled by the same tractor. By far the most soil conserving of the cropping practices is *no-till*. The crop is planted directly in the residue of the previous crop with no prior disturbance. In no-till systems, weed control must be accomplished by chemical means. No-till systems can result in considerable ground cover, depending on the type and yields of the previous crop. This increased cover can be a disadvantage for wet soils, because they are much slower to dry and to warm up in the spring. Reduced- and no-till planting result in fuel savings for the farmer, but at least part, if not most of that savings is required for increased chemicals to control weeds. Examples of the effect of different crop sequences, tillage practices, and management styles on the *C* factor are shown in

Figure 7-5   Examples of tillage implements and tillage systems. (a) Drawn rigid moldboard plow. (b) Chisel plow. (c) Light disking to prepare seedbed for winter wheat in Palouse region of Washington. (d) Minnesota farmer applying fertilizer, herbicide, and planting in same operation. (e) Zero-till corn planted in wheat stubble. (f) Zero-till soybeans in corn residue. (a and b Courtesy of John Deere Co.; c to e Courtesy of USDA Soil Conservation Service.)

TABLE 7-4 Crop management C-factor values for Central Illinois

	Crop sequences[a]				
	C-Sb	C-C-Sb	C-C	C-Sb-G_x	C-Sb-G-M
Conventional tillage					
Fall plow conventional till (residue left)	0.47	0.44	0.40	0.33	0.18
Fall plow conventional till (residue removed)	0.53	0.51	0.51	0.36	0.20
Plow-plant[b]	0.33	0.30	0.26	0.24	0.12
Conservation tillage					
Minimum-till[c] residue level (kg/ha)					
1100–2200	0.32	0.31	0.31	0.24	0.15
2200–3400	0.22	0.21	0.20	0.17	0.13
3400–4500	0.16	0.16	0.12	0.13	0.11
4500–6700	0.12	0.12	0.07	0.10	0.09
6700+	0.10	0.10	0.04	0.07	0.08
No-till[d] residue level (kg/ha)					
1100–2200	0.29	0.29	0.29	0.15	0.11
2200–3400	0.16	0.16	0.16	0.11	0.08
3400–4500	0.11	0.11	0.11	0.09	0.06
4500–6700	0.06	0.06	0.06	0.08	0.05
6700+	0.04	0.04	0.03	0.07	0.04

[a]C, corn; C-C, continuous corn; Sb, soybeans; G, small grain; G_x, small grain with catch crop; M, meadow.
[b]Crops are planted immediately after plowing in the spring.
[c]Systems such as chisel planting, till planting, or strip till. Level of corn residue or equivalent on the surface after planting.
[d]Systems such as zero-till, no-till, or slot plant.

Tables 7-4 and 7-5. For the various management practices in the tables, note that the residue level or the amount of cover in contact with the ground has a significant effect on soil loss in both tilled soils and in permanent pasture and rangeland. Since management practices are relatively easy to manipulate, the C factor is one of the first factors to be considered, when excess erosion is a problem.

Support practice factor (P). In addition to the type of crop or crop rotation chosen, the amount of residue left on the soil surface, or other considerations already covered by the cropping or C factor, there are also support practices that will slow the runoff of water and reduce the amount of soil it can carry off-site. The three most important are contour tillage, contour plus stripcropping, and terracing, and each will be discussed briefly. The effect of these erosion-reducing practices is accounted for in the USLE by the P factor, which is the ratio of loss with the support practice chosen compared to the loss if the soil were cultivated up and down the slope.

TABLE 7-5   Factor *C* for *Permanent Pasture, Range, and Idle Land*[a]

Vegetative canopy			Cover that contacts the soil surface: ground cover (%)					
Type and height	*Percent cover*[b]	*Type*[c]	*0*	*20*	*40*	*60*	*80*	*95+*
No appreciable		G	0.45	0.20	0.10	0.042	0.013	0.003
canopy		W	0.45	0.24	0.15	0.091	0.043	0.011
Tall weeds or	25	G	0.36	0.17	0.09	0.038	0.013	0.003
short brush		W	0.26	0.20	0.13	0.083	0.041	0.011
with average								
drop fall height	50	G	0.26	0.13	0.07	0.035	0.012	0.003
of 20 in.		W	0.26	0.16	0.11	0.076	0.039	0.011
	75	G	0.17	0.10	0.06	0.032	0.011	0.003
		W	0.17	0.12	0.09	0.068	0.038	0.011
Appreciable brush	25	G	0.40	0.18	0.09	0.040	0.013	0.003
or brushes,		W	0.40	0.22	0.14	0.087	0.042	0.011
with average								
drop fall height	50	G	0.34	0.16	0.08	0.038	0.012	0.003
of 6.5 ft		W	0.34	0.19	0.13	0.082	0.041	0.011
	75	G	0.28	0.14	0.08	0.036	0.012	0.003
		W	0.28	0.17	0.12	0.078	0.040	0.011
Trees, but no	25	G	0.42	0.19	0.10	0.041	0.013	0.003
appreciable low		W	0.42	0.23	0.14	0.089	0.042	0.011
brush; average								
drop fall height	50	G	0.39	0.18	0.09	0.040	0.013	0.003
of 13 ft		W	0.39	0.21	0.14	0.087	0.042	0.011
	75	G	0.36	0.17	0.09	0.039	0.012	0.003
		W	0.36	0.20	0.13	0.084	0.041	0.011

[a]The listed *C* values assume that the vegetation and mulch are randomly distributed over the entire area.

[b]Portion of total-area surface that would be hidden from view by canopy in a vertical projection (a bird's-eye view).

[c]G: Cover at surface is grass, grasslike plants, decaying compacted duff, or litter at least 2 in. deep. W: Cover at surface is mostly broadleaf herbaceous plants (as weeds with little lateral-root network near the surface) or undecayed residues, or both.

Source: After W. H. Wischmeier and D. D. Smith, *Predicting Rainfall Erosion Losses: a Guide to Conservation Planning*, USDA Agricultural Handbook No. 537 (1978).

Figure 7-6   (a) Contour planted early tomatoes San Diego Country, California. (b) Contour strip cropping of corn and alfalfa in northwestern Illinois. (c) No-till planting of corn in soybean residue on broad based terraces. (d) Well-maintained grass waterway. (Courtesy of USDA Soil Conservation Service.)

Contouring.   Contouring (Figure 7-6a) is the practice of operating tillage implements and planting the crop in a direction perpendicular to the slope of the land. Each row serves as a miniterrace to resist the movement of runoff, thus greatly reducing its load-carrying capacity. Contouring works well on slopes between about 3 and 8%. On slopes less than 3%, the *P* factor for contouring approaches 1, meaning that there is no significant benefit to contouring of soils with these slopes. When slopes are in excess of about 8%, the rather small barriers created by the contoured rows do not significantly reduce the rate of runoff compared to tillage up and down the slope, and rills or gullies form very rapidly. The effectiveness of contouring is dependent not only on slope steepness, but also on the length of the slope. A steeper slope must have a shorter maximum length for contouring to be effective. For example, for a parcel of land with a 21% slope, contouring would reduce predicted soil loss by only 10%, even if the slope length were 15 m or less. For soils with this much slope, contouring would

TABLE 7-6   Erosion-control practice (*P*) factor for contouring and contour strip cropping

Slope (%)	Contouring		P value contour strip cropping
	P value	Maximum slope length (m)[a]	
1–2	0.60	120	0.30
3–5	0.50	90	0.25
6–8	0.50	60	0.25
9–12	0.60	35	0.30
13–16	0.70	25	0.35
17–20	0.80	20	0.40
21–25	0.90	15	0.45

[a]Length limit may be increased 25% if residue cover after crop seeding exceeds 50%.

not be an acceptable support practice factor. *P* values, as functions of both slope steepness and slope length, are given in Table 7-6.

Contour Strip Cropping. Alternating strips of small grain or row crop and sod can be planted perpendicular to the slope to increase the effectiveness of contouring (Figure 7-6b). This choice of support practice requires that sod or hay crops be a part of the cropping rotation. The sod or hay crop acts as a receiver of soil moving off the row crop area. By keeping the strips relatively narrow (15 to 30 m), the devastating action of raindrop impact and splash is greatly reduced. While soil may be moved somewhat from the areas planted to row crops or small grains to the hay or grass strips, the loss of soil from the total field is much smaller. The percentage reductions in soil loss for various strip widths and slope lengths attributed to contour strip cropping are given in Table 7-6.

Terracing. A very effective erosion control measure is to reshape the land so as to divide the slope into segments so that water does not flow from one segment to another but is diverted to a channel or tile outlet at the downhill portion of each segment. The length of the slope then becomes the distance between the terraces. Terraces can be built as broadbased (Figure 7-6c), quite wide and usually able to be cultivated, or narrow with steep backslopes planted permanently to a grass crop. Terraces are constructed perpendicular to the slope, thus terracing is almost always combined with contour cropping practices. Terracing requires a large capital input but allows row crop production on steep or long slopes that otherwise could not be cultivated while maintaining tolerable soil losses.

Calculating soil loss using the USLE. Once the appropriate values for the USLE are known, this equation can be used both to calculate the estimated soil loss and to examine the effect of different erosion control practices on soil loss. Two examples using the USLE illustrate how it can estimate the extent of soil loss under present land use practices and how agronomically and environmentally sound management decisions can be achieved.

Example 1

Estimate the soil loss from an Ipava silt loam located in central Illinois. The field in question has a 4% slope 150 m in length that is utilized by a cash grain farmer growing continuous corn. The farmer has incorporated the practice of minimum tillage as a soil conservation practice and has estimated that this leaves the equivalent of approximately 2500 kg/ha corn residue on the soil surface. The appropriate factors for this situation are

$R = 300$ (Figure 7–3)

$K = 0.36$ (Table 7–1), $T = 11$ mt/ha/yr

$LS = 0.75$ (Table 7–3)

$C = 0.20$ (Table 7–4)

$P = 1$ (no support practice is followed)

$A = 300 \times 0.36 \times 0.75 \times 0.20 \times 1 = 16.2$ mt/ha/yr   (see Figure 7–7a)

Note: Erosion expressed in mt/ha/yr can be converted to tons/acre/yr by multiplying by 0.447.

$$16.2 \text{ mt/ha/yr} \times 0.447 = 7.23 \text{ tons/acre/yr}$$

Despite of the fact that minimum tillage is employed in this example, predicted soil erosion exceeds tolerable soil loss and should prompt a management decision to further reduce the loss.

Solution   A logical possibility might be contour farming. This would change the P factor to 0.5 and predicted soil loss would now be

$A = 300 \times 0.36 \times 0.75 \times 0.20 \times 0.5 = 8.1$ mt/ha/yr   (see Figure 7-7b)

Predicted soil loss no longer exceeds the tolerance level, *but* the maximum slope length, where contour farming is acceptable for a 4% slope, is 90 m. So contouring alone is not permissible. The farmer could consider a major investment of terracing and contouring. By constructing a terrace every 50 m, the effective slope length is reduced to 50 m and the LS factor becomes 0.45. The predicted soil loss is now

$A = 300 \times 0.36 \times 0.45 \times 0.20 \times 0.5 = 5.2$ mt/ha/yr   (See Figure 7-7c)

This meets the T value (11 mr/ha/yr) for this soil.

Another consideration would be contour strip cropping with alternate strips of corn and grasses or legumes instead of building terraces. This implies a change in the cropping system from continuous corn to a corn–forage rotation. This practice would be feasible only if the farmer had an economic use for the forages. With this management scenario, the predicted soil loss would be

$$A = 300 \times 0.36 \times 0.75 \times 0.20 \times 0.25 = 4.1 \text{ mt/ha/yr} \qquad \text{(see Figure 7-7d)}$$

Example 2

Next consider the case of an Ida silt loam in the southwest corner of Iowa with a 6% slope and 250 m in length. The land has been in permanent pasture with no appreciable canopy and is essentially 100% covered with grass. The landowner wants to know what the present predicted soil loss is and if this land can be converted to a corn–soybean rotation that would be expected to produce approximately 2500 kg/ha of corn equivalent residue without exceeding tolerance limits. Relevant factors for current soil loss from the Ida soil are $R = 300$, $K = 0.43$, $LS = 1.91$, $C = 0.003$, and $P = 1.0$.

$$A = 300 \times 0.43 \times 1.91 \times .003 \times 1 = 1 \text{ mt/ha}$$

The landowner can rest assured that permanent pasture is very effective in terms of controlling soil erosion.

Solution Converting to a corn–soybean rotation would mean a C factor of

Figure 7-7 Calculated soil loss based on a variety of production practices.

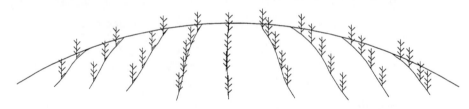

Up and down the hill

$$A = (300 \times 0.36 \times 0.75 \times 0.20 \times 1) = 16.2 \text{ mt/ha/yr}$$

(a)

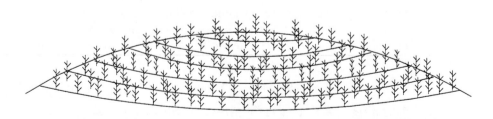

Contour

$$A = (300 \times 0.36 \times 0.75 \times 0.20 \times 0.5) = 8.1 \text{ mt/ha/yr}$$

(b)

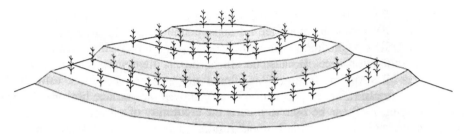

Terrace and contour

A = (300 x 0.36 x 0.45 x 0.20 x 0.5) = 5.2 mt/ha/yr

(c)

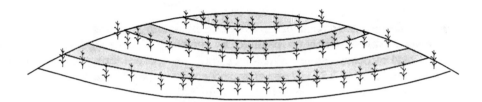

Strip cropping and contour

A = (300 x 0.36 x 0.75 x 0.20 x 0.25) = 4.1 mt/ha/yr

(d)

Figure 7-7   (cont.)

0.47 for conventional tillage, 0.22 for minimum tillage, and 0.16 for no-till. The resulting A values would be 116, 54, and 39 mt/ha, respectively. Additional support practices would be necessary in order that soil loss not exceed tolerance levels. If the landowner were willing to invest substantial capital in terraces with 25-m spacing and the tenant were to farm on the contour and use minimum tillage, soil loss would be within T values as shown:

$$A = 300 \times 0.43 \times 0.60 \times 0.22 \times 0.5 = 8.5 \text{ mt/ha/yr}$$

7.1.3 Erosion from Construction Sites

Erosion is not only the product of agricultural activities but also of urban activities. Large quantities of soil can be lost from bare soils associated with highway and road construction as well as from residential and commercial building sites. The USLE can be adapted to conditions found in urban areas. For example, the USLE can be used to compare potential soil loss from different development plans and to predict the need for sediment basins to prevent sediment movement into streams and reservoirs.

R factors from Figure 7-3 can be used for construction projects lasting several years but should be modified for construction projects involving shorter periods. See the USDA Agricultural Handbook No. 537[3] for the procedures to correct R factors for short-term projects. Because the soil surface is often unprotected during construction, the soil erodibility factor (K) is often of greater importance for construction sites than for cropland. Since construction often removes the surface horizons, the nomograph (Figure 7-4) can be used to predict the erodibility of subsoil materials. Some subsoil materials are substantially more erodible than the original surface horizons, while others are less erodible. The planner can usually obtain a detailed description of the subsoil materials from published soil survey data allowing K values to be calculated for these materials. Soil losses from subsoil materials, if exposed on similar slopes, would be directly proportional to the subsoil K values. Information on subsoil K values not only shows the depth of cut that would result in the least soil erosion, but also indicates whether return of stockpiled topsoil on exposed subsoil would be desirable for erosion control on a particular site.

Development planning could reduce erosion by exposing subsoil materials with lower K values whenever possible, by shortening slope lengths or creating convex instead of concave slopes. Erosion can also be reduced through the use of anchored mulches. Use of 1.5 tons/acre of an anchored straw or hay mulch on a 6 to 10% slope 200 ft long can reduce the C factor in the USLE from 1.0 to 0.06, a 94% reduction in erosion. Wood chips and crushed stone mulches can also be used to reduce erosion; these are especially useful on highly sloped areas. Seven tons of wood chips applied on a 16 to 20% slope 50 ft long will reduce the C factor from 1.0 to 0.02[4].

7.1.4 Control of Gully Erosion

As already mentioned, the USLE does not apply to soil loss from gully erosion. Gullies are rills that have become too large to be crossed and smoothed by normal farming equipment or practices. Gullies carry large amounts of water and soil from a watershed and if unchecked, rapidly become large ravines. Where gullies are not too large, they can be plowed or graded in with appropriate machinery, so vegetation can be established. Successful revegetation is much more likely to succeed where a diversion dam or ditch can be constructed to redirect the flow of water away from the gully. Where slopes are too steep or erosion damage too great, the gully can be stabilized by building stairstep dams or by creating temporary check dams using logs, brush, or rock to slow the flow of water and prevent further erosion. For extreme gully damage, permanent check dams of concrete or other material may be necessary, but these are very expensive and from an agricultural standpoint, uneconomical. By far the best solution is to incorporate soil management practices that will prevent the formation of gullies in the first place.

[3]*Predicting Rainfall Erosion Losses: A Guide to Conservation Planning*, USDA Agriculture Handbook No 537, 1978, p. 44.
[4]Ibid.

7.2 WIND EROSION

Wind erosion is a natural phenomenon and was the basis for the formation of loess deposits (see Chapter 3) which formed some of the most productive soils in the world. Today, however, wind erosion is a serious problem in many parts of the United States and the world. It has been estimated that 2 million hectares of land are moderately to severely damaged each year in the United States and that wind erosion is a dominant problem on 30 million hectares of land.

Conditions conducive for wind erosion are (1) loose, dry, finely divided soil; (2) absence of vegetation and a smooth soil surface; (3) large contiguous areas of soil subject to wind erosion; and (4) persistent, strong winds (Figure 7-8). The dustbowl days of the 1930s are a prime example of conducive conditions in the southwestern United States that resulted in huge dust clouds being carried to the

Figure 7-8 (a) Severe dust caused by wind erosion, visibility was so poor that motorist used headlights, even though it was midday. (b) Soil drifted into road ditches from strong winds. (c) Blowouts near Imperial, Nebraska, caused by a combination of cattle traffic, overgrazing, and wind erosion. (d) Improper tillage resulted in wind blown sand and silt filling a concrete irrigation ditch. (Courtesy of USDA Soil Conservation Service.)

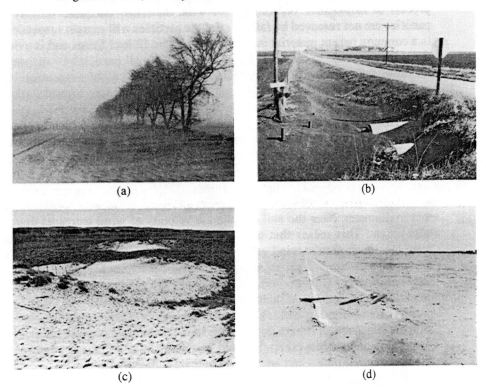

(a)

(b)

(c)

(d)

Atlantic Ocean. While wind erosion is usually considered to be a major problem mainly in regions of low rainfall, it also has detrimental consequences in semiarid and even humid regions like the midwest, where blowing soil reduces visibility on highways, reduces seedling survival, fills road ditches, and covers lawns. Where wind erosion is significant, it removes the most fertile portion of the soil. The very fine material removed by wind may also contain the highest concentrations of herbicides, insecticides, or other pesticides or agricultural chemicals, which can result in pollution.

7.2.1 Mechanics of Wind Erosion

Like water erosion, wind erosion involves the basic processes of detachment, transportation, and deposition. Detachment of soil particles with wind takes place as a result of the direct energy of the wind or as a result of energy expended when soil particles are struck by other particles already in motion as a result of wind. These forces are in contrast to the detachment forces of the falling raindrop in water erosion.

Once detached from the soil surface, the particles are transported by wind in three distinct ways, depending primarily on the size of the soil particles. Particles 0.05 mm or less in diameter, once detached from the soil surface, are moved by the process of *suspension*. As long as the wind maintains adequate velocity and the particles are not removed by falling rain, the particles will remain suspended. Loess is a common parent material in many parts of the United States and is evidence of the ability of wind to transport large quantities of soil particles by suspension. As the wind looses velocity, the particles tend to settle out of suspension, with the smallest-diameter particles settling out last and hence being carried farthest from the source.

The second transportation process is that of *saltation*, which applies to soil particles 0.05 to 0.5 mm in diameter. Saltation results in temporary suspension by the wind with the particles soon falling to the soil surface again. As they fall, they may dislodge other particles of similar size or may "bounce" again into the air. Thus, by a series of short bounces, particles of this size class are transported along the surface. Wind velocities alone are not sufficient to suspend particles up to 0.5 mm in diameter. Near the soil surface the velocity of wind increases dramatically with height. This means that as a soil particle is spinning along the soil surface, there is a much greater wind velocity above the particle than below it, resulting in decreased pressure or a partial vacuum above the particle. This phenomenon, called the *Venturi effect* (Figure 7-9), aids in lifting the soil particles. Under many soil conditions, the process of saltation can account for one-half to three-fourths of particles transported by wind. The abrasive energy of saltated particles, upon impact with the soil surface, results in detachment and movement of other particles by both suspension and surface creep. Saltated particles can also damage young plants; for example, cotton plants can be "shot holed" by saltated particles.

Soil particles larger than 0.5 mm but smaller than about 1 mm can also be

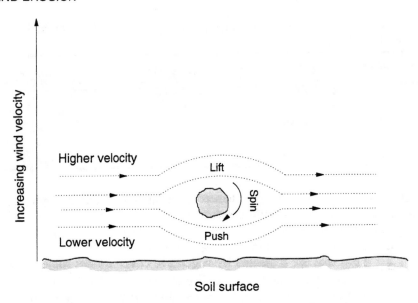

Figure 7-9 Venturi effect of the wind near the soil surface as it moves a spinning sand grain.

transported by wind or by the abrasive action of other wind-suspended or wind-moved particles by an active bumping process. This process, termed *surface creep*, generally involves fairly short distances. If the particles are significantly larger than about 1 mm, the energy directly from wind or even from collisions with other suspended or temporarily suspended particles is not great enough to result in wind transportation. If enough large particles or aggregates are present to result in considerable surface coverage, wind erosion will be kept in check. The "desert pavement," often found in the southwestern United States, is an example of coarse particles protecting the remaining soil from wind erosion.

7.2.2 Wind Factors

The magnitude of the effects of wind erosion on soils is determined by both wind factors and soil factors.

Wind velocity. The velocity of the wind is an important factor in the amount of wind erosion. Wind velocities are usually measured at a standard height of 10 m above the soil surface. The relation between wind velocities at 10 m and the effect of such winds on soil erosion is given in Table 7-7. Wind velocities less than 12 km/h (7.45 mph) have little or no soil-erosive effects. Velocities of 12 to 20 km/h result in some erosion hazard to dry organic soils (Histosols) because of their very low bulk densities. Some dust is visible and slight soil erosion hazard exists at wind velocities of 20 to 40 km/h. Winds of 40 to 75 km/h represent a significant

TABLE 7-7   Wind velocities measured at a height of 10 m and their relation to wind erosion

Description of wind	Velocity (km/h)	Wind erosion hazard
Calm to light breeze	<1.5–12	None
Gentle breeze	12–20	Some for Histosols
Moderate to fresh breeze	20–40	Slight for mineral soil
Strong breeze to fresh gale	40–75	Considerable
Strong gale to hurricane	75–120+	Severe

soil erosion hazard, and the hazard is considered severe at winds above 75 km/h. Table 7-7 lists the potential erosion hazards of different wind velocities, while the actual amount of erosion experienced is dependent on many other factors, including soil properties and the amount of vegetative cover.

Soils that are protected by vegetative cover are not subject to wind erosion, even at very high wind velocities. For example, wind velocity approaches zero at the soil surface under a corn crop (Figure 7-10). The erosive power of wind is related more to frictional velocity than to speed of the air movement. Frictional velocity is influenced by the velocity profile and by the shape and composition of the soil surface. Both wind velocity and frictional velocity are reduced as the soil surface is approached and as nonerodible clods or vegetation are encountered.

Wind turbulence.   Wind turbulence is also a factor in suspension and movement of soil particles. Wind turbulence or mixing is greater over a rough surface than a smooth one and is increased by temperature differentials above the soil surface. In general, turbulence is greater closer to the soil surface than higher in the main windstream. Increased wind turbulence results in greater suspension and movement of soil particles.

Wind gustiness.   Winds are not constant forces that very gradually build up or slow down, but rather, winds fluctuate broadly and frequently. Soil surfaces that are temporarily stabilized under one wind condition are, once again, eroded when a large gust of wind strikes them. Research has shown that constant winds would result in much less soil erosion than gusty winds because the soil surface eventually becomes covered with larger less-erodible particles. The gusty nature of wind tends to stir up the surface, moving these larger particles and exposing smaller, more erodible soil particles.

Predominant wind direction.   Just as changes in wind velocity or gustiness can result in increased soil erosion, so can changes in wind direction increase soil erosion. If wind is blowing from the same general direction, a somewhat stabilized surface can result. As wind direction changes, a temporarily stabilized surface will start to erode again. For each set of conditions, such as surface cover,

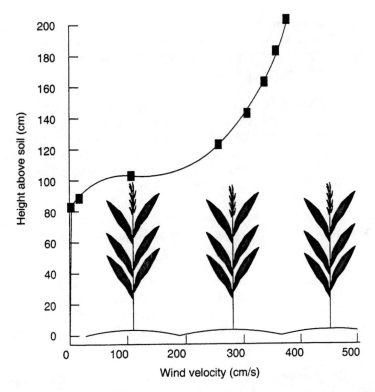

Figure 7-10 Wind speed with height above the soil surface as influenced by the canopy of a corn crop.

roughness and wind direction, there is a threshold wind velocity below which a soil particle will not move. However, a change in one of these factors, such as wind direction, can change the threshold velocity. If the predominant wind direction is predictable for a given region, cropping practices such as stripcropping perpendicular to the prevailing wind direction can significantly lower soil losses from wind erosion.

7.2.3 Soil Factors

Aggregate size. Soil particles larger than 1 mm in diameter are nonerodible by wind. The size does not apply only to individual particles, but can also refer to clods or soil aggregates. Thus a well-aggregated soil will not be as susceptible to severe wind erosion as a soil with poor surface structure. Not only are large aggregates, clods, or particles less subject to erosion, but they also serve as a temporary surface barrier and protect finer particles beneath them. Aggregate formation and stability is a function of soil properties, such as texture, organic matter content, exchangeable cations, and free calcium carbonate. Approximately 27% clay is optimal for aggregation. Soils with lower clay contents lack binding ability, and soils with higher clay contents are more subject to weathering. Exchangeable cations (Ca^{2+}, Mg^{2+}, etc.), higher organic matter contents, and free calcium carbonate all foster stronger soil structure.

Surface geometry. Tillage operations can shape the soil surface into ridges, reducing the effect of wind. Ridges serve as mini wind barriers, reducing wind velocity near the ground, especially between the ridges, provided that they are not parallel to the wind direction. The depression between ridges also serves as a natural "trap" for particles that might otherwise be subject to saltation or soil creep.

Soils with short steep slopes or knolls in a repeating fashion also influence soil erosion. With a slope length of less than 200 m and a slope steepness of 5 to 10%, the wind speed at the crests of the slopes will be much greater than in the valleys between slopes. Erosion can be 50 to 100 percent greater on the crests of such slopes than farther down the slopes.

Surface water content. Moist soil is not subject to wind erosion. Thus timely rainfall can effectively prevent wind erosion of soils. However, it should be noted that wind (convective energy) is very effective in drying a thin layer of soil, which is then subject to wind erosion. Moisture indirectly protects soils from erosion by encouraging plant cover.

Crop cover. The most effective way to reduce wind erosion is to cover the soil. Plant cover, either living or dead, serves to dramatically reduce, if not eliminate, wind erosion. The percentage of cover, geometry of cover, and stability of the cover are all important factors in reducing wind erosion. In general, standing residues are more effective than chopped, flat residues.

7.2.4 Prediction and Control of Wind Erosion

The amount of soil loss by wind erosion is a function of five factors and is expressed by the following equation:

$$E = f(I',K',C',L',V)$$

where E = predicted annual soil loss (mt/ha)

I' = soil–erodibility factor (mt/ha)

K' = soil–ridge–roughness factor

C' = climatic factor

L' = width–of–field factor

V = vegetative–cover factor

In contrast to the USLE for water erosion, the equation for calculating wind erosion is not a simple straightforward calculation. The calculation of soil loss by wind erosion (E) is a five-step process. Each step involves obtaining a separate E_s value by a calculation or tabular interpolation, where s is the number of the step.

Soil-erodibility factor. (I') The soil-erodibility factor represents the potential annual soil loss due to wind erosion and is based on empirical measurements from wind tunnel experiments. The major variable affecting potential erodibility is the percentage of the surface particles that are of nonerodible size, that is, particles greater than 0.84 mm in diamter. A graphic representation of the effect of the percentage of nonerodible clods on the value for I' is given in Figure 7-11. As shown in the figure, the relationship is exponential, resulting in very large values of I' when the percentage of nonerodible clods is less than 5% and very small I' values when the percentage of nonerodible clods exceeds 80%. The potential erosion for a soil with 10% nonerodible clods is 300 mt/ha/yr. Estimates of the I' factor can also be obtained from soil texture and other soil properties and are given in Table 7-8. This table illustrates that potential wind erosion is nullified if the soil remains wet.

The soil-erodibility value must be modified when dealing with rolling topography. Wind is more erosive over the crests of knolls than over level land. In this

Figure 7-11 Soil erodibility (I') in mt/ha-yr as related to percentage of nonerodible clods (>0.84 mm in diameter).

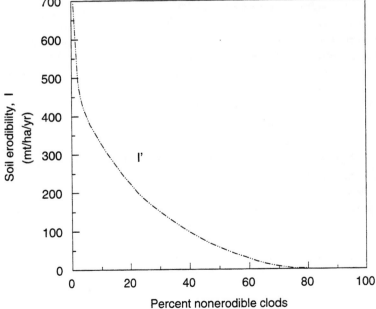

TABLE 7-8   Relationship of soil characteristics to the soil-erodibility factor (I')

Texture and other characteristics	*Dry soil aggregates >0.84 mm (%)*	*Soil erodibility factor, I' (mt/ha-yr)*
Very fine, fine, and medium sands; dune sands	1–7	694–356
Loamy sands; loamy fine sands	10	300
Very fine sandy loams; fine sandy loams; sandy loams	25	190
Clays; silty clays; noncalcareous clay loams, silty clay loams with >35% clay	25	190
Calcareous loams, silt loams; noncalcareous clay loams and silty clay loams with <35% clay	25	190
Noncalcareous loams and silt loams with <20% clay; sandy clay loams; sandy clays	40	126
Noncalcareous loams and silt loams with >20% clay; noncalcareous clay loams with <35% clay	45	108
Silts; noncalcareous silty clay loams with <25% clay	50	84
Very wet or stony soils, usually not erodible	—	—

Source: USDA Soil Conservation Service.

case, the I' value is multiplied by I_s, which represents the increased (percent) amount of wind erosion at the crest or windward side of the knoll. The relationship between I_s and slope is given in Figure 7-12. For example, as shown in Figure 7-12, potential wind erosion from the crest of a knoll with a 6% slope would be approximately 50% more than the potential erosion on level land.

$$Step\ 1: \quad E_1 = I' \times I_s \qquad\qquad [7\text{-}3]$$

Soil-ridge-roughness factor (K').   While soil texture and vegetative cover dramatically affect potential erosion, the soil-ridge-roughness factor is an expression of the effect of tillage and planting equipment in creating nonerodible ridges, very coarse material, that reduce wind erosion. The numerical value for K' is derived from a ridge roughness equivalent (K_r) based on the height of the ridge and the spacing between ridges. Note that ridge spacing is measured from the crest of one ridge to the crest of the other ridge. The ridge roughness equivalent is given by

$$K_r = \text{measured ridge height} \times \frac{R_m}{R_s} \qquad\qquad [7\text{-}4]$$

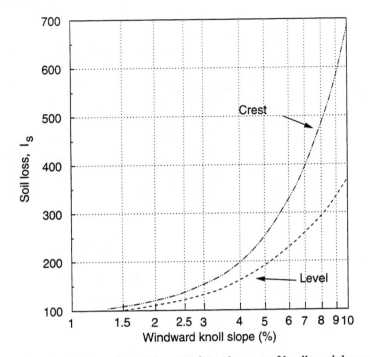

Figure 7-12 Potential soil loss (I_s) from the crest of knolls and the upper third or level portion of the windward slope. For slopes less than 150 m long. I_s expressed as a percentage of I' for level ground. [After *Soil Sci. Soc. Amer. Proc.* 29:602–608 (1965).]

where R_m is the field measured ratio of ridge height to ridge spacing and R_s is the standard ratio (1/4) of ridge height to ridge spacing. Thus for a tillage operation that left ridges 5 cm in height with a spacing of 20 cm, $R_m = 5/20 = 1/4$ and the ratio R_m/R_s = (1/4)/(1/4) = 1, and hence the resultant K_r would be 5 cm. This soil-ridge-roughness equivalent is converted to a unitless K' value, as shown in Figure 7-13. Remember that the K' factor has a value of 1; that is, it will not reduce the potential wind erosion unless ridges are formed with mechanical equipment.

$$Step\ 2:\quad E_2 = E_1 \times K' \qquad\qquad [7\text{-}5]$$

Climatic factor. The climatic factor used for calculating wind erosion is a function of wind velocity and rainfall and is given by the following equation:

$$C' = 34\frac{\mu^3}{(P{-}E)^2} \qquad\qquad [7\text{-}6]$$

where μ(km/h) is wind velocity and P-E is a moisture index. Both high winds and low rainfall will contribute to greater C' factors; hence the C' values for the desert southwest are higher than for the Pacific Northwest. Equation [7-6] emphasizes that

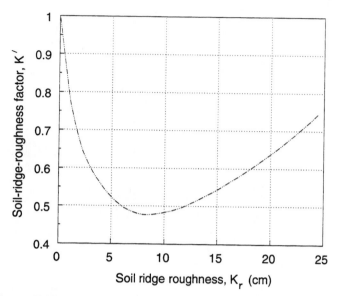

Figure 7-13 Relationship of equivalent soil-ridge-roughness in centimeters, K_r, to the soil ridge roughness factor, K'. [After *Soil Sci. Soc. Amer. Proc.* 29:602–608, (1965).]

the amount of wind erosion is a function of the cube of the wind velocity. It is based on the fact that the detaching power of wind is a function of the square of wind velocity, while the ability of wind to transport detached soil particles is a function of the fifth power of wind velocity. The moisture index (P-E) is the sum over 12 months of a function of monthly precipitation (p) and monthly temperature (T) using the following equation:

$$P\text{-}E = 115 \sum \frac{p}{T - 10} \qquad [7\text{-}7]$$

The minimal values for temperature and precipitation are -2°C (28.4°F) and 13 mm (0.5 in.) of rainfall regardless of the weather statistics for an individual local. Climatic values will vary by location and time of the year. Climatic isopleths for various parts of the United States for the month of April are given in Figure 7-14. C' values calculated using the equations above or obtained from a local soil conservation office are used in step 3 of the prediction process as follows:

$$Step\ 3: E_3 = E_2 \times C' \qquad [7\text{-}8]$$

Width-of-field factor. The width-of-field factor L' is an adjustment based on the relative unsheltered distance downwind from the prevailing wind direction. The actual unsheltered field width (D) is multiplied by a factor K_{50} to obtain an equivalent field width. The factor K_{50} is a function of both the weighted or preponderance of erosion forces in the prevailing wind direction and the degrees of

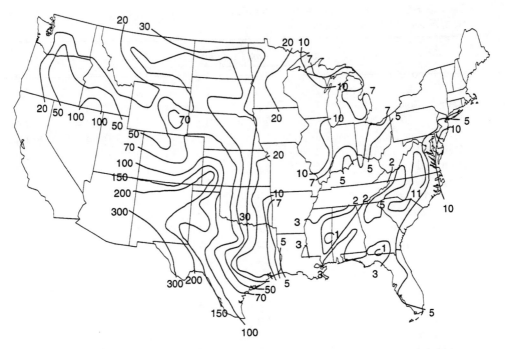

Figure 7-14 Wind climatic factors, C', for April. Note that the C' factor will vary from month to month. (After E. L. Skidmore and N.P. Woodruff, in *Wind Erosion Forces in the United States and Their Use in Prediction Soil Loss*. USDA Agricultural Handbook No. 346, 1968.)

deviation of prevailing wind direction from perpendicular to the field in question. Values for preponderance of erosion forces are obtained by subjecting to vector analysis the cube of the mean wind speed and a duration factor for all wind speed groups. A value for preponderance, R_p, of 1.0 means that there is no prevailing wind-erosion direction. Where R_p equals 2.0 there is a prevailing wind-erosion direction, and erosion forces parallel to it are twice as great as those perpendicular to it. R_p values for selected months for 25 locations in the United States are given in Table 7-9. The data in this table indicate that R_p values vary considerably from month to month and from location to location. Such values can be obtained from local soil conservation offices. The degrees of deviation of prevailing wind direction from perpendicular to the field in question can be used along with R_p to obtain a value for K_{50} using Table 7-10. To satisfy the calculations of wind erosion for a given situation, a value for L' (width of field) is needed. The value for L' is given by the equation

$$L' = D \times K_{50} - 10B \qquad [7\text{-}9]$$

where D is the actual measured field width, K_{50} is obtained as described above, and

TABLE 7-9 Preponderance of wind erosive forces in prevailing wind erosion direction, R_p, for selected months

Location	Feb.	Apr.	Jun.	Aug.	Oct.	Dec.
San Diego, Cal.	1.5	1.3	1.2	1.1	1.4	1.3
Stockton, Cal.	2.2	2.0	2.4	2.9	3.7	2.3
Denver, Colo.	1.1	1.2	1.2	1.3	1.1	1.3
Hartford, Conn.	1.8	2.0	1.6	1.5	1.4	1.6
Wilmington, Del.	2.0	1.3	1.2	1.3	1.1	1.4
Daytona Beach, Fla.	1.2	1.3	1.7	1.7	2.0	1.3
Athens, Ga.	1.4	1.4	1.3	1.5	1.8	1.4
Boise, Idaho	4.9	3.1	2.7	1.7	2.3	3.0
Peoria, Ill.	1.4	1.3	1.3	1.5	1.5	1.2
South Bend, Ind.	1.2	1.1	1.3	1.3	1.2	1.2
Des Moines, Iowa	1.1	1.6	1.1	1.2	1.5	1.3
Topeka, Kan.	1.5	1.4	1.7	1.2	2.2	1.6
Battle Creek, Mich.	1.4	1.4	1.5	1.3	1.2	1.5
Minneapolis, Minn.	1.7	1.5	1.3	1.2	1.6	1.4
Great Falls, Mont.	3.3	1.6	1.6	1.5	2.1	3.6
Lincoln, Neb.	1.5	1.6	1.6	1.5	1.7	2.1
Las Vegas, Nev.	1.4	1.9	2.6	2.0	1.9	1.8
New York, N.Y.	1.9	1.5	1.4	1.2	1.1	1.7
Fargo, N.D.	2.0	1.7	1.2	1.6	1.9	2.2
Cleveland, Ohio	1.2	1.3	1.4	1.4	1.4	1.2
Tulsa, Okla.	2.8	2.2	2.7	1.9	2.7	2.1
Austin, Tex.	1.8	1.8	1.9	1.9	2.1	1.8
Spokane, Wash.	2.8	3.1	2.1	1.6	3.4	3.5
Madison, Wis.	1.3	1.5	1.3	1.3	1.1	1.2
Cheyenne, Wyo.	1.7	1.5	1.3	1.2	1.7	2.3

TABLE 7-10 Table to obtain multiplier (K_{50}) used to calculate the median travel distance of erosive forces across a field from the degrees of deviation of prevailing wind direction from perpendicular to the field in question (A) and the preponderance of erosion forces in the prevailing direction

R_p	K_{50} value at A (deg):				
	0	10	20	30	40
1.0	1.90	1.90	1.90	1.90	1.90
1.2	1.55	1.62	1.69	1.77	1.84
1.5	1.33	1.42	1.50	1.60	1.77
2.0	1.18	1.29	1.35	1.04	1.68
2.5	1.13	1.23	1.31	1.40	1.67
3.0	1.10	1.19	1.30	1.44	1.77
3.5	1.06	1.17	1.32	1.51	1.95
4.0	1.04	1.16	1.36	1.63	2.23

B is 10 times the height of any barriers, such as trees or hedges, that shelter the field in question.

Step 4: A soil loss estimate (E_4) can be approximated using the values for E_2 and E_3 and L' as shown in Table 7-11.

Vegetative factor. The final factor needed to predict the amount of soil loss from wind erosion is the vegetative factor (V), which accounts for the protec-

TABLE 7-11 Selected values for soil loss, $E_4 = F(I', K', C', L')$ for specific values of $E_2 = I'K'$, $E_3 = I'K'C'$, and the unsheltered distance across a field[a]

		E_4 value at unsheltered distance L (m):						
E_2	E_3	20	40	80	200	400	800	2000
100	10	0	0	1	4	6	8	9
100	40	0	7	14	23	29	34	38
100	75	4	20	34	48	58	65	72
100	200	36	83	116	147	166	181	195
100	400	114	206	263	315	346	371	392
100	800	314	481	578	665	714	753	788
200	10	0	1	3	6	8	9	10
200	40	6	13	20	29	35	38	40
200	75	19	32	44	58	67	72	75
200	200	80	112	139	167	184	195	200
200	400	201	257	301	348	375	392	399
200	800	472	568	642	717	760	788	799
400	10	1	3	6	9	10	10	10
400	40	14	22	29	37	39	40	40
400	75	34	47	58	70	74	75	75
400	200	116	144	168	191	198	200	200
400	400	263	310	349	385	397	400	400
400	800	579	655	719	777	796	799	800
600	10	3	6	8	10	10	10	10
600	40	21	29	36	40	40	40	40
600	75	46	58	68	74	75	75	75
600	200	142	168	187	199	200	200	200
600	400	306	349	380	398	400	400	400
600	800	650	718	768	797	799	800	800

[a]Values other than those given can be interpolated or specific values obtained using equations given by R. P. C. Morgan, *Soil Erosion and Conservation*, Longman Scientific and Technical/John Wiley & Sons, New York (1986), or by a nomogram method given in *Soil Sci. Soc. Amer. Proc.* 29:602–608 (1965).

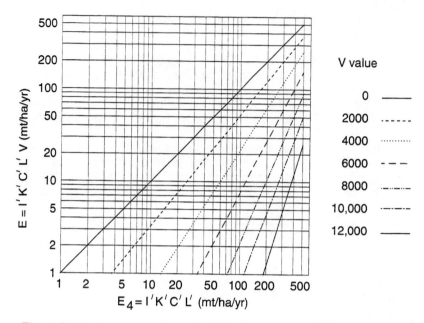

Figure 7-15 Graph to convert values of E_4 and V to predicted rates of wind erosion (E). The E value is determined by entering the figure at the E_4 value on the bottom scale, following upward to the V value and then horizontally across to the final E value. [After *Soil Sci. Soc. Amer. Proc.* 29:602–608, (1965).]

tion from surface cover by dry matter. The V factor is entered into calculations by converting the vegetative matter present, live or residual, to the equivalent of flattened wheat straw. Table 7-12 can be used to convert small grain in the seedling stages, small grain stubble, or sorghum stubble, to equivalent amounts of flattened wheat straw (V). Figure 7-15 can then be used to convert E_4 values previously obtained to E (predicted rates of soil erosion in metric tons/ha/yr).

Step 5: The final predicted value for soil loss by wind erosion can be obtained by using E_4 and the vegetative factor, V, using the relationship shown in Figure 7-15.

The final E value is determined by entering the figure at the E_4 value on the bottom scale, following upward to the V value, and then across to the E value.

The effect of the amount of vegetative cover on the final predicted soil loss is easily seen. For example, with an E_4 of 20 and barren soil, the predicted soil loss would be 20 mt/ha/yr, while an equivalent vegetative cover of 4000 mt/ha would result in a predicted soil loss of only 2 mt/ha/yr.

Example: Calculating soil loss from wind erosion

Consider a Mollisol near Denver, Colorado. This soil is found on level terrain with no knolly features and the field in question is 473 m long in the east-west

TABLE 7-12   Equivalent amounts of flattened wheat straw (*V*) from various vegetative residues or cover

Vegetative cover	Weight (kg/ha)	Flattened straw equivalent, V
Living or dead small grain in stooling stages (smooth soil surface)	200	850
	400	2,400
	600	3,700
	800	5,600
	1,000	8,100
	1,200	11,000
	1,400	14,000
Small grain stubble, standing	500	3,600
	1,000	8,400
	1,500	15,200
	2,000	21,600
	2,500	28,400
	3,000	35,200
Sorghum stubble flat	1,000	600
	3,000	2,800
	5,000	6,400
	7,000	10,400
Sorghum stubble standing 20 cm tall	1,000	800
	3,000	4,000
	5,000	8,400
	7,000	13,600
30 cm tall	1,000	1,100
	3,000	5,000
	5,000	10,800
	7,000	17,400
50 cm tall	1,000	2,200
	3,000	8,400
	5,000	17,600
	7,000	27,400

direction and prevailing winds are at 110°. The field in April contains standing small grain stubble with the equivalent weight of 500 kg/ha. The ridge roughness equivalent from planting and cultivating is 10 cm. An analysis of the soil, by sieving, revealed 20% nonerodible clods. In this example we calculate the annual erosion loss based on values for the month of April. In practice, it would be necessary to make the same calculations with data specific for each month of the year and give each month appropriate weight.

Solution

Step 1: $E_1 = I' = I \cdot I_s$, $I = 220$ (see Figure 7-11)

$I_s = 1$ (no knolly features); therefore,

$E_1 = 220 \times 1 = 220$ mt/ha/yr

Step 2: $E_2 = E_1 \times K'$, $K' = 0.49$ (see Figure 7-13)

$= 220 \times 0.49 = 108$ mt/ha/yr

Step 3. $E_3 = E_2 \times C'$, $C' = 70\%$ (see Figure 7-14)

$= 108 \times \dfrac{70}{100} = 76$ mt/ha/yr

Step 4: E_4 is a function of E_2, E_3, and L' $L' = K_{50} D$

$R_p = 1.2$ (Table 7-9)   and $K_{50} = 1.69$ (Table 7-10)

$L' = 1.69 \times 473 = 800$

Using the information above and Table 7-11, we find that $E_4 = 65$

Step 5:   Using the value for E_4 (step 4) and a value for $V = 3600$ (Table 7-12), the value for soil loss $E = I' \, K' \, C' \, L' \, V = 15$ mt/ha/yr (Figure 7-15).

Figure 7-16   Windbreaks to protect a sandy soil. Windbreak in foreground is to protect town road from drifting snow. (Courtesy of USDA Soil Conservation Service.)

Thus soil loss for this situation based on data for April would be 15 mt/ha/yr. Similar calculations would be made for each month and averaged to obtain the expected soil loss for 1 year. The effect of cover is dramatic. For example, if the cover in this example were small, grain stubble with the equivalent of 750 kg/ha, the resulting V would be 6000 (Table 7-12) and E would be approximately 3.5 mt/ha/yr.

Windbreaks are established by planting trees to serve as barriers to wind movement. Windbreaks alone or in combination with cover crops or conservation tillage systems offer the most efficient method of controlling wind erosion. Windbreaks slow wind velocities and shorten the effective length of the field that wind can act upon. Windbreaks offer additional environmental advantages in that they provide shelter and nesting habitat for wildlife. Windbreaks also serve as snow fences to prevent snow from closing rural roads. Note the large windbreak next to the road in Figure 7-16. The purpose of this windbreak is to prevent snow from drifting and closing the adjacent road in the winter.

7.3 LAND CAPABILITY CLASSES

The erodibility of soil is one of the factors that determines the intensity of land utilization. Other factors that determine long-term land use are drainage, rock outcrops, restricted climatic factors, physical conditions, and chemical status of the soils. The Soil Conservation Service has established a system of land capability classes that are numbered from I (the least limited) to VIII (land with severely restricted use). Land uses allowed for each of the eight classes are shown in Figure 7-17, and a brief description for the soil classes (Figure 7-18) follows.

Class I lands have few limitations that would restrict their use. They are generally deep, well-drained soils with little or no erosion hazard and they can be continuously cropped. They are usually naturally fertile and high yields are maintained by standard fertilizer and liming practices. The water-holding capacity is high and the soils receive adequate rainfall or are irrigated.

Class II lands may be used for the same crops as class I soils but not quite as intensively. Moderate erosion hazards may require conservation practices such as terracing, strip cropping, or grassed waterways. Soils may not be as deep as class I soils or may have less than ideal soil structure. Soils can be classified as class II soils because of slight chemical restrictions of alkalinity or salinity or they may exhibit somewhat restricted drainage.

Class III lands are restricted to moderate use as cultivated lands. Soils with this classification have severe limitations such as moderately steep slopes resulting in erosion hazards, very slow water permeability, restricted root zones resulting in shallow soils, chemical limitations such as moderated alkalinity or salinity, or unstable soil structure. Class III land benefits from special management practices

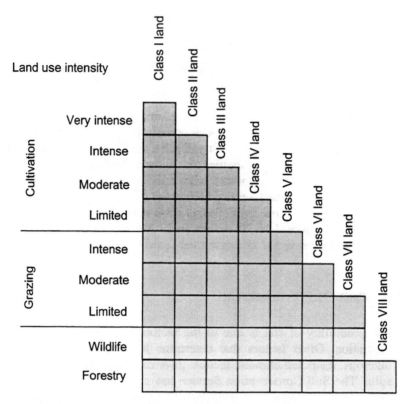

Figure 7-17 Land use capability classes indicating suitable land use intensities.

such as restricted choice of crops, special conservation practices, tile drainage, and more grasses and legumes in the rotation.

Class IV lands can be cultivated only with very restricted cropping choices and careful management practices. The most limiting factors for class IV lands are severe erosion potential (steep slopes), severe erosion damage from past mismanagement, very shallow soils, low water-holding capacity, poor drainage, or severe chemical problems such as alkalinity or salinity. As with the other land classes, one or more of the restrictions mentioned may result in soil being classified as class IV.

Class V lands are not suitable for cultivation. They are also unique in that erosion is not a hazard for class V lands. Soils are limited to class V lands because of outcropping of large stones, frequent stream overflow, ponding, or climatic restrictions, such as a short growing season. Usually, pastures on class V land can be improved with proper management practices.

Class VI lands can accommodate moderate grazing or can be used for wildlife

Figure 7-18 Photograph showing land use capability classes. Class I land is being used for row crop production. The class II, III, and some IV land is being used to produce grapes. Other class IV land and VI land is in permanent pasture. The class VII and VIII land is covered with natural tree and brush and serves as a watershed for the agricultural land. (Courtesy of USDA Soil Conservation Service.)

habitat or forestry purposes. The limitations on class VI lands are similar to those of class IV lands but are more severe.

Class VII lands are restricted to permanent pasture, range, woodland, or wildlife habitat. The physical limitations arc similar to class VI lands, but pasture improvement is not feasible.

Class VIII lands are best used for recreation, permanent wildlife habitat, water supplies, or simply aesthetic purposes. They cannot support commercial plant life. These would include sandy beaches and active floodplains (river wash), severely restricted wetlands, and very dense rock outcropping.

Within the land capability classes described, there are capability subclasses that designate the specific limitation for the soil described. The subclasses are designated by the letters e (erosion hazard), w (water in or on the soil restricting plant growth or cultivation), s (shallow, droughty, or stony soils), and c (climatic limitations of low rainfall or cold temperatures). A soil with a moderate drainage limitation, for example, would be classified as IIw.

Approximately 248 million hectares (57%) of the land in the United States is suitable for at some intensity of cultivation (classes I to IV). The remainder of the land (classes V to VIII) has severe restrictions and can be used only for permanent pasture, grazing, or for forests and wildlife purposes. Land capability classifications for individual soils are provided in the county soil survey reports available from the USDA Soil Conservation Service.

STUDY QUESTIONS

1. What are some of the costs to society of water erosion?
2. What are the mechanisms by which no-till or minimum-till farming reduces soil loss caused by water erosion?
3. What is the relative contribution of length of slope to soil erosion as percent slope increases from <2 to as much as 20%?
4. What is meant by *tolerable soil loss*? What factors are considered when a value of tolerance (*T*) is determined?
5. What are the three major types of soil movement caused by wind?
6. How does soil texture affect wind erosion?
7. What is meant by *conservation tillage*, and how does it affect wind and water erosion?
8. What are the two most important climatic factors affecting wind erosion, and how are they related to wind erosion?
9. What makes the land classification capability of class V lands unique?
10. Approximately how much of the land area in the United States is suitable for at least limited cultivation?
11. What are the limitations on the use of the USLE?
12. When considering the *K* factor in the USLE, what is the influence of organic matter content, drainage, and texture?
13. Which factors does a grower have the greatest control over in the USLE?

8

Properties of Soil Colloids

Colloidal particles are particles with effective diameters less than or equal to 1μm (10^{-6} m). These particles exhibit unique properties associated with their small sizes and subsequent large surface area/mass ratios. Figure 8-1 illustrates the dramatic increase in surface area of a constant mass of material, as the material is subdivided into smaller and smaller spheres. One major consequence of the high surface areas of colloidal particles is that many important soil processes occur on colloidal surfaces. The adsorption of water, inorganic cations and anions, and organic molecules occurs primarily on the surface of colloidal particles. Other soil fractions are involved in these type of reactions, but because of their low surface-area-to-mass, ratios their contributions are minor or negligible. The colloidal fraction of the

Figure 8-1 Relationship between total surface area of a given mass of material and particle size.

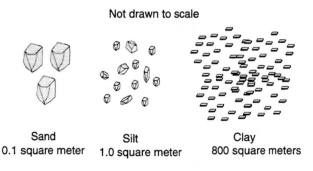

Not drawn to scale

Sand	Silt	Clay
0.1 square meter	1.0 square meter	800 square meters

soil is dominated by phyllosilicate minerals, humic substances, and in more highly weathered soils, by the hydrous oxides of iron, aluminum, and manganese.

8.1 CLAY MINERALS

The structure of the clay minerals (phyllosilicates) and their relationship to other silicate minerals was discussed in Chapter 3. As was pointed out in that chapter, clay minerals are negatively charged due to isomorphous substitution that occurred during mineral formation and/or due to dissociation of weak-acid groups at surface irregularities or mineral edges. The effect of this negative charge on soil processes is basically dependent upon the ability of the clay minerals to swell, allowing internal (interlayer) surfaces of the clay to interact with the soil solution.

8.1.1 Swelling of Clay Minerals

Swelling results when water enters between the layers of a clay mineral, causing the layers to move apart a few additional nanometers. This results in an increased volume of a given mass of clay (see Figure 8-2). For swelling to occur, water must be able to enter between two adjacent layers into the interlayer. Swelling means that

Figure 8-2   Swelling of a 2:1 clay mineral.

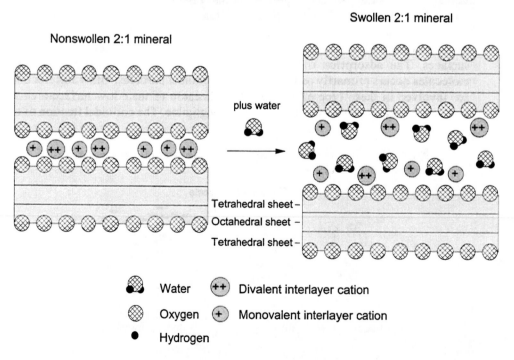

the interlayer is available for processes such as cation exchange, water adsorption, and other colloidal phenomena. Swelling also means that the clay will expose a much greater surface area/unit weight to the soil solution and hence be potentially more chemically reactive. Compare the potential surface area of a textbook when it is glued tightly shut to when it is opened page by page. Swelling is dependent on the type of clay (1:1, 2:1, etc.), on the unit-layer charge of the clay, and on the nature of the interlayer cation.

1:1 minerals. Adjoining layers of these minerals are oriented in such a manner that the bottom of the octahedral sheet of one layer is above the top of a tetrahedral sheet of the adjacent layer. This results in a surface of hydroxyls of one 1:1 layer being adjacent to a surface of oxygens in the next 1:1 layer, as shown in Figure 8-3.

Oxygen is a very electrophilic element. In covalent bonds with elements such as hydrogen, the electrons are pulled closer to the oxygen, resulting in a slight residual negative charge on the oxygen and a slight residual positive charge on the hydrogen. A similar unequal sharing of electrons between silicon and oxygen also occurs. This creates a layer of positive charges due to the surface of hydroxyls of the octahedral sheet over a layer of negative charges due to the oxygen surface of the tetrahedral sheet of the underlying 1:1 layer. The interaction of these slightly electron-rich and electron-poor layers results in the formation of hydrogen bonds between the adjacent hydroxyls and oxygens. While a single hydrogen bond is weak, the collective effect of the formation of many hydrogen bonds is that 1:1 layers are strongly bonded together. The bonding is strong enough to prevent the

Figure 8-3 Hydrogen bonding of adjacent 1:1 layers.

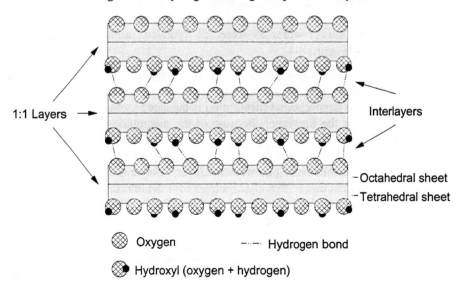

entry of water between the layers so that 1:1 minerals are nonswelling minerals. This bonding between the 1:1 layers can be likened to gluing the pages of a book together, making it virtually impossible to access the individual pages.

2:1 minerals.   Unlike the 1:1 minerals, the bottom of a 2:1 layer is composed of oxygens and the top of the adjacent 2:1 is also a surface of oxygens (Figure 8-2). This prevents the formation of multiple hydrogen bonds as in the 1:1 case. In *unsubstituted* 2:1 minerals, the adjacent layers are bound together by weak van der Waals bonds. In *substituted* 2:1 minerals, adjacent layers are bound together by their mutual attraction for interlayer cations in addition to the weak van der Waals bonding.

Swelling in 2:1 minerals is dependent on the bonding of the two adjacent layers to each other. For the unsubstituted 2:1 layers the bonding of adjacent layers is weak. There is, however, no real driving force causing water to enter into the interlayer. The unsubstituted silicon–oxygen surface is basically hydrophobic (water repelling) in nature. Because there are no interlayer cations subject to hydration, the only hydrophilic (water-loving) sites in the interlayer of unsubstituted 2:1 minerals are associated with >SiOH sites due to crystal irregularities.

For substituted 2:1 minerals, the affinity of one 2:1 layer for adjacent 2:1 layers is dependent on the degree of attraction of the two negatively charged adjacent layers to mutual interlayer cations. The degree of bonding is a function of both the amount of isomorphous substitution and the size of the hydrated interlayer cations.

If the affinity of the layers for the interlayer cations is strong, water is prevented from entering into the interlayer, hydrating the interlayer cations, and bonding to other hydrophilic sites. If the affinity of the adjacent layers for the interlayer cations is weaker, the mineral swells because water can enter between the layers, increasing the hydration of the interlayer cations and wetting other hydrophilic sites. The main hydrophilic entities in the interlayer are interlayer cations that attract water as water of hydration and >SiOH sites at crystal irregularities (Figure 8-4).

Micas.   The micas have a high unit-layer charge ($X \approx 2$) due to the great amount of isomorphous substitution that occurred during mineral formation. The negative charge due to isomorphous substitution is balanced by the presence of cations such as K^+ or Ca^{2+} in the interlayer. Because of the magnitude of the unit-layer charge, the interlayer cations are held very tightly by the two adjacent 2:1 layers. Due to the strength of attraction of the two adjacent layers for the same interlayer cations, water is prevented from entering between the layers. Hence micas are nonswelling minerals. The strong attraction of the adjacent layers for the interlayer cations also results in the interlayer cations being nonexchangeable (fixed). The interlayer cations are only released as the mineral is altered by weathering.

Illites and Vermiculites.   The unit-layer charge of these minerals is lower (X = 1.0 to 1.5) than for the micas. Because of the lower charge, these minerals will only fix certain-size cations, thus holding the layers tightly together and preventing

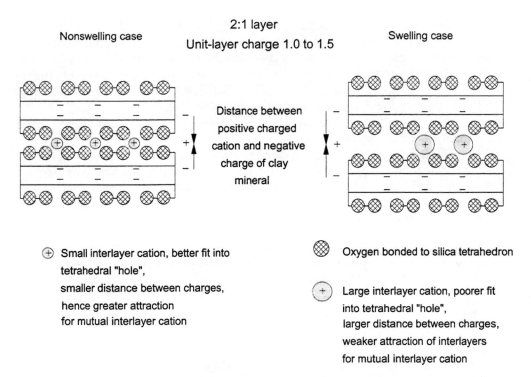

Nonswelling case

2:1 layer
Unit-layer charge 1.0 to 1.5

Swelling case

Distance between
positive charged
cation and negative
charge of clay
mineral

⊕ Small interlayer cation, better fit into
tetrahedral "hole",
smaller distance between charges,
hence greater attraction
for mutual interlayer cation

⊗ Oxygen bonded to silica tetrahedron

(+) Large interlayer cation, poorer fit
into tetrahedral "hole",
larger distance between charges,
weaker attraction of interlayers
for mutual interlayer cation

Figure 8-4   Mutual attraction of 2:1 layers for interlayer cations.

swelling. Cations such as K^+ and NH_4^+, due to their smaller hydrated sizes, are thought to have a favorable fit into the "holes" of the tetrahedral sheets. The holes are a result of the ring pattern of the tetrahedrons in the tetrahedral sheets (see Figure 3-10). A favorable fit allows the cations to approach closer to the source of the negative charge. Since the strength of attraction between charges increases as the distance between the charges decreases, this results in a much stronger attraction of the adjacent layers for the interlayer cations and prevents swelling.

When the dominant interlayer cations are Ca^{2+} or Mg^{2+} or some other cation which, due to the size of the hydrated cation, is too large to fit into the "hole," the distance between the layers is increased, the attraction for interlayer cations is lower, and the mineral will swell. This means that not only can water enter the interlayer, but also that the interlayer cations are exchangeable.

Illite, as found in soils, tends to have K^+ as its major interlayer cation and therefore occurs as a nonswelling mineral. Vermiculite, on the other hand, is formed under conditions where Ca^{2+} and Mg^{2+} are more abundant and thus tends to have Ca^{2+} or Mg^{2+} as dominant interlayer cations. Therefore, vermiculite occurs as a swelling mineral. If the interlayer cations of a naturally occurring vermiculite were replaced with K^+ and the mineral allowed to dry collapsing the interlayers, the mineral would not swell upon rewetting.

Smectites. These minerals have a low unit-layer charge ($X = 0.5$ to 0.9). The strength of attraction for the interlayer cations is less than for the illites, vermiculites, or micas. Because of this lower charge, montmorillonite and other smectites have very low capacities to "fix" cations, even cations with favorable fits, in the interlayer "holes." Hence water can enter between the layers and cause swelling. In addition, all interlayer cations are exchangeable.

Note: Smectites at the higher end of unit-layer charge may exhibit some capacity to fix K^+ or NH_4^+ ions.

Table 8-1 compares some of the properties of three major clay minerals, a 1:1 mineral (kaolinite), a swelling 2:1 type (montmorillonite), and a nonswelling 2:1 type mineral (illite). The specific surface area and cation-exchange capacity, as well as other physical properties of the minerals, are primarily functions of particle size, charge, and the ability of the clay minerals to swell.

TABLE 8-1 Comparative properties of clay minerals

Property	Montmorillonite	Illite	Kaolinite
Size (μm)	0.01–1.0	0.1–2.0	0.1–5.0
Shape	Irregular flakes	Irregular flakes	Hexagonal crystals
Total surface area (m²/g)	700–800	100–200	5–20
External surface area	High	Medium	Low
Internal surface area	Very high	Low to none	None
Plasticity	High	Medium	Low
Cohesiveness	High	Medium	Low
Swelling capacity	High	Low to none	Low
CEC	80–100	15–25	3–15
Unit-layer charge	0.5–0.9	1.0–1.5	0

8.2 ORGANIC COLLOIDS

Soil organic matter is a collection of plant, animal, and microbial residues in various stages of decomposition. Humus is the more or less stable fraction of soil organic matter. Humus is a collection of complex organic compounds of microbial and higher plant origin that are resistant to decay, and/or compounds protected from decay due to adsorption onto mineral surfaces. The humus content of a soil is a function of the soil formation factors. Jenny[1] found that for loamy soils in the United States the effects of soil formation factors on humus content were in the

[1]H. Jenny, *A Study on the Influence of Climate upon Nitrogen and Organic Matter Content of Soil*, Missouri Agr. Exp. Sta. Res. Bull. No. 52, University of Missouri, Columbia, Mo., 1930; see also H. Jenny, The nitrogen content of soil as related to the precipitation–evaporation ratio, *Soil Sci. 29:193–206 (1930)*.

order climate > vegetation > topography = parent material > age. Soil organic matter is generally considered to be in equilibrium or in a steady state with the environment. This implies that new humus is produced at approximately the same rate as old humus is decomposed. Anthropogenic activities such as soil drainage (i.e., improved soil aeration), fertilization, liming, tillage practices, and erosion can dramatically alter the equilibrium humus content of a soil.

Humus is composed of two major classes of compounds. The first class is nonhumic substances, consisting of well-defined groups of compounds such as amino acids, lipids, and carbohydrates. The second major class, the humic substances, is a series of high-molecular-weight amorphous compounds that are brown to black in color and that were formed by secondary synthesis reactions associated with the decay processes.

Humic substances can be separated into several components based on the methods used to extract the material from the soil.

Humic acid. Dark-colored amorphous organic materials that can be extracted from the soil by a variety of reagents, such as strong bases, neutral salts, organic chelates, or organic solvents and that are *insoluble* in dilute acid are called humic acid. This implies that humic acid contains primarily acidic functional groups, such as phenolic or carboxylic groups.

Fulvic acid. The colored organic materials that are extracted with humic acid but remain in solution upon acidification with dilute acid are called fulvic acid. This implies that fulvic acid contains acidic functional groups since it is soluble in strong bases and extracted with humic acid and that fulvic acid also contains basic groups since it remains in solution upon acidification.

Humin. The strong base insoluble fraction of humus is called humin.

Humic and fulvic acids are the two most widely studied humic substances. Humic acid is composed of molecules with molecular weights in the range 20,000 to 1,360,000, while fulvic acid is composed of molecules with molecular weights in the range of 275 to 2110. The elemental composition of the two substances is given in Table 8-2. Both humic and fulvic acids contain a variety of functional

TABLE 8-2 Composition of humic and fulvic acids (percent)

Element	Humic acid	Fulvic acid
C	50–60	40–50
O	30–35	44–50
H	4–6	4–6
N	2–6	<2–6
S	0–2	0–2

Source: F. J. Stevenson, *Humus Chemistry: Genesis, Composition, Reactions,* Wiley-Interscience, New York, 1982, p. 222.

groups. Carboxyl, phenolic, enolic, hydroxyquinone, lactone, ether, and alcoholic hydroxyl groups have been identified on different humic and fulvic acids. In general, fulvic acids are considered to be decay products of higher plant and microbial residues, while humic acids are considered to be polymerization or condensation products of fulvic acids and other decay products.

Humic and fulvic acids are considered to be hydrophilic colloids. As such, they have a high affinity for water and are solvated in aqueous solutions. These colloids are electrically charged, usually with a net negative charge due to the dissociation of carboxyl and phenolic functional groups (Figure 8-5). The resulting negative charge is neutralized by cations such as Ca^{2+} and Mg^{2+}. These cations are generally considered to be a part of the reservoir of exchangeable cations in the soil. Humic acids are viewed as being coiled long-chain molecules, which are cross-linked from one portion of the coil to another. The cross-linking is probably due to the bonding of hydrophobic (i.e., hydrocarbon) portions of the molecule to other hydrophobic sites, with the hydrophilic sites (polar functional groups) oriented out into the soil solution or toward mineral surfaces (Figure 8-6).

Humic substances are active in many soil processes. They contribute to soil

Figure 8-5   Dissociation of pH-dependent functional groups.

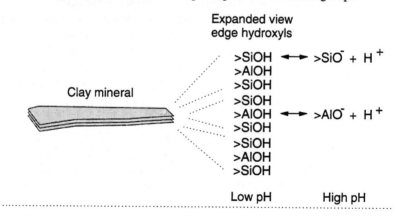

Edge hydroxyls of clay minerals and functional groups of humic substances dissociate in response to shift in soil pH

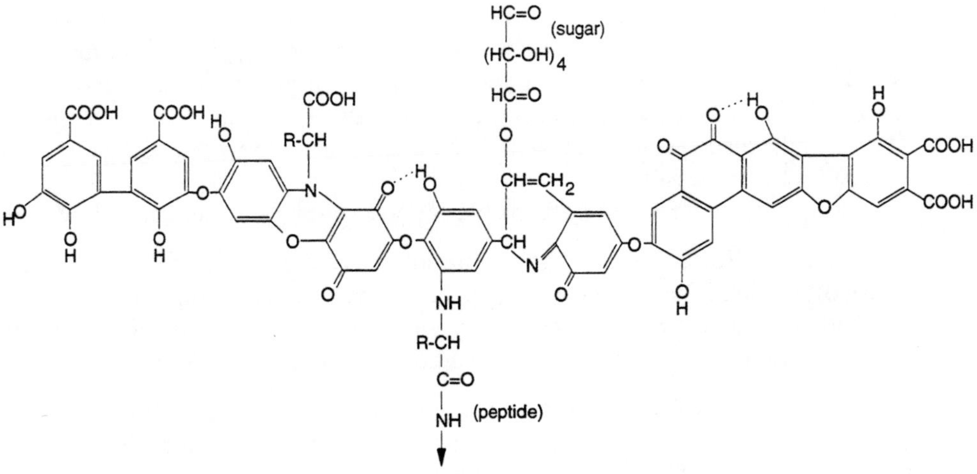

Figure 8-6 Diagrammatic structure of humic acid. (After F. J. Stevenson, *Humus Chemistry: Genesis, Composition, Reactions*, Wiley- Interscience, New York, 1982, p. 259.)

cation and anion exchange, complex or chelate many metal ions, serve as important pH buffers, and promote the formation of soil horizons and structural units through their cementing action on soil particles. Humic substances coat, or partially coat, mineral particles, often protecting the coated particles from the extremes of weathering. Many of these important properties of humus are discussed in other sections.

8.3 ADSORPTION

Adsorption results when a solution component is concentrated at an interface, that is, at a boundary between two phases. For soil-water systems we are primarily interested in the solid–solution interface. Adsorption occurs when the attractive forces between the solid soil surface (adsorbent) and the solution component (adsorbate, for example a dissolved herbicide) overcomes both the attractive forces between the solution component (solute or dissolved herbicide) and the soil solution (solvent), as well as any repulsive forces between the soil surface and the adsorbing species (herbicide). That is, the net adsorbate–adsorbent interaction overcomes the net solute–solvent interaction. Note that *solute* refers to the adsorbing species in solution, *adsorbate* to the adsorbing species upon adsorption on the surface. Since colloids present a much greater surface area than do other size fractions, most adsorption occurs on colloidal surfaces.

Adsorption is important because how a chemical species is distributed between the soil solution and colloidal surface is often the major factor controlling the movement and bioavailability of the chemical. Nutrient availability, as well as correct application rates of herbicides and other chemicals, is often a function of the degree of adsorption. The potential contamination of groundwater is much greater for weakly adsorbed soil amendments and pollutants than it is for more strongly adsorbed species. Cation and anion exchange are two important examples of adsorption processes.

A variety of forces are responsible for adsorption, as described below.

van der Waals Forces. These are attractive forces that arise from the in-phase oscillation of electron clouds of adjacent molecules or atoms. These in-phase oscillations induce small dipoles of opposite sign in neighboring atoms, resulting in a weak attraction of the neighboring atoms or molecules for each other.

Coulombic or Electrostatic Attraction. These are electrostatic forces resulting from the attraction of ions to oppositely charged surfaces or ions. The attraction of cations to the negatively charged surfaces of clay minerals is an example.

Charge Transfer. This is the formation of a donor–acceptor complex between an electron-donor molecule and an electron-acceptor molecule. It involves the partial overlap of molecular orbitals and a partial exchange of electron densities. Hydrogen and π bonds are examples of charge transfer forces.

Dipole-Dipole and Dipole-Induced Dipole. These are the attractions of permanent dipoles for other dipoles, or for molecules that can be polarized to produce a dipole. Note that a dipole is a molecule that contains both a positive and a negative charge separated by a finite distance. This is often the result of unequal sharing of electrons: for example, water molecules.

Hydrophobic. These forces are related to the exclusion of the molecule from the solvent rather than to forces related to the bonding the molecule to the surface. Hydrophobic forces are related to the destruction of cavities in water that nonpolar organic molecules occupy and to the destruction of water structure surrounding poorly solvated nonpolar organic molecules (Figure 8-7). Hydrophobic forces can be compared to other adsorption forces using the following analogy. Consider two strong men both holding onto a bag of gold. Both men, one the solvent, the other the sorbing surface, pull on the bag; the one that is the strongest will get the gold. If the solvent has greater attraction for the solute (bag of gold), no adsorption occurs. If the sorbing surface has greater attraction for the solute, adsorption occurs. Now consider the same two men (solvent and surface) holding onto a skunk. Neither of these men want the skunk, so both are pushing it away toward the other person. If the surface is the strongest, the solute (skunk) stays in solution. If the solvent is the strongest, the skunk (solute) is pushed onto the surface and adsorption occurs.

Adsorption that occurs when there is a strong attraction both of the solvent and surface for the solute is driven by differences in bonding energies (enthalpy) of the solute for the solvent and the surface. Adsorption that occurs due to little affinity

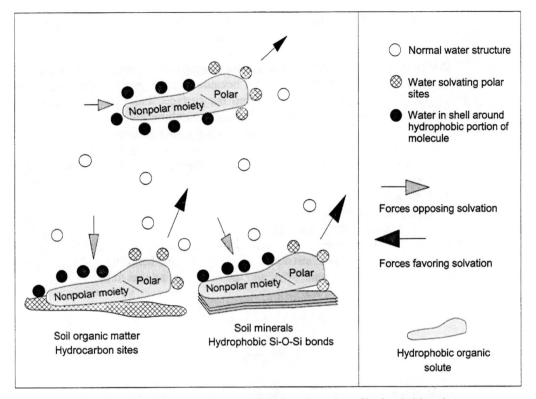

Figure 8-7   Forces responsible for adsorption of hydrophobic solutes.

for the solvent (both pushing) is driven by increases in entropy due to the removal of the solute from solution.

8.3.1 Cation Exchange

Cation exchange is the reversible, low-energy transfer of ions between solid and liquid phases. Cation exchange affects many soil processes, including:

1. Weathering of soil minerals
2. Nutrient absorption by plants
3. Leaching of electrolytes
4. Buffering of soil pH

Cation exchange is the result of the neutralization of the negative charge on soil colloids by cations. The cations are held to the colloidal surface by Coulombic attraction with van der Waals forces and induced dipoles increasing the strength of bonding of some types of cations. Along with the attraction to, and the subsequent concentration of, cations at the colloidal surface is the repulsion of anions, which are negatively charged, like the colloidal surface.

Conceptual model of the cation-exchange process. Cations are not held rigidly on the colloidal surface, but because of their thermal energies, have some degree of motion on and away from the surface. This movement near the surface defines a hemisphere of motion for each combination of ion and colloidal surface. Consider two cases: the first case, a tightly held nonexchangeable cation, and the second case, a less tightly held exchangeable cation (Figure 8-8).

What is cation exchange? Cation exchange occurs (Figure 8-9a) when ions in the soil solution, because of their thermal energies, move into the hemisphere of motion of a cation on the surface at a time when the exchangeable cation is far from the surface. The ion initially in solution becomes "trapped" within the hemisphere of motion on the surface by the negative charge, and the ion initially on the surface moves into the soil solution. If an ion that is initially associated with the surface is close to the charged colloidal surface when a solute ion moves randomly into the hemisphere of motion, ion exchange does not occur (Figure 8-9b) and the solute ion returns to the solution.

Figure 8-8   Fixed versus exchangeable cations.

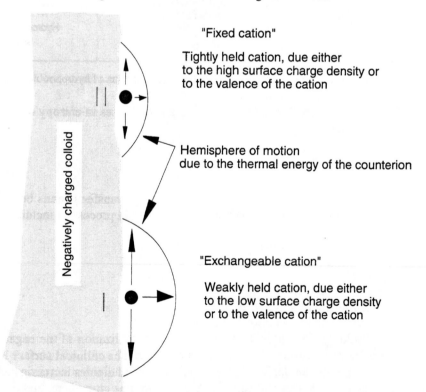

"Fixed cation"

Tightly held cation, due either to the high surface charge density or to the valence of the cation

Hemisphere of motion due to the thermal energy of the counterion

"Exchangeable cation"

Weakly held cation, due either to the low surface charge density or to the valence of the cation

Negatively charged colloid

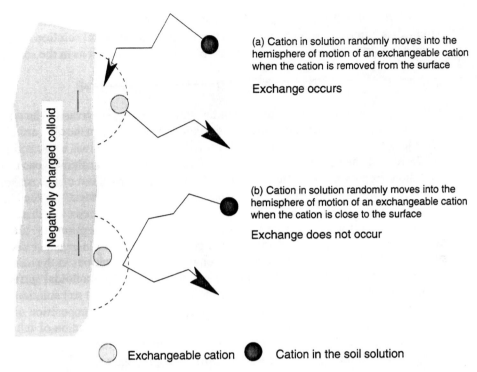

(a) Cation in solution randomly moves into the hemisphere of motion of an exchangeable cation when the cation is removed from the surface

Exchange occurs

(b) Cation in solution randomly moves into the hemisphere of motion of an exchangeable cation when the cation is close to the surface

Exchange does not occur

Negatively charged colloid

○ Exchangeable cation   ● Cation in the soil solution

Figure 8-9 Example of cation exchange between cations in the soil solution and on the colloidal surface.

Many factors affect the distribution of cations between the soil solution and the colloidal surface. An intuitive sense for the factors affecting cation exchange can be developed using the conceptual model. First, as the concentration of a particular cation in the soil solution increases, the probability of that type of cation exchanging with the surface cation increases. As the concentration of a cation in solution increases, there is a corresponding increase in the ratio of that cation to other cations on the colloidal surface (exchange sites). Second, as the valence of an exchangeable cation increases, so does the affinity of the cation for the surface resulting in a smaller hemisphere of motion. The higher the valence of an adsorbed cation, the lower the exchangeability of the ion, and the cation's concentration on the colloidal surface increases relative to cations of lower valence.

Other factors, such as the hydrated size of the cation and the density of charge on the colloidal surface, also affect the degree of attraction of the cation to the colloidal surface. The hydrated size of the cation determines how closely the cation can approach the negatively charged surface, while the density of charge on the surface affects the strength of attraction of the cation. Both factors affect the hemisphere of motion of the cation and therefore its exchangeability.

Mass-action concept of cation exchange. Cation exchange can be treated as a mass-action chemical reaction; cations in the soil solution compete for exchange sites on the soil colloids based on their concentrations in the soil solution.

$$2Na\text{-clay} + Ca^{2+}(aq) \leftrightarrow Ca\text{-clay} + 2Na^+(aq) \qquad [8\text{-}1]$$

When the concentration of $Ca^{2+}(aq)$ in the soil solution is increased, the reaction is shifted to the right, increasing the concentration of calcium on the clay and releasing Na^+ ions to the soil solution. If the concentration of calcium ions in the soil solution is decreased, then the reaction will shift to the left releasing calcium ions to the soil solution. Equation [8-1] illustrates that the ionic composition of the soil solution and the colloids are coupled and cannot be varied independently. When the concentration of a cation in the soil solution is increased by processes such as fertilization, or decreased by processes such as nutrient absorption by plants, the concentration of the cation on the exchange sites must also increase or decrease.

Analysis of the soil shows that the concentration of cations in the soil solution is low compared to the concentration of cations held on the colloidal surfaces. The actual ratio of total exchangeable cations to total cations in the soil solution depends primarily on the exchange capacity of the soil. The ionic composition of the soil solution is reasonably constant from soil to soil, with the exception of salt-affected soils. The amount of exchangeable cations is a function of the amount and type of clay as well as the humus (organic colloid) content of the soil.

Percentage Base Saturation. The concentration of a cation in the soil solution is controlled by the ratio of the concentration of that cation to the concentration of all other cations on the exchange complex. For example, the concentration of Ca^{2+} in the soil solution of neutral and acid soils is a function of the ratio of the concentration Ca^{2+} to the concentration of all other types of cations on the exchange complex. It also holds that the concentration of H^+ in the soil solution is a function of the ratio of H^+ on the exchange complex to the concentration of all other types of cations on the exchange complex.

Note: H^+ ions can be produced by the hydrolysis of water by Al^{3+}. This most commonly occurs in soils with pH values below 5.5 (see Chapter 9). Hence in soils with pH values below 5.5 the concentration of H is a function of both exchangeable H^+ and exchangeable Al^{3+}, while in soils with pH values above 5.5 the concentration of H^+ in the soil solution is only a function of exchangeable H^+. *Because of this* H^+ *and* Al^{3+} *cations are known as acidic cations, whereas other cations, such as* Ca^{2+}, Mg^{2+}, K^+, Na^+, *or* NH_4^+, *are known as basic cations or bases.*

The concentration of H^+ ion in the soil solution can be shown to be a direct function of the percentage of the CEC of a soil that is occupied by acidic cations or an inverse function of the percentage of the CEC of a soil that is occupied by basic cations. The latter is known as *percentage base saturation*:

$$\% \text{ base saturation} = \frac{\text{concentration of basic cations}}{\text{CEC}} \times 100 \qquad [8\text{-}2]$$

Both the concentration of basic exchangeable cations and the CEC are expressed in cmol of positive charge per kilogram of clay or soil ($cmol_c$ kg^{-1}). The importance of percent base saturation is that soils with similar types of colloids will have comparable pH values if they have corresponding percent base saturations (see Chapter 9).

This concept can be expanded and generalized. The concentration of any cation in the soil solution that is not controlled by the solubility of some soil mineral will be a function of the percentage of the CEC occupied by that cation. For example, the concentration of K^+ in the soil solution and its availability to plants is a function of the percentage of the soil's CEC occupied by K^+ ions. In fact, most soil tests used to estimate the availability of K^+ measure exchangeable K^+.

Electrical double layer.   As discussed earlier, the negative charge of the soil colloids is balanced by cations (counterions) attracted to the colloidal surface. The negatively charged surface of the colloid, the counterions and associated water, and even a few anions, are collectively known as the *electrical double layer* (Figure 8-10). The electrical double layer extends for a distance from the colloidal surface into the soil solution and represents the volume of water and ions that is controlled by colloidal phenomena.

The volume or size of the double layer affects many important soil properties. The flocculation or dispersion of soil colloids in sodic and saline-sodic soils is controlled by the size of the double layer. The permeability of soil horizons to air and water is impaired by the dispersion of soil colloids. To explain these phenomena it is necessary to understand how the double layer is influenced by ion valence and by the concentration of ions in the soil solution.

Short-range van der Waals forces are considered to be responsible for the bonding of colloidal particles to each other. When two colloidal particles approach closely enough for these short-range forces to function, the particles bond together, exceeding colloidal size limits, flocculate, and settle out of suspension. For these short-range forces to be effective, it is necessary for the colloids to be close together. As colloidal particles approach each other, their double layers begin to overlap. This overlapping of the double layers results in a repulsive force due to the repulsion of the negative charges of the colloids (Figure 8-11). If the repulsive forces become too large before the particles are close enough together for the van der Waals forces to function, the colloid particles separate and remain in a dispersed state.

If the sizes of the double layers surrounding colloidal particles are large, it is impossible for the particles to approach closely enough together for van der Waals forces to function and the particles will remain dispersed. If the sizes of the double layers are smaller, the particles can approach closely enough for the attractive forces to function and the colloidal particles will flocculate. The size of a colloidal particle's double layer is primarily a function of the valences of the counterions and the total concentration of ions in the soil solution.

Effect of Valence.   The valence of the counterions in the double layer has a major effect on the size of the double layer and hence whether the colloids will be

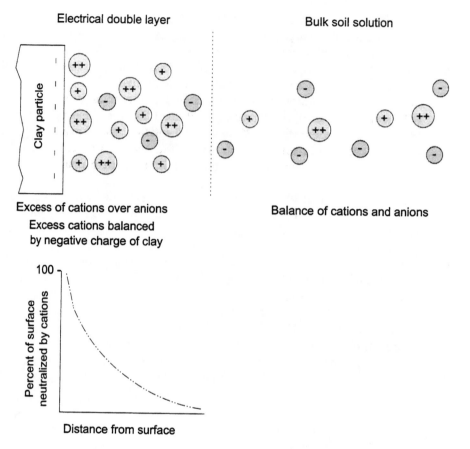

Figure 8-10 Electrical double layer (Gouy–Chapman model).

dispersed or flocculated. The colloidal particles have a certain magnitude of negative charge that must be neutralized to maintain electrical neutrality. If this charge is balanced by monovalent cations, it takes twice as many ions as it would if it were balanced by divalent cations and three times as many cations as it would if it were balanced by trivalent cations.

In "normal" soils Ca^{2+} is the major exchangeable cation with smaller amounts of Mg^{2+}, K^+, Na^+, and NH_4^+. When this is the case the double layers of the colloids are compressed and there is little or no tendency for the colloidal particles to be dispersed, that is, to be pushed apart by their double layers. If some process, such as irrigation or groundwater flow, removes Ca^{2+} and other divalent cations from the exchange complex and replaces them with monovalent cations such as Na^+, the double layers will be expanded. When the double layer contains about 15% exchangeable Na^+, the double layers will have expanded enough to push apart the flocculated colloids, causing them to disperse. Sodic soils are soils that contain enough exchan-

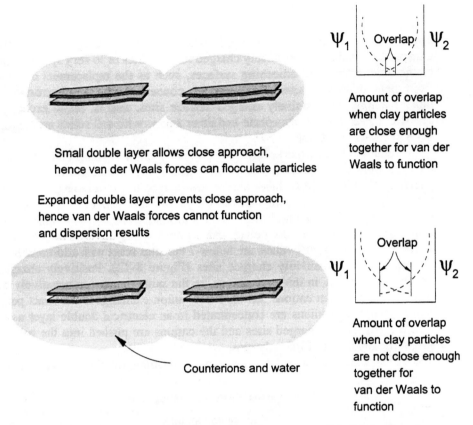

Small double layer allows close approach,
hence van der Waals forces can flocculate particles

Amount of overlap
when clay particles
are close enough
together for van der
Waals to function

Expanded double layer prevents close approach,
hence van der Waals forces cannot function
and dispersion results

Counterions and water

Amount of overlap
when clay particles
are not close enough
together for
van der Waals to
function

Figure 8-11   Attractive and repulsive forces responsible for the flocculation and dispersion of colloids.

geable Na^+ to result in dispersion of some of the colloidal particles, resulting in poor
physical structure, as well as other problems related to the resulting low permeabilities.

Effect of salt concentration.   The size of the colloidal double layer is also
affected by the concentration of dissolved salts in the soil solution. As the concentration of salts is increased, more ions are compressed into a given volume of
the soil solution. As the ions are crowded together in the soil solution, the ions in
the double layer are also packed closer together. This results in smaller double
layers and favors the flocculated state. For example, both sodic and saline-sodic
soils contain >15% exchangeable Na^+. The saline-sodic soil contains high concentrations of salts in the soil solution, while the sodic soil contains normal amounts
of dissolved salts. Because of the high salt concentration the double layers of the
colloids are compressed in the saline-sodic soils and these soils are flocculated,
whereas colloids in the sodic soils are in a dispersed state.

8.3.2 Anion Exchange

The adsorption of anions by soil solids can be due to relatively simple electrostatic attractions of anions to positively charged surface sites or to very specific reactions of the anions with the adsorbing surfaces, such as the replacement of hydroxyls from metal hydroxide sites. This section is concerned with the electrostatic attraction of anions by positively charged surface sites, that is, anion exchange. The specific interactions of phosphate and other anions with soil solids are the topics of other sections (see Chapter 14).

Metal oxide-hydroxide surfaces, such as Al and Fe oxides and hydroxides, as well as some pH-dependent charge sites associated with clay minerals, are amphoteric. That is, these surfaces and/or sites can be negatively charged, neutral, or positively charged, depending on the pH of the soil. For example, hematite (α-Fe_2O_3) is electrically neutral at pH values around 7. When pH values are greater than 7 the >FeOH sites dissociate and produce a negatively charged site and a hydrogen ion. When pH values are below 7 the sites react with additional hydrogen ions to produce positively charged sites (Figure 8-12). Positively charged sites attract counterions, in this case anions, in the same manner as negatively charged clay minerals attract cations. In this sense, anion exchange is the exact parallel of cation exchange. Anions are concentrated in an electrical double layer associated with the positively charged sites and the cations are pushed into the soil solution due to the repulsion of like charges.

Metal oxides/hydroxides often exist as coatings on clay surfaces and in the

Figure 8-12   Amphoteric nature of soil oxide-hydroxide minerals.

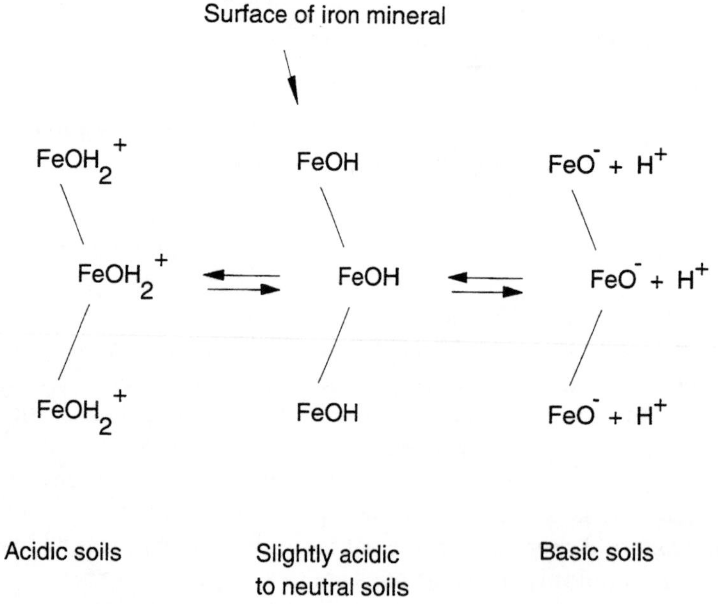

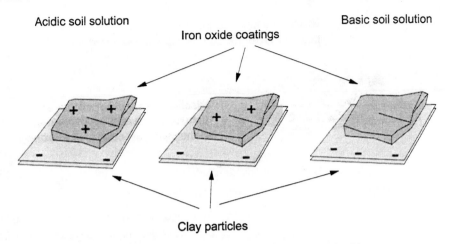

Acidic soil solution

Iron oxide coatings

Basic soil solution

Clay particles

In acidic soils the clay iron oxide complex has a net positive
charge, in neutral soils a net charge of zero, and in
basic soils a net negative charge

Figure 8-13   Effect of pH on oxide coated clay particles.

interlayers of clay minerals. This can result in ion exchange systems that are
intermediate between the clay minerals and the oxides. The noncoated portions of
the clay are negatively charged and contribute to the CEC of the soil. Depending
on pH, the oxide coatings can be negatively or positively charged or be neutral.
Figure 8-13 illustrates the effects of oxide-hydroxide coatings. At a pH of 2 the soil
has a net positive charge, at a pH of 4 the soil is electrically neutral, although it still
has cation and anion exchange sites, and at a pH of 7 the soil contains only
negatively charged sites, that is, cation-exchange sites.

8.3.3 Adsorption of Organic Compounds

Organic chemicals such as herbicides and other pesticides are often applied to the
soil during normal agricultural operations. In addition, these and other organic
chemicals can be spilled, or disposed of, on soils. The fate of these chemicals, that
is, their biological activity, their potential to leach into groundwaters, their introduction into food chains, even their eventual degradation by soil organisms, is often
a function of adsorption processes.

A large volume of research has shown that the two most active adsorbers of
organic compounds in soils are the various humic substances and clay minerals. The
organic chemicals, independent of their physiological activity, can be divided into
three chemical classes: organic cations, anions, and neutral molecules.

Reaction with clay minerals.   The ability of clay minerals to adsorb
organic compounds is dependent first, on the ability of the clay minerals to swell,

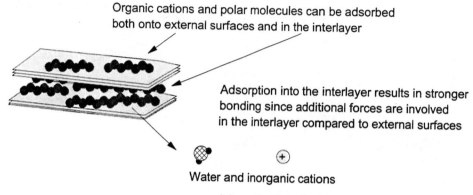

Organic cations and polar molecules can be adsorbed
both onto external surfaces and in the interlayer

Adsorption into the interlayer results in stronger
bonding since additional forces are involved
in the interlayer compared to external surfaces

Water and inorganic cations

Adsorption into the interlayer may result in
the loss of water and inorganic cations from the
interlayer

Figure 8-14   Interaction of organic chemicals in the interlayer of swelling clays.

and second, on the class of the chemical. Both swelling and nonswelling clay minerals have been shown to adsorb strongly organic cations, such as paraquat and diaquat, up to their cation-exchange capacities.

The adsorption of organic cations on external surfaces generally involves only coulombic attraction, whereas adsorption into the interlayer can involve the same forces, as well as additional forces such as dipole-induced dipole forces, because the molecule is surrounded by the negative charge of the clay (Figure 8-14). The adsorption of organic cations by swelling clays has been shown to be irreversible except by exchange for other like-sized organic cations. Adsorption into the inter-layer often displaces the inorganic cations and interlayer water. Adsorption of organic cations on the external surfaces of nonswelling clays is not as strong as adsorption in the interlayer, and the adsorbed cations can be replaced by inorganic as well as organic cations.

Swelling clay minerals have also been shown to be capable of adsorbing neutral organic molecules. For this to occur, the neutral organic compound must have some degree of polarity. The greater the polarity of the compound, the greater the potential for the compound to enter into the interlayer of the clay. Neither swelling nor nonswelling clays have significant capacities to adsorb organic anions.

Reaction with soil organic matter.   Soil organic matter is a very active adsorber of all classes of organic chemicals. Organic cations can react with the pH-dependent cation-exchange sites of soil organic matter. Since no interlayer is involved in the adsorption of organic cations by soil organic matter, the adsorption is reversible and similar to the adsorption of organic cations on external surfaces of clay minerals. Soil humic substances also contain a variety of functional groups that can potentially interact with the functional groups of the organic chemicals. Adsorption involving the interaction of functional groups is very specific and difficult to generalize.

Soil organic matter has also been shown to be a very good adsorber of neutral organic species. Unlike swelling clay minerals, adsorption of neutral organic chemicals has been shown to be greatest for the nonpolar (hydrophobic) chemicals. Adsorption of nonpolar organic molecules has been shown to increase predictably with decreasing water solubility of the organic compounds.[2]

STUDY QUESTIONS

1. Why are colloidal particles considered to control the chemical and physical properties of a soil?
2. What are the dominant minerals in the colloidal-size fraction?
3. What mineral properties determine whether a clay mineral will swell?
4. Discuss the nature of bonding between two sheets of the same layer of a phyllosilicate and between 1:1 and 2:1 layers of swelling and nonswelling minerals.
5. Define *humic acid*, *fulvic acid*, and *humin*.
6. What is *adsorption*?
7. Discuss the process of cation exchange. What colloidal, solute, and solution properties determine the ionic composition of the exchange complex?
8. Why does an accumulation of Na^+ on the exchange complex cause colloids to disperse?
9. Discuss the process of anion exchange.
10. If quartz, a tectosilicate, were ground into clay-size material, would it exhibit colloidal properties, such as cation or anion exchange?
11. Let montmorillonite, a smectite, have a CEC of 100 $cmol_c$ kg^{-1}. Let kaolinite have a CEC of 20 $cmol_c$ kg^{-1}. Assume that you have a clay mixture in the soil that is 20% kaolinite and 80% montmorillonite. Ignore the contribution of soil organic matter. What is the CEC of soil with 20% total clay (i.e., kaolinite + montmorillonite)? With 50% total clay?

[2]J. J. Hassett and W. L. Banwart, The sorption of nonpolar organics by soils and sediments, in *Reactions and Movements of Organic Chemicals in Soils*, B. L. Sawhney and K. Brown (eds.), Soil Sci. Soc. Amer. Spec. Publ. No. 22, 1989.

9

Soil Reaction: Soil pH

Soil reaction is one of the most interesting and informative soil properties. *Soil reaction* is a measure of the degree of acidity or alkalinity of the soil solution. Soil reaction is expressed as *pH*, which is the negative logarithm of the hydrogen ion concentration in the soil solution.

$$pH = -\log (H^+) \hspace{3cm} [9\text{-}1]$$

Soil pH is an indicator of the degree of soil weathering. Soil pH values reflect the mineral content of the parent material, the length of time and severity of weathering, and especially, the leaching of basic materials, such as Ca^{2+} and Mg^{2+} ions, from the soil profile. Factors such as the type of vegetation, amount of annual rainfall, internal soil drainage, and the activities of people influence soil pH.

The plant availabilities of iron, copper, phosphorus, zinc, and other nutrients, as well as that of toxic substances are controlled in a large part by soil pH. Some potentially toxic substances in soils, such as aluminum (Al^{3+}) and lead (Pb^{2+}), have little effect on plant growth in alkaline calcareous soils but can be serious concerns in acid soils. Many nutrients, such as phosphorus, are most available to plants in soils with slightly acid to neutral pH values, showing marked decreases in plant availability above or below this optimum pH range. Nutrient availability and/or toxicities often restrict species of plants to certain ranges of soil pH (Table 9-1).

Soil pH is also an indicator of serious soil problems. Soil pH values above 8.5 are indicative of sodic soils. Soil pH values below 4 suggest that the oxidation of reduced sulfur compounds has occurred or is occurring in the soil. The specific effects of pH on nutrient availability and nutrient uptake are covered in later chapters.

TABLE 9-1   Desirable pH ranges of some common plants and crops

Crop	pH range
Alfalfa	6.0–6.7
Alsike clover	5.8–6.2
Azaleas and camellias	5.0–5.5[a]
Barley	5.8–6.2
Bent grass (except creeping)	4.8–5.5
Blueberries	4.8–5.2[a]
Buckwheat	5.8–6.2
Carrots	5.8–6.2
Cassava	4.4–5.0[b]
Cauliflower	5.8–7.2
Corn	5.8–6.7
Crimson clover	5.8–6.2
Irish potatoes	4.8–5.5[a]
Kentucky bluegrass	5.8–6.2
Lespedeza	5.8–6.2
Peanuts	5.8–6.2
Rice	5.8–6.2
Soybean	5.8–6.7
Sweet and red clover	6.3–6.7
Tobacco	5.0–6.2[a]
General garden crops	5.8–6.2
General legume crops	5.8–6.7

Source: C. D. Sopher and J. V. Baird, *Soils and Soil Management*, 2nd ed., Reston Publishing Co., Reston, Va.
[a]May become susceptible to disease at higher pH values.
[b]Will respond to higher pH values if zinc levels are adequate.

9.1   GENERAL PROCESSES

9.1.1   Role of Water

Water is one of the most important components of aqueous systems, not only because it is an excellent solvent, but also because of its role in acid-base reactions. Water will autohydrolyze into a hydronium ion, H_3O^+, or as more commonly written the hydrogen ion, H^+, and a hydroxyl ion, OH^-.

$$2H_2O \leftrightarrow H_3O^+ + OH^- \qquad [9\text{-}2]$$

or

$$H_2O \leftrightarrow H^+ + OH^-, \qquad K_w = 10^{-14} \text{ at } 25°C \qquad [9\text{-}3]$$

where K_w is the equilibrium constant for the autohydrolysis of water and is given by

$$K_w = (H^+)(OH^-) = 10^{-14} \quad\quad [9\text{-}4]$$

This equation says the concentration of hydrogen ions multiplied by the concentration of hydroxyl ions is a constant and always equal to 1×10^{-14} at 25°C. The equilibrium constant expression for water has special significance for aqueous systems. This expression, and its logarithmic form, not only establishes the pH scale and the definition of acidic, basic, and neutral solutions, but also illustrates the interdependence of (H^+) and (OH^-) concentrations.

In logarithmic form:

$$\log (H^+) + \log (OH^-) = -14 \quad\quad [9\text{-}5]$$

In terms of negative logarithms, $p = -\log$ and $pH = -\log (H^+)$:

$$pH + pOH = 14 \quad\quad [9\text{-}6]$$

Equations [9-4] and [9-6] illustrate that in aqueous solutions the concentrations of the H^+ and OH^- ions cannot be varied independently. When one species is increased there must be a corresponding decrease in the concentration of the other, such that the product of their concentrations is a constant (K_w). It must also be remembered that the pH scale is a negative logarithmic scale. A change in pH from 5 to 4 is a tenfold increase in H^+ ion concentration, and a pH change from 4 to 8 is a 10,000-fold decrease in hydrogen ion concentration.

9.1.2 Hard and Soft Acids and Bases and Their Salts

Many soil components are acids, bases, or salts of acids and bases. The role of these materials in the control of soil pH, as well as the effect of soil pH on the concentration of their ions and molecules in the soil solution, is determined by the nature of the acid, base, or salt.

1. Strong (hard) acids and bases are those acids and bases that are completely dissociated into their respective cations and anions under the conditions found in the soil.

2. Weak (soft) acids and bases are those acids and bases that are not completely dissociated into cations and anions under the conditions found in the soil.

For example, consider two acids: the strong acid H_2SO_4 (sulfuric acid) and the weak acid H_2CO_3 (carbonic acid). For the strong acid:

$$H_2SO_4 \leftrightarrow HSO_4^- + H^+, \quad Ka_1 = 10^{1.98} \quad\quad [9\text{-}7]$$

$$HSO_4^- \leftrightarrow SO_4^{2-} + H^+, \quad Ka_2 = 10^{-1.98} \quad\quad [9\text{-}8]$$

where Ka_1 and Ka_2 are the first and second dissociation constants for the acid. The numerical values of the dissociation constants provide information about the conditions under which the acid would be completely dissociated.

$$Ka_2 = \frac{(SO_4^{2-})(H^+)}{(HSO_4^-)} = 10^{-1.98}$$ [9-9]

When $(H^+) = Ka_2$, that is, when pH = 1.98, the concentrations of (SO_4^{2-}) and (HSO_4^-) are equal. At soil pH values above 1.98, the strong acid (H_2SO_4) is completely dissociated into (SO_4^{2-}) and (H^+) ions. This means that if sulfuric acid is added to a soil with a pH greater than 1.98, the sulfuric acid molecule will completely dissociate into sulfate and hydrogen ions. It also means that if sulfate ions are added to a soil with a pH greater than 1.98, there will be little or no tendency for the sulfate ions to hydrolyze water, re-forming sulfuric acid. If sulfate ions are added to a soil with a pH less than 1.98, there would be a tendency for the sulfate ions to react with water, re-forming sulfuric acid.

For the weak acid,

$$H_2CO_3 \leftrightarrow HCO_3^- + H^+, \qquad Ka_1 = 10^{-3.76}$$ [9-10]

$$HCO_3^- \leftrightarrow CO_3^{2-} + H^+, \qquad Ka_2 = 10^{-10.25}$$ [9-11]

Examination of equations [9-10] and [9-11] show that for pH values below pK_1, that is, pH <3.76, H_2CO_3 will have little tendency to dissociate. For pH values between 3.76 and 10.25, H_2CO_3 will be dissociated into HCO_3^- (bicarbonate ion) and (H^+) ions. Soil pH values must approach 10.25 before H_2CO_3 is dissociated to any extent into CO_3^{2-} (carbonate ion) and H^+ ions. For the extreme range of pH values found in soils, that is, pH values from 2 to 11, H_2SO_4, which is a strong acid, will be found in significant concentrations only as the SO_4^{2-} and H^+ ions, while H_2CO_3, which is a weak acid, will exist in the soil solution as H^+, HCO_3^-, and CO_3^{2-} ions and the undissociated acid, H_2CO_3.

An additional important aspect of strong and weak acids and bases is related to the reactions that take place when the anion of an acid or the cation of a base is added, as a component of a salt, to the soil. A salt will affect soil pH only when either the cation of the base or the anion of the acid, which is produced upon dissolution of the salt, hydrolyzes water.

Consider the dissolution of calcium sulfate $[CaSO_4(s)]$, a salt of a strong base $[Ca(OH)_2]$ and a strong acid (H_2SO_4), upon addition to water:

$$CaSO_4(s) \leftrightarrow Ca^{2+} + SO_4^{2-}$$ [9-12]

For this salt to affect pH, either the cation must hydrolyze water, making the solution acidic,

$$Ca^{2+} + H_2O \leftrightarrow CaOH^+ + H^+, \qquad K = 10^{-12.70}$$ [9-13]

or the anion must hydrolyze water, making the solution basic,

$$SO_4^{2-} + H_2O \leftrightarrow HSO_4^- + OH^-, \qquad K = 10^{-12.09}$$ [9-14]

The calcium ion (Ca^{2+}), which is a cation of a strong base, will hydrolyze water, that is, form the base ($CaOH^+$), only when pH is close to or above 12.7. The sulfate ion (SO_4^-), which is the anion of a strong acid, will hydrolyze water only when the pOH is greater than 12.02, that is, when the pH is less than 1.98. Hence $CaSO_4$ and other salts of strong acids and strong bases form neutral solutions when added to water.

Now consider the addition of the salt of a weak acid and a strong base [$CaCO_3$(s)] to water.

$$CaCO_3(s) \leftrightarrow Ca^{2+} + CO_3^{2-} \tag{9-15}$$

For this salt to affect soil pH, either the cation or the anion must hydrolyze water. Since the calcium ion is the cation of a strong base, it has little or no tendency to hydrolyze water at pH values below 12.70 (equation [9-13]). Since the carbonate ion is the anion of a weak acid, it will have a strong tendency to hydrolyze water.

$$CO_3^{2-} + H_2O \leftrightarrow HCO_3^- + OH^-, \quad K = 10^{-3.75} \tag{9-16}$$

Reaction [9-16] shows that the addition of the carbonate ion to water with pOH values greater than 3.75, that is, at pH values less than 10.25, results in the hydrolysis of the water. The H^+ ion released by the hydrolysis combines with the carbonate ion to form the bicarbonate ion. The OH^- ion released by the hydrolysis stays in solution, increasing the pH of the solution. It can be generalized that salts of strong bases and weak acids form basic solutions.

The addition of a salt of a strong acid and a weak base will affect pH only when either the anion of the strong acid or the cation of the weak base hydrolyses water. Aluminum sulfate [$Al_2(SO_4)_3$(s),] is the salt of a strong acid (H_2SO_4) and a weak base [$Al(OH)_3$(s)]. Upon addition to the soil, aluminum sulfate will dissolve producing Al^{3+} and SO_4^{2-} ions.

$$Al_2(SO_4)_3(s) \leftrightarrow 2Al^{3+} + 3SO_4^{2-} \tag{9-17}$$

Since SO_4^{2-} is the anion of a strong acid, it has little tendency to hydrolyze water (see equation [9-14]). Since Al^{3+} is the cation of a weak base, it has a marked tendency to hydrolyze water.

$$Al^{3+} + H_2O \leftrightarrow AlOH^{2+} + H^+, \quad K = 10^{-5.0} \tag{9-18}$$

Equation [9-18] shows that the trivalent aluminum ion results in hydrolysis of water at pH values close to 5 or above. In fact, second and third hydrolysis reactions producing $Al(OH)_2^+$ and $Al(OH)_3$(amorphous) are possible, resulting in up to three hydrogen ions for each Al^{3+} added to soil solution. As a general rule, salts of strong acids and weak bases form acidic solutions. In soil solutions covering a normal range of concentrations and pH values, HCl, H_2SO_4, and HNO_3 can be considered to be strong acids, while NaOH, KOH, and the first dissociations of $Ca(OH)_2$ and $Mg(OH)_2$ can be considered to be strong bases.

9.2 MECHANISMS THAT CONTROL SOIL pH

Soil formation in subhumid and humid regions is, in general, an acidifying process. When a geologic process such as erosion, sedimentation, loess deposition, or glaciation creates a parent material, the parent material normally has a pH around 7 (neutrality) and often has pH values above 8 if the parent material contains carbonate minerals, such as $CaCO_3$(calcite) or $CaMg(CO_3)_2$(dolomite). During soil formation, the parent material and the subsequent soil are subject to a variety of processes that modify and usually lower soil pH.

Processes resulting in increased acidity

1. The soil is constantly leached by precipitation containing carbonic acid, H_2CO_3. Carbon dioxide gas [CO_2(gas)] dissolves in water to produce an aqueous form of the gas, which reacts with water molecules to produce carbonic acid.

$$CO_2(gas) \leftrightarrow CO_2(aq) \qquad K_1 = 10^{-1.41} \qquad [9\text{-}19]$$

$$CO_2(aq) + H_2O \leftrightarrow H_2CO_3, \qquad K_2 = 10^{-2.62} \qquad [9\text{-}20]$$

For water in equilibrium with atmospheric carbon dioxide gas [PCO_2(gas) = 3×10^{-4} atm), the concentration of $H_2CO_3 = 2.83 \times 10^{-8}$ M. For water in equilibrium with the upper limit of carbon dioxide found in the soil, due to microbial and root respiration, [PCO_2(gas) = 0.01 atm], $H_2CO_3 = 9.5 \times 10^{-7}$ M. During soil formation, parent materials are subjected to constant leaching by a dilute solution of this weak acid. This process, accumulated over the length of time involved in soil formation, can add large quantities of hydrogen ions to the soil profile.

2. Organic acids are added to the soil upon decomposition of organic residues by soil microbes. Organic acids acidify the soil and often complex metallic cations, increasing their movement and loss from the soil.

3. H^+ ions are released by plant roots and other organisms during nutrient uptake. The principle of electroneutrality means that the uptake of cations, by roots, must be balanced either by an equivalent uptake of negatively charged anions or by the release of hydrogen ions or other cations.

4. Anions, such as sulfate, move downward and are eventually lost into the groundwater. Electroneutrality requires that equivalent amounts of cations and anions be lost into the groundwater. The inputs of hydrogen ions at the surface of the soil due to CO_2(gas) charged rain, decomposition of organic residues, and plant uptake result in the downward displacement of the original soil cations, usually Ca^{2+}, Mg^{2+}, and K^+. It is these cations that are usually lost with the anions.

5. The oxidation of reduced substances, such as sulfide minerals, organic matter,

or ammonium-containing fertilizers, is an acidifying process. These are discussed in later sections of this and other chapters.

Processes resulting in increased alkalinity

1. The reduction of ferric iron, manganese, and other oxidized substances consumes H^+ ions or releases OH^- ions and increases soil pH. This is usually due
   to the creation of poorly aerated conditions associated with flooding of the
   soil. For example:

$$Fe(OH)_3(\text{amorphous}) + e^- \leftrightarrow Fe(OH)_2(\text{amorphous}) + OH^- \qquad [9\text{-}21]$$

   Upon reduction, the valence of iron changes from +3 to +2, resulting in the
   release of one OH^- group.
2. Deep-rooted plants recycle basic cations. The uptake of Ca^{2+}, Mg^{2+}, K^+, and
   other basic cations by roots deep within the soil profile, with the eventual
   deposition on and in surface horizons upon the death of the plants, tends to
   offset the downward movement and loss of cations due to leaching. This
   process helps maintain base saturation and, as will be discussed in future
   sections, increases soil pH or at least prevents or slows the acidification of the
   upper soil horizons.

Note: Some activities of humans, such as liming, tend to offset the natural
acidification associated with soil formation, while other human activities, such as
nitrogen fertilization and the release of sulfur from smokestacks, resulting in the
formation of "acid rain," tend to accelerate the process.

Soil pH is controlled by many mechanisms. Some mechanisms are direct
sources of H^+ and/or OH^- ions, while others operate by reacting with the H^+ and/or
OH^- ions to buffer the soil solution. Major mechanisms that potentially dictate soil
pH include:

1. Oxidation and reduction of iron, manganese, or sulfur compounds
2. The dissolution and precipitation of soil minerals
3. The reactions of gases such as $CO_2(\text{gas})$ with the soil solution
4. The dissociation of weak acid groups on the edges of silicate clays, hydrous
   oxides, or humic substances
5. Ion-exchange reactions

Soil pH values vary from as low as 2 in some acid-sulfate soils or strip-mine
spoil piles, to as high as 10.5 in severe sodic soils. Humid-region soils, which have
been subjected to greater amounts of leaching than arid region soils, generally have
pH values between 4.5 and 7.2, while arid region soils generally have pH values in
the range 6.8 to 8.5.

TABLE 9-2 Mechanisms that control soil pH

Soil pH range	Major mechanism(s) controlling soil pH
2–4	Oxidation of pyrite and other reduced sulfur minerals; dissolution of soil minerals
4–5.5	Exchangeable Al^{3+} and its associated hydroxy ions; exchangeable H^+
5.5–6.8	Exchangeable H^+; weak acid groups associated with soil minerals and humic substances; dissolved CO_2(gas) and other aqueous species of dissolved CO_2(gas)
6.8–7.2	Weak acid groups on humic substances and soil minerals
7.2–8.5	Dissolution of solid divalent carbonates, such as $CaCO_3$(calcite)
8.5–10.5	Exchangeable Na^+ under normal salt conditions; dissolution of solid Na_2CO_3(s)

Table 9-2 depicts the major mechanisms that produce H^+ or OH^- ions or react with the H^+ or OH^- to buffer soil pH. The mechanisms are identified with the soil pH ranges where they are the dominant processes controlling pH. The mechanisms may function over a wider range than shown, but are not the dominant pH controlling mechanisms for that pH range. In the following sections we examine, in detail, the mechanisms that control and buffer soil pH and explain why the mechanisms are restricted to their respective pH range. Sulfur and aluminum are discussed, in greater detail, in other chapters. Only their effects on soil pH and soil buffering are examined in this chapter.

9.2.1 Oxidation of Reduced Sulfur Compounds

When soils or deposits that contain reduced forms of sulfur, such as iron sulfide, are exposed to aerobic environments, conditions are created that favor the oxidation of the reduced sulfur compounds. Associated with the oxidation of the reduced sulfur minerals is the production of H^+ ions and their release to the soil solution.

$$2FeS_2(\text{pyrite}) + 7.5O_2(\text{gas}) + 4H_2O \leftrightarrow \alpha Fe_2O_3(\text{hematite}) + 8H^+ + 4SO_4^{2-} \quad [9\text{-}22]$$

The reduced sulfur compounds are oxidized by chemoautotrophic bacteria, such as *Thiobacillus ferroxidans*. The chemical energy released by these oxidations is used by the bacteria to drive their metabolic processes. The initial process of sulfur oxidation in neutral soils is carried out by bacteria, but once the soil has been acidified to pH values around 3.5 or less, the oxidation of the sulfide ions can be coupled with the reduction of ferric iron and the process becomes chemical (abiotic) in nature.

$$8Fe^{3+} + S^{2-} + 4H_2O \leftrightarrow 8Fe^{2+} + SO_4^{2-} + 8H^+ \qquad [9\text{-}23]$$

The source of the acidity in these soils is the oxidation of reduced sulfur. The total amount of acidity that can potentially be produced is determined by the quantity of reduced sulfur minerals in the soil. The rate at which the acidity is produced is governed by the rates of the biological and chemical oxidation mechanisms.

If the pH values of these soils, in which sulfur oxidation is occurring, are determined and compared to the amount of acid being produced, it is often apparent that some buffering mechanism in the soil is operating and reacting with the H^+ because extremely low pH values are not always present. In some acid sulfate soils, it has been shown that more that 98% of the acid released during pyrite oxidation reacts with soil minerals and is neutralized, 1 to 2% reacts with dissolved alkalinity (HCO_3^-) and less than 1% remains in the soil solution as free acid. Equation [9-24] is an example of how a soil mineral can neutralize the H^+ ions produced by sulfur oxidation. Soil minerals, in addition to gibbsite, are also dissolved in response to the elevated H^+ ion levels and contribute to overall soil buffering.

$$Al(OH)_3(gibbsite) + H^+ \leftrightarrow Al(OH)_2^+ + H_2O \qquad [9-24]$$

Two major situations exist in which the oxidation of reduced sulfur compounds is the major source of soil acidity.

Cat clays.   Cat clays occur in low-lying coastal areas. These soils were at one time shallow marine sediments covered with seawater containing sulfate (SO_4^{2-}) as a major anion. Since the sulfate concentration in seawater is higher than the sulfate concentrations in the pore waters of the sediments, sulfate diffused into the underlying sediments. Once in the sediments, the sulfate served as a terminal electron acceptor for the anaerobic oxidation of the organic materials contained in the sediments and was reduced to sulfide (S^{2-}). Since sulfide forms very insoluble compounds with iron and other metals, it precipitated as FeS_2(pyrite) or other solid sulfides and accumulated in the sediments. The precipitation of sulfide, coupled with the influx of additional organic materials, provided the driving force for the continued diffusion of sulfate out of the seawater into the sediments, as well as its subsequent reduction and accumulation. When geologic processes or the actions of people raise or drain the sediments, this creates well-aerated conditions favorable to the oxidation of the FeS_2(pyrite) and the formation of strongly acidic soils (equations [9-22] and [9-23]).

Prelaw acid-strip-mine spoils.   The process of strip mining coal generally involves removal of large volumes of overburden (Figure 9-1). Often the shale or rock materials that lie directly over the coal seam contain appreciable quantities of sulfide minerals. In their natural position, these materials are isolated from the oxygen in the atmosphere by layers of earth material and are stable as the reduced minerals. Upon strip mining, the shale and its reduced sulfide minerals are exposed to the atmosphere. This creates conditions favorable for the oxidation of the sulfide minerals to sulfate (equation [9-22]) and the production of strongly acidic conditions. Present federal mining laws require that sufficient sulfide-free soil material be placed on top of the spoil to isolate it from the atmosphere and prevent the

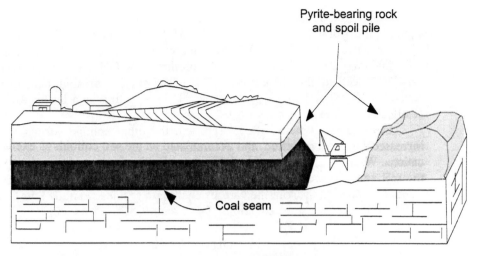

Figure 9-1   Strip mining for coal. Residual materials create the possibility of the formation of acid mine spoils.

oxidation of the sulfide minerals and formation of soils too acidic to support vegetation.

9.2.2 Exchangeable Aluminum

The trivalent aluminum ion (Al^{3+}) is the cation of a weak base and, as such, has the potential to hydrolyze water, producing hydrogen ions. The combination of the Al^{3+} ion, its hydrolysis species, and the insoluble weak base $Al(OH)_3s$ represents a vigorous buffer system for soils in this pH range (4–5.5). Consider the following species of aluminum and their chemical equilibria. Additional species of aluminum exist, but these species will be sufficient to illustrate how pH is controlled in the pH range 4 to 5.5.

Soluble species:

$$Al^{3+}, AlOH^{2+}, Al(OH)_2^+$$

Solid species:

$$Al(OH)_3(amorphous)$$

Reactions:

$$Al(OH)_3(amorphous) + H^+ \leftrightarrow Al(OH)_2^+ + H_2O, \qquad K_1 = 10^{-0.081} \qquad [9\text{-}25]$$
$$Al(OH)_2^+ + H^+ \leftrightarrow AlOH^{2+} + H_2O, \qquad K_2 = 10^{4.7} \qquad [9\text{-}26]$$
$$AlOH^{2+} + H^+ \leftrightarrow Al^{3+} + H_2O, \qquad K_3 = 10^{5.0} \qquad [9\text{-}27]$$

In neutral soils the solubility of the solid phase aluminum species is very low.

For example, the concentration of $Al(OH)_2^+$ in equilibrium with $Al(OH)_3$(amorphous) at a pH of 7 is equal to 8.3×10^{-7} M. By the time the soil has acidified to pH 5, the concentration of $Al(OH)_2^+$ has increased to 8.3×10^{-5} M. This represents a 100-fold increase in solubility. The dissolution of $Al(OH)_3$(amorphous) (equation [9-25]), as well as the shift from one aqueous species of aluminum to another (equations [9-26] and [9-27]), consumes H^+ ions, acting as a buffer to slow the acidification process.

Once the soil has acidified, Al^{3+}, and the other soluble species will have increased in the soil solution, and accumulated on the soil colloids as exchangeable cations. When the soil has reached this state, the aluminum system not only buffers the soil against decreases in soil pH, but also against increases in pH. If a basic material is added to the soil, the various aluminum species neutralize the OH^- ions, as shown by the reverse of equations [9-25] and [9-27].

$$Al^{3+} + OH^- \leftrightarrow AlOH^{2+}, \qquad K = \frac{1}{K_w K_3} \qquad\qquad [9\text{-}28]$$

or

$$AlOH^{2+} + OH^- \leftrightarrow Al(OH)_2^+, \qquad K = \frac{1}{K_w K_2} \qquad\qquad [9\text{-}29]$$

The pH of the soil is buffered not only by the aluminum species in the soil solution, but also by the aluminum species on the cation-exchange complex. The aluminum species, associated with both phases, must be converted back to insoluble solid phase $[Al(OH)_3]$ to increase pH up to the level (pH ≈ 6.2) usually recommended for crop production. Aluminum species on the cation-exchange complex, along with exchangeable H^+, constitute what is known as exchangeable or reserve acidity. The control of soil pH by exchangeable acidity is discussed in the next section.

Soluble aluminum, in particular Al^{3+}, is toxic to many plant and animal species. The lower production found with many crop species on soils with pH values <5.5 is due not only to decreased nutrient availability, but also to aluminum and, in some cases, manganese toxicity. In many tropical regions, where soil acidity is a severe problem, recommended pH values are set just high enough to decrease the level of soluble aluminum (pH ≈ 5.5), but not as high as suggested in many temperate regions. The lower recommended pH values are related to the pH at the zero point of charge of the hydrous oxides in the tropical soils. Much of the damage caused by acid rain is related to increased aluminum levels, in particular the Al^{3+} species, found in surface runoff and groundwaters coming from affected soils. The increased aluminum levels in these runoff and groundwaters are in part the result of the interaction of the soil buffer mechanisms with the acid rain.

9.2.3 Exchangeable Hydrogen Ions (Reserve Acidity)

The concentration of a particular type of cation on the cation-exchange complex relative to the concentration of all other types of cations on the exchange complex governs the concentration of that type of cation in the soil solution. There is an equilibrium established between the cations on the exchange complex and the cations in the soil solution.

$$[\text{exchangeable cation}] \leftrightarrow [\text{cation in soil solution}] \quad K \quad [9\text{-}30]$$

$$[\text{exchangeable H}^+] \leftrightarrow [\text{H}^+ \text{ in the soil solution}] \quad K \quad [9\text{-}31]$$

where K is the equilibrium or exchange constant for the reaction. Another way of expressing equations [9-30] and [9-31] is that the percentage of the exchange complex that is occupied by a particular cation determines the concentration of that cation in the soil solution.

Figure 9-2 diagrammatically illustrates the relationship between hydrogen ions on the exchange complex (exchangeable acidity) and the concentration of hydrogen ions in the soil solution (active acidity). When liquid is withdrawn from the standpipe, the level in the standpipe decreases (A) and stays low until the valve is shut. Once the valve is shut, the level in the standpipe will increase back to the level in the large tank. Since the volume of the tank is much larger than the standpipe, it is necessary to remove many standpipe volumes before significant change is reflected in the level of liquid in the tank (B). The volume of the large tank is a model of the amount of H^+ ions on the exchange complex, while the volume of the standpipe is a model of the amount of H^+ ions in the soil solution. The level of H^+ ions in the soil solution is controlled by the level of H^+ ions on the exchange complex. When H^+ ions are added to or removed from the soil solution, the exchangeable acidity is affected very little and the level of H^+ ions in the soil solution is restored to nearly the original level by transfer of H^+ in or out of

Figure 9-2 Relationship between exchangeable and active acidity.

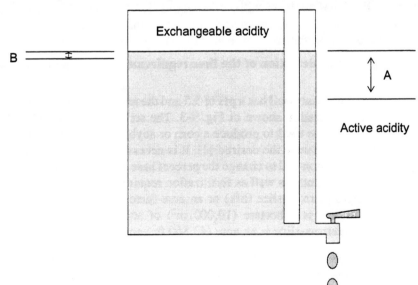

the large exchangeable pool. The exchangeable H^+ ions buffer the level of H^+ ions in the soil solution.

Percentage base saturation is the ratio of basic (nonacidic) cations on the exchange complex to the total cation exchange capacity of the soil.

$$\% \text{ base sat.} = \frac{\text{conc. basic cations } [\text{cmol}_c/\text{kg}]}{\text{CEC } [\text{cmol}_c/\text{kg}]} \times 100 \qquad [9\text{-}32]$$

When percentage base saturation is high, the concentration of basic cations (i.e., Ca^{2+}, K^+, Mg^{2+}, etc.) on the exchange sites is high, and the concentration of acidic cations (H^+, Al^{3+}, and the related hydrolysis species) is low.

Remember, the level of ions on the exchange complex governs the level of ions in the soil solution. Therefore, when percent base saturation is high, the relative concentration of basic cations in the soil solution is high and the relative concentration of acidic cations is low. When base saturation is low, the relative concentration of basic cations in the soil solution is low and the relative concentration of acidic cations in the soil solution is high.

Figure 9-3 is a titration curve for a soil. This figure is generated by adding acid or base to a soil to raise or lower soil pH and by measuring the percentage base saturation at each pH value. The figure shows that when the soil solution has a pH of 5.5, the corresponding percent base saturation is equal to 50%. When the pH of the soil is 6.5, the corresponding percent base saturation is 90%. To change the pH of the soil from 5.5 to 6.5 as would be the case in a liming operation, not only must the hydrogen ions in the soil solution be neutralized, but the percent base saturation must also be changed from 50% to 90%. That is, the amount of hydrogen equal to 40% (90% − 50%) of the CEC must be removed from the exchange complex, neutralized, and replaced with basic cations. To change the H^+ ion concentration in the soil solution (active acidity) permanently, the ratio of H^+ ions to other cations on the exchange complex (exchangeable acidity) must be changed.

Example 1: Calculation of the lime requirement of a cash grain cropping system

Assume that a soil has a pH of 5.5 and the relationship between pH and percent base saturation shown in Fig. 9-3. The soil has a CEC of 20 cmol_C kg^{-1} soil and will be used to produce a corn or soybean crop that requires a soil pH of 6.5. To achieve the desired pH, it is necessary to neutralize the acidity in the soil solution and to change the percent base saturation from 50% to 90%. Lime requirements, as well as fertilization requirements are usually calculated on a hectare-furrow-slice (hfs) or an acre-furrow-slice (afs) basis. A hectare-furrow-slice is a hectare (10,000 m^2) of soil approximately 16 cm deep; an acre-furrow-slice is an acre (43,560 ft^2) approximately $6\frac{2}{3}$ in. deep.

Solution: (a) *Hydrogen ions in the soil solution* (active acidity):   Assume that the soil contains 35% water (W %) and has a bulk density of 1.3 g/cm^3. Since the soil has a pH of 5.5 and pH = −log(H^+),

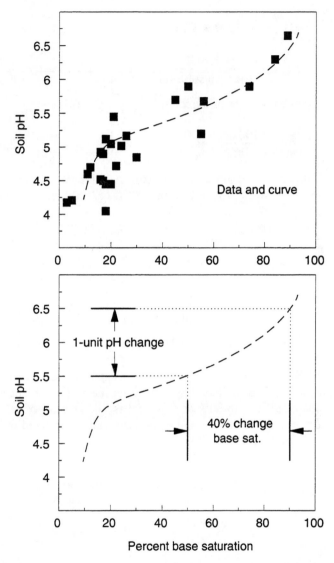

Figure 9-3   Titration curve for an ideal soil, showing the relationship between soil pH and percent base saturation. Top curve shows the theoretical curve drawn through actual data for Piedmont soils. Lower curve illustrates the relationship between on unit change in pH and the change in percentage base saturation required to facilitate the change in pH. [Data from A. Mehlich, *Soil Sci. Soc. Amer. Proc.* 6:150–156 (1941).]

$$(H^+) = 10^{-5.5} M = 3.16 \times 10^{-6} M = 3.16 \times 10^{-6} \frac{\text{moles } (H^+)}{\text{liter soil solution}}$$

$$3.16 \times 10^{-6} \frac{\text{moles } (H^+)}{\text{liter soil solution}} \times \frac{1 \text{ g } (H^+)}{\text{mole}} = 3.16 \times 10^{-6} \frac{\text{g } H^+}{\text{liter soil solution}}$$

Oven-dry weight of a hectare-furrow-slice (hfs).

$$1.3 \frac{g}{cm^3} \times \left(\frac{100 \text{ cm}}{m}\right)^3 \times \frac{10,000 \text{ m}^2}{\text{hectare}} \times \frac{1 \text{ kg}}{1000g} \times \frac{0.16 \text{ m}}{\text{hfs}} = 2,080,000 \frac{\text{kg soil}}{\text{hfs}}$$

Amount (liters) of water in the soil. Assume that the soil solution has the same density as water: $\rho_w = 1$ g/cm^3 = 1000 g/1000 cm^3 =1 kg/liter.

$$2,080,000 \frac{kg}{hfs} \times 0.35\% \text{ (W \%)} = 728,000 \frac{\text{kg soil solution}}{\text{hfs}}$$

$$728,000 \frac{\text{kg soil solution}}{\text{hfs}} \times 1 \frac{kg}{\text{liter}} = 728,000 \frac{\text{liter soil solution}}{\text{hfs}}$$

$$728,000 \frac{\text{liter soil solution}}{\text{hfs}} \times 3.16 \times 10^{-6} \frac{\text{g } H^+}{\text{liter}} = 2.3 \frac{\text{g } H^+}{\text{hfs}}$$

The soil contains 2.3 g of H$^+$/hfs in the soil solution at a pH of 5.5 and 35% water.

(b) *Hydrogen ions on the exchange complex* (reserve acidity):   To achieve the desired pH, it is necessary to change the soil's percent base saturation from 50% at pH 5.5 to 90% at 6.5. That is, the amount of H$^+$ ions that have to be removed from the exchange complex and neutralized is equivalent to 90% − 50%, or 40% of the CEC.
CEC = 20 cmol$_c$ kg $^{-1}$ of soil

$$20 \frac{cmol_c}{\text{kg soil}} \times 0.40 = 8 \frac{\text{cmol } H^+}{\text{kg soil}}$$

$$8 \frac{\text{cmol } H^+}{\text{kg soil}} \times 0.01 \frac{\text{g } H^+}{\text{cmol } H^+} = 0.08 \frac{\text{g } H^+}{\text{kg soil}}$$

$$0.08 \frac{\text{g } H^+}{\text{kg soil}} \times 2,080,000 \frac{\text{kg soil}}{\text{hfs}} = 166,400 \frac{\text{g } H^+}{\text{hfs}} = 166.4 \frac{\text{kg } H^+}{\text{hfs}} = 0.166 \frac{\text{mt } H^+}{\text{hfs}}$$

Clearly, the 2.3 g of H$^+$/hfs in the soil solution example 1(a), while responsible for measured soil pH, is insignificant compared to 166,400 g of H$^+$/hfs on the colloids example 1(b) that must be neutralized to change the measured pH to the desired pH.

(c) *Neutralization of hydgrogen ions:* A liming material, such as calcium carbonate, can be used to neutralize H$^+$ ions.

$$CaCO_3(s) + 2H^+ \leftrightarrow Ca^{2+} + CO_2(gas) + H_2O \qquad [9\text{-}33]$$

Since the molecular weight of calcium carbonate is 100 g and it neutralizes 2 g of hydrogen, it requires 50 units of $CaCO_3$ to neutralize 1 unit of H^+ ions. (1) For the H^+ ions in the soil solution:

$$2.3 \text{ g } H^+/hfs \times 50 \text{ g } CaCO_3(s)/g \text{ } H^+ = 115 \text{ g of } CaCO_3(s)/hfs$$

(2) For the H^+ ions on the exchange complex:

$$166{,}400 \text{ g } H^+/hfs \times 50 \text{ g } CaCO_3(s)/g \text{ } H^+ = 8{,}320{,}000 \text{ g } CaCO_3(s)/hfs$$
$$= 8.32 \text{ mt } CaCO_3(s)/hfs$$

The example illustrates two points: first, the soil contains 208,000 g of H^+ ions on the exchange complex ($166{,}400 \times 50/40$) compared to 2.3 g of H^+ ions in the soil solution, and second, that it requires 8.32 mt $CaCO_3(s)/hfs$ to change the soil's pH from 5.5 to 6.5. This is a good illustration of the soil's buffering ability in this pH range.

Example 2

How much sulfur (S mol. wt. = 32) would it require to change the pH of this same soil from 6.5 to 5.5?

Solution: In this case it is necessary to calculate the amount of H^+ ions that must be added to the soil, which is equal to the amount of H^+ ions removed in example 1b. Sulfur will be oxidized to sulfuric acid in the soil by soil organisms, such as *Thiobacillus ferroxidans*. Each mole of sulfur produces 1 mol of sulfuric acid, which produces 2 mol of H^+ ions.

$$S + 1\tfrac{1}{2}O_2 + H_2O \rightarrow H_2SO_4 \rightarrow 2H^+ + SO_4^{2-}$$

$$166{,}400 \text{ g } H^+ \text{ needed}/hfs \times 32/2 = 2{,}662{,}400 \text{ g } S/hfs = 2.66 \text{ mt } S/hfs$$

To grow plants such as azaleas or rhododendrons, which require acid soil, 2.66 mt/hfs of sulfur would have to be added. This again illustrates the magnitude of the buffer capacity of this soil.

For soils in the pH range 5.5 to 6.8, the source of H^+ ions is the exchangeable H^+ ions associated with soil clays and humic substances. The buffer mechanism that controls pH in this range is the relationship between pH and percent base saturation. The strength of the buffer is illustrated by the large change in percent base saturation that is required to change the soil's pH.

9.2.4 Weak Acid Groups on Soil Clays, Hydrous Oxides, and Soil Organic Matter

Important sources of H^+ ions that contribute to the overall buffering of soils include the dissociation of weak acid groups, such as >AlOH and >SiOH groups, at the edges of soil clays, >FeOH and >AlOH groups associated with hydrous oxides, and phenolic and carboxyl groups of soil organic matter. In addition, these groups are responsible for the pH-dependent charge of soils. These functional groups are tightly bound to, and very specific for, H^+ ions in acid soils.

$$>AlOH \leftrightarrow >AlO^- + H^+ \tag{9-34}$$

$$>SiOH \leftrightarrow >SiO^- + H^+ \tag{9-35}$$

$$\text{low pH} \quad >FeOH \leftrightarrow >FeO^- + H^+ \quad \text{high pH} \tag{9-36}$$

$$R{-}OH \leftrightarrow R{-}O^- + H^+ \tag{9-37}$$

$$R{-}COOH \leftrightarrow R{-}COO^- + H^+ \tag{9-38}$$

R represents the organic molecule and > represents the soil mineral or silicate clay to which the functional group is attached. In acid soils the reactions are shifted to the left, toward the undissociated groups, because of the higher concentration of H^+ ions in the soil solution. As the soil pH is raised, the groups dissociate, becoming an important source of H^+ ions in near-neutral soils. If H^+ ions are added to the soil, the weak acid groups buffer the soil by reacting with the H^+ ions, and if OH^- ions are added, additional weak acid groups can dissociate, producing H^+ ions to react with the added OH^- ions. The >AlOH, >SiOH, and >FeOH groups tend to dissociate, over a fairly narrow pH range, while the groups associated with soil organic matter begin to dissociate at lower pH values and continue over a wider pH range.

9.2.5 Carbonic Acid and Aqueous Carbon Dioxide

Carbon dioxide gas and the associated aqueous species comprise a very important source of H^+ ions and an important buffer system in soils and natural waters. The atmosphere above the soil contains a concentration of CO_2(gas), expressed as the partial pressure of CO_2(gas) equal to 0.0003 atm. This gas dissolves in precipitation, eventually to produce carbonic acid (H_2CO_3). Although carbonic acid is a weak acid, it is a constant source of H^+ ions. The following equations detail the aqueous chemistry of CO_2(gas):

$$CO_2(\text{gas}) \leftrightarrow CO_2(\text{aq}), \qquad K_1 = 10^{-1.41} \tag{9-19}$$

$$CO_2(\text{aq}) + H_2O \leftrightarrow H_2CO_3, \qquad K_2 = 10^{-2.62} \tag{9-20}$$

$$H_2CO_3 \leftrightarrow HCO_3^- + H^+, \qquad Ka_1 = 10^{-3.76} \tag{9-10}$$

$$HCO_3^- \leftrightarrow CO_3^{2-} + H^+, \qquad Ka_2 = 10^{-10.25} \tag{9-11}$$

For a soil in equilibrium, with atmospheric CO_2(gas), the concentration of CO_2(aq) and H_2CO_3 are constant and can be found from equations [9-19] and [9-20].

$$CO_2(\text{aq}) = K_1 PCO_2(\text{gas}) = (3.93 \times 10^{-2})(3 \times 10^{-4}) = 1.179 \times 10^{-5} \, M \tag{9-39}$$

$$(H_2CO_3) = K_1 K_2 PCO_2(\text{gas}) = (3.93 \times 10^{-2})(2.42 \times 10^{-3})(3 \times 10^{-4})$$

$$= 2.83 \times 10^{-8} \, M \tag{9-40}$$

Since CO_2(gas) is a waste product of aerobic respiration, levels in the soil atmosphere have been shown to reach a maximum of about 0.01 atm. Hence H_2CO_3 levels can vary from about $2.8 \times 10^{-8} M$ to about $9.5 \times 10^{-7} M$.

The carbonic acid–bicarbonate ion (HCO_3^-) pair contributes to soil buffering, especially in acid soils.

$$H_2CO_3 \leftrightarrow HCO_3^- + H^+, \qquad Ka_1 = 10^{-3.76} \qquad\qquad [9\text{-}10]$$

When H^+ ions are added to the soil solution, whose H^+ concentration is close to Ka_1, a portion of the added H^+ ions combines with the HCO_3^- ions to form the undissociated H_2CO_3 molecule. When OH^- ions are added or H^+ ions are removed, additional H_2CO_3 molecules dissociate to produce the HCO_3^- and H^+ ions.

9.2.6 Calcareous Soils

The pH of calcareous soils is due to the presence of salts of a weak acid and a strong base. The salts, or solid phases, may be calcite or aragonite, different crystal arrangements of $CaCO_3$s, Mg-containing calcium carbonates, magnesite $MgCO_3$ or dolomite $CaMg(CO_3)_2$. Regardless of the solid phase present, when the salt dissolves the anion (CO_3^{2-}) of a weak acid, carbonic acid is released into the soil solution. For a soil containing calcite,

$$CaCO_3(\text{calcite}) \leftrightarrow Ca^{2+} + CO_3^{2-}, \qquad K_{sp} = 10^{-8.3} \qquad\qquad [9\text{-}41]$$

The carbonate ion will hydrolyze water, as shown by the equation

$$CO_3^{2-} + H_2O \leftrightarrow HCO_3^- + OH^-, \qquad K = K_w/Ka_2 \qquad\qquad [9\text{-}42]$$

If equations [9-41] and [9-42] are combined with the equations relating CO_2(gas) to CO_3^{2-}, the following equation relating the H^+ ion concentration of the soil solution to the solubility of the carbonate solid phase and the partial pressure of CO_2(gas) in the soil atmosphere can be derived.

$$(H^+)^2 = \frac{Ka_2Ka_1K_2K_1PCO_2(\text{gas})\,(Ca^{2+})}{K_{sp}} \qquad\qquad [9\text{-}43]$$

When the partial pressure of CO_2(gas) equals 3×10^{-4} atm and the concentration of Ca^{2+} equals $10^{-2} M$, the H^+ concentration is equal to $10^{-7.63} M$ or pH equals 7.63. When the level of Ca^{2+} is reduced to $10^{-4} M$, pH is increased to 8.63 because of the increased solubility of the $CaCO_3$ solid phase. In general, processes that increase Ca^{2+} decrease pH, and processes that decrease Ca^{2+} increase pH. Examples of processes that decrease pH include the presence of a calcium salts, such as gypsum, that are more soluble than calcite. The more soluble gypsum maintains high levels of Ca^{2+} ions that shift the calcite equilibria (equation [9-41]) toward the solid phase—this is an example of the common ion effect. This lowers the concentration of CO_3^{2-} ions by mass action and results in lower pH values.

Example 3: Acidification of a calcareous soil

Assume that a soil that contains 3% $CaCO_3(s)$ has a pH of 8.0. The soil has the same pH–percent base saturation relationship and CEC as shown in Figure 9-3 and Examples 1 and 2. A grower wants to raise plants, such as azaleas, rhododendrons, or pin oaks, which require acid soils with pH values around 5.5. How much sulfur must be added per hectare-furrow-slice (hfs) to change the pH from 8.0 to 5.5?

Solution   To solve this problem, the amount of sulfur needed to dissolve the 3% $CaCO_3$ must first be calculated, and then the amount of sulfur needed to change the percent base saturation from 100% to 50% must be calculated. The second calculation will be very similar, in procedure and in amount of sulfur needed, to Example 2.

The amount of $CaCO_3$ in the soil can be found by multiplying the ODwt by the percentage of $CaCO_3$ in the soil.

$$2{,}080{,}000 \text{ kg soil/hfs} \times 0.03 = 62{,}400 \text{ kg of } CaCO_3(s)/hfs$$

As discussed in Example 2, sulfur will be oxidized to sulfuric acid. Each mole of sulfur will produce 1 mol of sulfuric acid, and each mole of sulfuric acid can react with 1 mol of $CaCO_3$:

$$S + 1\tfrac{1}{2}O_2 + H_2O \rightarrow H_2SO_4$$

$$CaCO_3(s) + H_2SO_4 \leftrightarrow Ca^{2+} + SO_4^{2-} + CO_2(gas) + H_2O \qquad [9\text{-}44]$$

Since the molecular weight of sulfur is 32 and the molecular weight of calcium carbonate is 100, 32 units of S will dissolve and neutralize 100 units of $CaCO_3(s)$. The weight of S that is needed will be equal to the weight of $CaCO_3(s)$ in the soil multiplied by the ratio 32/100.

$$62{,}400 \text{ kg } CaCO_3(s) \times \frac{32}{100} = 19{,}968 \text{ kg of sulfur is needed}$$

In this example, 19,968 kg or 19.9 mt of sulfur acid is needed to remove the solid calcium carbonate. Once the solid $CaCO_3$ is removed, the soil will be at 100% base saturation and have a neutral pH. An additional amount ($\approx$ 2.66 mt) is needed to lower the pH to 5.5 (i.e., 50% base saturation). This is an unrealistic and uneconomical amount of S to add to the soil. The grower should not attempt to raise on this soil azaleas or other plants that require acid soil pH values, but to select species and cultivars that are adapted to the higher pH values of calcareous soils.

9.2.7 Sodic Soils

The high pH of sodic soils is due to a complex combination of exchange reactions and the autohydrolysis of water. The sodium ion (Na^+) is a monovalent ion that can easily be replaced on the exchange complex by H^+ ions. This exchange of H^+ for

Na$^+$ can occur readily, under certain conditions, and result in increased pH values in the soil.

$$H_2O \leftrightarrow H^+ + OH^- \qquad \text{[9-45]}$$

$$\text{clay–Na}^+ + H^+ + OH^- \leftrightarrow \text{clay–H}^+ + Na^+ + OH^- \qquad \text{[9-46]}$$

Consider two situations: first, a soil solution containing a high concentration of dissolved salts, and second, a soil solution with low level of dissolved salts such that the H$^+$ and OH$^-$ ions, resulting from the autohydrolysis of water, are significant ions. In the first case, when exchange of Na$^+$ occurs, it most probably would be for a basic cation such as K$^+$, Ca^{2+}, or Mg^{2+}, since they would be the major cationic species present in the soil solution. This would have no effect on the pH of the soil solution, since there would be no change in the balance of H$^+$ and OH$^-$ ions in the soil solution. In the second case, H$^+$, because of the low salt environment of the soil solution, is a major cationic species. When exchange for the Na$^+$ ion occurs, the exchange will often be with the H$^+$ ion. The exchange of H$^+$ for Na$^+$ results in a decrease in the H$^+$ ion concentration relative to the OH$^-$ ion concentration, and hence an increase in the pH of the soil solution. In a sense the clay is acting as an anion of a weak acid.

The Na$^+$ ion is not readily exchanged by the H$^+$ ion until there is a fairly large portion of the exchange capacity occupied by the Na$^+$ ion. Research has shown that the exchange of H$^+$ for Na$^+$ does not become a problem until more than 15% of the exchange complex is occupied by Na$^+$ and unless the dissolved salt content of the soil solution is low enough that the electrical conductivity of the soil's saturated extract is less than 4 dS m^{-1}.

If the pH of a sodic soil becomes high enough, a second mechanism becomes involved and buffers the soil. As the pH of the soil increases, CO$_2$(gas) in the atmosphere dissolves in the soil solution. Because of the high pH of the soil, the aqueous CO$_2$ system is shifted toward the carbonate species (CO$_3^{2-}$), resulting in a high concentration of the carbonate ion. When the CO$_3^{2-}$ ion concentration multiplied by the concentration of the Na$^+$ ion squared exceeds the K_{sp} (solubility product) of sodium carbonate (Na$_2$CO$_3$s), sodium carbonate will precipitate and buffer the pH of the soil solution.

$$2Na^+ + CO_3^{2-} \leftrightarrow Na_2CO_3(\text{solid}) \qquad \text{[9-47]}$$

$$K_{sp} = (Na^+)^2(CO_3^{2-}) \qquad \text{[9-48]}$$

Once the solid phase has formed, it will buffer the soil at pH values close to 10.5 by a mechanism very similar to that controlling pH in calcareous soils.

9.3   MEASUREMENT OF SOIL pH

Soil pH, as used in this book, refers to the negative logarithm of the hydrogen ion concentration (H$^+$) in the soil solution. Predictions about the chemistry of a soil constituent are often based on the pH of the soil solution. In actual practice, these predictions are based on measured pH values, and measured pH values may or may not be reliable estimates of the actual pH of the soil solution. McLean[1] identified the following factors that may influence the accuracy of the measured pH:

1. The nature and type of inorganic and organic constituents that contribute to soil acidity
2. The soil-to-solution ratio used to measure the pH
3. The salt content of the diluting solution used to achieve the desired soil-to-solution ratio
4. The CO_2(gas) content of the soil and solution
5. Errors associated with the standardization of the equipment used to measure pH and the liquid junction potential

The measurement of soil pH generally requires that additional water or salt solution be added to the soil. The most common soil-to-solution ratios, used to measure pH with a pH electrode, are a saturated paste and a 1:1 (water/soil) dilution, although 2:1 and 10:1 dilutions are often used in soil testing.

The addition of water or salt solution can result in the H^+ concentration in the final diluted solution being different from the H^+ concentration in the original soil solution. Upon dilution, weak acid groups associated with soil organic matter and soil minerals may associate or dissociate. The salt content of the diluting solution can result in the release of H^+ ions from the exchange complex. Depending on the nature and type of colloids and weak acid groups present, the effect of dilution may differ from one soil to another, making consistent corrections difficult, if not impossible. The use of 0.01 M $CaCl_2$ has been recommended to minimize the effects of dilution, since this concentration of salt is generally considered to represent the salinity of a "normal" soil solution in the field.

Dilution of the soil solution may also result in differences in the type and amount of dissolved gases. Gases such as CO_2(gas) are present in the soil in higher concentrations than in the aboveground atmosphere. Dilution and stirring, as used in the actual measurement of soil pH, can result in the loss of such gases, and in differences between the measured and actual pH of the soil.

Soil pH measurements are normally made using either electrometric or colorimetric techniques. Colorimetric techniques are based on structural changes of chromophore groups of organic compounds in response to the pH of the soil solution. Normally, the pH indicator is added to the soil and the resulting color is compared with a color chart. Problems are encountered relating to the interference of the native color of the soil with the color of the pH indicator and the error in matching the resultant color with a standard color chart.

[1]E. O. McClean, Soil pH and lime requirement, in *Methods of Soil Analysis: Part 2, Chemical and Microbiological Properties*, A. L. Page, R. H. Miller, and D. R. Keeney (eds.), American Society of Agronomy, Milwaukee, Wis., 1982.

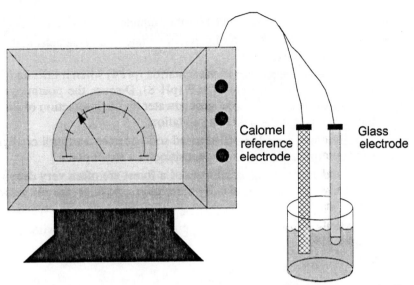

Figure 9-4   Typical electrometric (pH meter) setup for measurement of soil pH.

Electrometric methods, based on pH-sensitive glass electrode, are generally accepted as the standard method for determining soil pH. The glass electrode consists of a thin glass membrane containing a solution of HCl and KCl in contact with an internal AgCl or Hg–Hg_2Cl_2 electrode, which is connected to the pH meter. The thin glass membrane is in contact with external soil solution (diluted soil solution) of unknown (H^+) ion concentration and in contact with the internal HCl–KCl solution of known (H^+) ion concentration.

The glass membrane has cation-exchange properties and is particularly sensitive to H^+ ions. When the glass membrane is placed in contact with an external solution, a potential is developed across the membrane and is sensed by the internal AgCl or Hg–Hg_2Cl_2 electrode.

The measurement of pH using the glass electrode requires that there be a complete conducting path for the electrons. The circuit is completed using a second external reference electrode, usually a calomel electrode, which is placed in the unknown external solution and connected to the pH meter. Figure 9-4 gives a typical arrangement for the electrometric measurement of pH.

S T U D Y   Q U E S T I O N S

1. Describe the buffer(s) mechanisms that control soil pH in the following pH ranges. (a) 2 to 4; (b) 4 to 5.5; (c) 5.5 to 6.8; (d) 6.8 to 7.2; (e) 7.2 to 8.5; (f) 8.5 to 10.5.

2. Explain why calcium sulfate is not a suitable material to increase soil pH.

3. Explain why the addition of calcium sulfate to a calcareous soil can result in lower pH values.

4. Discuss the effects of acid precipitation (pH 4) when it falls on (a) an acid soil (pH 5) or (b) a calcareous soil (pH 8). Discuss the potential effects of the materials released into the groundwater by the interaction of the soil's buffer mechanisms and the acid precipitation.

5. Explain how an acid, well-drained soil becomes a neutral or slightly alkaline sediment upon erosion and deposition in a small pond.

6. The understory shrubs and trees of a forest are often very deep rooted. What effect will nutrient uptake, by these plants, have on the pH values of surface horizons?

7. Calculate a lime requirement in terms of an acre-furrow-slice for Example 1 (page 214).

10

Salt-Affected Soils

10.1 CLASSIFICATION OF SALT-AFFECTED SOILS

Salt-affected soils are classified into three categories: saline, saline-sodic, and sodic soils (Figure 10-1). Salt-affected soils are found predominately in arid regions, but they are also found in the humid midwest (sodic soils), in areas that are influenced by marine environments or impacted by human activities. Soils in close proximity to marine environments can be salinized by salt spray from the ocean or by saltwater incursion in the groundwater as the fresh water table is lowered. Human activity that can result in the formation of salt-affected soils include irrigating with surface waters or groundwaters that contain dissolved salts, deicing of roads and walkways, and the pumping of oil and associated salt water out of oil wells.

Saline Soils.   Saline soils are soils that contain high enough levels of dissolved salts in the soil solution to affect the growth of sensitive plants. Saline soils do not have a sodic or a potential sodic problem. Specifically, saline soils are soils that have a conductivity of their saturated extracts of 4 dS m^{-1} or greater. A saturated extract is obtained by saturating a soil sample with distilled water and then using vacuum to remove the gravitational water. Pure water is a poor conductor of electricity, but as dissolved salts are added to pure water its conductivity increases. The conductivity of a soil's saturated extract is a quantitative measure of the amount of dissolved salts in the soil solution. Saline soils have less than 15% exchangeable sodium and pH values less than 8.5.

Saline-Sodic Soils.   Saline-sodic soils contain high enough levels of soluble salts to interfere with the growth of sensitive plants. These soils have a potential

(a)

(b)

(c)

Figure 10-1 Salt-affected soils. (a) The soil material shows the normal soil color (dark) and the salt crust (white) coating the surface of the soil. These crusts are formed when evaporation removes the water and leaves the dissolved salts on the soil surface. (b) Salt-affected soil along Chalk Creek, Utah, illustrates a common landscape position that is depressional to other sites in the landscape. (c) Cotton crop failure, Imperial Valley, California, caused by high salinity levels in the soil. Note the white salt crusts near the top of the furrow ridges. (Courtesy of USDA Soil Conservation Service.)

sodic problem, which will be manifested if improper reclamation procedures are used. Specifically, saline-sodic soils have conductivities of their saturated extracts that are greater than 4 dS m^{-1} and they contain greater than 15% exchangeable sodium with pH values less than 8.5. The high concentration of dissolved salts prevents the dispersion of the soil colloids and helps maintain soil porosity and permeability. The high dissolved salt concentrations also prevents the development of the elevated pH values associated with sodic soils.

Sodic Soils. Sodic soils are soils that *do not* have excess levels of dissolved salts, but *do* have high levels of exchangeable sodium. Specifically, these soils have saturated extract conductivities less than 4 dS m^{-1}. In addition, they contain greater

TABLE 10-1   Salt tolerance of crops

Crop	Salinity at initial yield decline (threshold) (dS m^{-1})	Yield decrease per unit increase in salinity beyond threshold[a] (%)
Alfalfa	2.0	7.3
Almond	1.5	19
Barley	8.0	5.0
Beet, garden[b]	4.0	9.0
Bermudagrass	6.9	6.4
Blackberry	1.5	22
Broccoli	2.8	9.2
Carrot	1.0	14
Clover, aslike, red	1.5	12
Corn, grain	1.7	12
Cotton	7.7	5.2
Grapefruit	1.8	16
Peanut	3.2	29
Rice, paddy	3.0	12
Ryegrass, perennial	5.6	7.6
Soybean	5.0	20
Sudangrass	2.8	4.3
Sugar beet[b]	7.0	5.9
Sugarcane	1.7	5.9
Tomato	2.5	9.9
Wheat	6.0	7.1
Wheatgrass	7.5	4.2

Source: E. V. Maas and G. J. Hoffman, Crop salt tolerance—current assesment, *J. Irrig. Drain.* 103:115–134, (1977).
[a] Relative yield for salinity exceeding threshold [$Y = 100$ - (value in this column)(dS m^{-1} - A)]. For alfalfa with a salinity of 5.4 dS m^{-1}, $Y = 100$ - $7.3(5.4 - 2.0) = 75\%$. Yield at a soil salinity of 5.4 dS is 75% of yield at threshold.
[b] Sensitive during germination. Salinity should not exceed 3 dS m^{-1}.

than 15% exchangeable sodium and have pH values greater than 8.5. The high level of exchangeable sodium results in partial dispersion of the soil colloids and high pH values. These soils are very difficult to reclaim, due to their very low permeabilities, which interfere with reclamation procedures.

The poor growth of plants on salt-affected soils is related primarily to three properties associated with these soils: (1) the effect of dissolved salts on the availability of water to plants, (2) the effect of Na^+ on the physical properties of the soil and on soil pH, and (3) specific toxicity due to the high concentration of certain ions, such as chloride (Table 10-1).

Effect of dissolved salts. Plants are energetically capable of using water that is held in the soil at matric potentials between $-\frac{1}{3}$ and -15 bar (see Chapter 5). The total potential that water is under in the soil is primarily the result of two major forces. The first force is due to the attraction of water by the soil solids and is called the matrix potential (ϕ_m); the second is due to the effect of salts dissolved in the soil solution and is called the osmotic potential (ϕ_o). In nonsaline soils, the matrix potential is approximately equal to the total moisture potential (ϕ_t), while in saline and saline-sodic soils the osmotic potential makes a substantial contribution to the total moisture potential of the soil:

$$\phi_t = \phi_m + \phi_o \qquad\qquad [10\text{-}1]$$

Crop yields begin to decrease when the total moisture potential falls below -2 to -3 bar, even though plants will survive potentials to about -15 bar. Since dissolved salts increase the total moisture potential, any increase in the amount of dissolved salts in the soil solution can decrease crop yields. The actual effect of a given salt addition would be greatest in soils that have low water-holding capacities and in drought-stressed soils.

Effect of sodium. Soil colloids in the pH ranges found in salt-affected soils are primarily negatively charged. The source of the negative charge is either isomorphous substitution or the disassociation of weak acid groups at the edge of the clay particles and at sites of crystal irregularities. The negative charge on the colloids results in oppositely charged ions, counterions, being attracted to the colloids to maintain electroneutrality. The volume occupied by the counterions and the associated water molecules dictates how closely the colloidal particles can approach each other and whether the colloids are dispersed or flocculated. When divalent cations dominate the counterion swarm (i.e., Na^+ is less than 15% of the total counterions), the double layer (i.e., the counterion swarm and the associated water molecules) is compact and the colloids are flocculated. When monovalent cations, such as Na^+, become a significant portion, approximately 15%, of the double layer, the double layer is thick enough to cause some of the colloids to disperse (see Chapter 8). The increased size of the double layer is due primarily to the need to have two monovalent cations to replace one divalent cation.

In addition, it has been shown[1] that in mixed calcium-sodium colloidal systems, Ca^{2+} occupies the interlayer regions of swelling clays until sufficient Na^+, from 10 to 15%, is present to initiate interlayer occupation. Once this level of exchangeable Na^+ is exceeded, Na^+ not only results in the dispersion of a portion of the colloidal particles, but also results in each swelling clay particle expanding and occupying a larger volume. Both processes result in a dramatic decrease in soil porosity and permeability.

---

[1]E. Breseler et al., *Saline and Sodic Soils: Priciples, Dynamics, Modeling*, Springer-Verlag, New York, 1982.

10.2   DETERMINATION OF SALINITY AND SODIUM HAZARDS IN SOILS AND WATERS

Measurement of salinity. Salinity in saturated-soil extracts or irrigation waters was first measured, quantitatively, by evaporating a known volume of solution to dryness and gravimetrically determining the quantity of dissolved materials. These determinations were time consuming and subject to major limitations. The method of choice today is to measure the electrical conductivity of the saturated soil solution or the irrigation water and to calculate salinity. This technique is simple, rapid, and requires only a modest investment in equipment. Electrical conductivity has been correlated with the osmotic potential of the water by the U.S. Salinity Laboratory staff using the following approximate equation.

$$\phi_o(\text{bar}) = -0.36 \times EC \ (\text{dS m}^{-1}) \qquad [10\text{-}2]$$

where ϕ_o is the osmotic potential in bars and EC is the electrical conductivity in dS m^{-1}. A saline or saline-sodic soil with a minimum EC of 4 dS m^{-1} would have an osmotic potential of −1.44 bar.

Sodium adsorption ratio. The sodium adsorption ratio (SAR) is used to describe ion exchange and is based on mass-action principles. The equation was developed by the U.S. Salinity Laboratory staff[2] to predict the sodium hazard of irrigation waters. The equation for SAR is based on saline systems containing predominantly calcium, magnesium and sodium cations:

$$SAR = \frac{[\text{Na}^+]}{\sqrt{\dfrac{[\text{Ca}^{2+}] + [\text{Mg}^{2+}]}{2}}} \qquad [10\text{-}3]$$

The SAR equation had been used extensively not only to predict the sodium hazard of irrigation waters (Figure 10-2), but also to determine the sodium status of soils. The determination of SAR on the saturated extract of a soil is less problematic than are measurements of CEC and of exchangeable cations.

The SAR of an irrigation water or a saturated-soil extract has been related to the exchangeable sodium percentage (ESP) of the soil.[3] The relationship is based on the establishment of an ion-exchange equilibrium between the soil and the irrigation water after repeated applications of water. The potential ESP of a soil can be predicted from the SAR of an irrigation water. An SAR of 13 is used to differentiate between sodic and nonsodic soils. If an SAR value of 13 is substituted into equation [10-4] an ESP value of 15 is predicted. When the SAR of a soil's saturated extract is used, the SAR is reflective of the present exchangeable sodium status of the soil.

$$ESP = \frac{(100\,(-0.0126 + 0.01475 \times SAR))}{(1 + (-0.0126 + 0.01475 \times SAR))} \qquad [10\text{-}4]$$

[2]*Diagnosis and Improvement of Saline and Alkali Soils*, U.S. Salinity Laboratory Staff, Agriculture Handbook No. 60, 1954.

[3]Ibid.

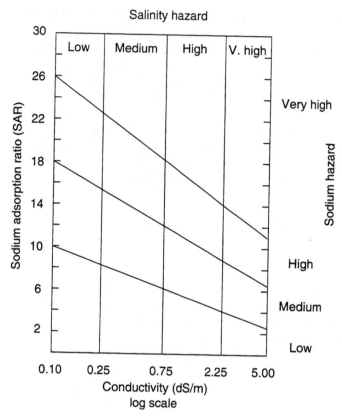

Figure 10-2 Sodium and salinity hazards of irrigation waters. (U.S. Salinity Laboratory Staff, *Diagnosis and Improvement of Saline and Alkali Soil*, Agricultural Handbook No. 60.)

10.3 RECLAMATION OF SALT-AFFECTED SOILS

The first step in reclaiming a salt-affected soil is to establish drainage. This may be difficult or impossible to do because of the landscape position of the soil (Figure 10-3). Salt-affect soils often are found in the low areas of closed basins in arid regions. In these site positions no outlets exist for the drainage water. The next step, which is often as difficult as the first, is to locate an adequate supply of good-quality water (i.e., containing low salt and low sodium) to be used to leach the soil.

Saline soils. Reclamation of saline soils involves leaching the soil with good-quality water to remove the excess salts. The amount of water needed depends on the salt content of the soil and the quality of water used to leach the soil. When the excess soluble salts are removed, the osmotic pressure of the soil solution will not make a significant contribution to the total soil water potential.

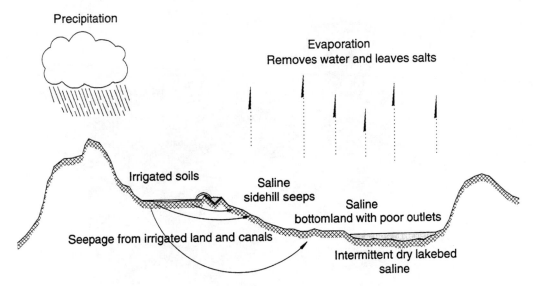

Figure 10-3   Relative landscape relationships of salt-affected soils.

Saline-sodic soils.   Reclamation of saline-sodic soils involves a two-step process, because these soils have both a salinity problem and a potential sodic problem. The soil must first be leached with a water containing Ca^{2+} ions to facilitate the exchange and removal of the sodium on the soil colloids. Then the soil must be leached with a good-quality irrigation water to remove the excess soluble salts. A good-quality irrigation water can be amended with a soluble source of calcium, such as gypsum, to achieve the first step.

If a saline-sodic soil is leached with a good-quality irrigation water before removing the exchangeable sodium, the level of salt in the soil solution will be lowered and a sodic soil could be formed. When the soluble salts are removed, the exchangeable sodium can cause the soil colloids to disperse and the pH value to increase. This would create a more serious soil problem than originally existed and one that is much more difficult to correct.

Sodic soils.   Because of the low permeabilities of sodic soils, it is very difficult to leach water through the soil profiles. Reclamation of these soils usually consists of mixing a soluble calcium source with the sodic horizons. Since the affected horizons often include the B horizons, this involves mixing the calcium source to a much greater depth of soil than is normally done for other soil amendments, such as fertilizers or limestone. Powdered gypsum is often used to accomplish the required Ca^{2+}–Na^{+} exchange. When the sodic (natric) horizons occur above calcareous materials, the calcareous materials can be mixed with the sodic horizons by plowing using very large plows (Figure 10-4). The calcareous materials slowly dissolve in the high pH sodic horizons and in time facilitate the required Ca^{2+}–Na^{+} exchange.

Figure 10-4   Large plows such as shown in this photograph can be used to mix soluble calcium sources with sodic horizons to facilitate reclamation. This plow is being used to mix sand deposited on a floodplain with the underlying soil. The depth of plowing is 42 in. (Courtesy of USDA Soil Conservation Service.)

10.4 MANAGEMENT OF SALT-AFFECTED SOILS

Salt-affected soils occur naturally, but they can also be formed by human activities, such as irrigation, waste disposal, and deicing roads and highways. Irrigation, especially in climates where rainfall in noncrop seasons is not sufficient to leach excess salt out of the soil profile, has the potential for the formation of salt-affected soils. Probably the single most critical issue in the management of irrigated soils is the prevention of soil degradation due to accumulation of salt or sodium.

Irrigation involves the application of surface waters or groundwaters to soils. These waters differ from natural precipitation in the amount and composition of dissolved salts. When surface water or groundwater is used for irrigation, water is lost by evapotranspiration and the dissolved salts can accumulate in the soil. The use of irrigation waters represents a potential salinity or sodium hazard to the soil. The actual hazard represented by an irrigation water is dependent on the amount and type of dissolved salts (Figure 10-2).

Water with conductivities in the range 0.1 to 0.25 dS m^{-1} have low salinity hazards, while water with conductivities in the range 0.25 to 0.75, 0.75 to 2.25, and 2.25 to 5.0 dS m^{-1} represent, respectively, medium, high, and very high salinity hazards. Water with low salinity hazards require little or no special management, although the conductivity of soil should regularly be monitored and excess water applied to remove any salt buildup, as indicated by increases in the conductivity of the soil's saturated extract. Waters with higher salinity hazards may be used for irrigation only if excess water is routinely applied to leach the excess salts. The amount and frequency of the required leaching is dependent on the conductivity of the irrigation water.

The leaching requirement (LR) is a function of the electrical conductivity of the irrigation water (EC_{iw}) and the electrical conductivity of either the drainage water (Ec_{dw}) or the maximum acceptable electrical conductivity of the soil solution. The maximum acceptable electrical conductivity is often set for individual crops as the level of salinity that will cause a 50% reduction in yield.

$$LR = \frac{EC_{iw}}{EC_{dw}} \times 100 \qquad [10\text{-}5]$$

If an irrigation water has a conductivity of 2 dS m^{-1} and the maximum acceptable conductivity of the soil solution is 4 dS m^{-1}, the leaching requirement would be 2/4 × 100, or 50%. This means that an excess of irrigation water amounting to 50% more than needed for consumptive use must be applied to prevent the accumulation of salts in the soil profile. Depending on the availability and cost of the irrigation water, this may be an unacceptable situation. Table 10-2 gives the leaching requirements as a function of both the conductivities of the irrigation water and the maximum acceptable conductivity of the soil solution.

Irrigation waters can also represent a sodium hazard to soils. The sodium hazard is dependent on the SAR (equation [10-3]) of the irrigation water. Waters with SAR values in the range 0 to 10 are low, while values in the range of 10 to 18, 18 to 26, and >26 are, respectively, medium, high, and very high. The sodium hazard of an irrigation water can be lowered by adding a soluble calcium source, such as gypsum, to the irrigation water or by mixing with higher-quality waters.

TABLE 10-2 Leaching requirements (percent)[a]

Conductivity of irrigation water (dS m^{-1})	Maximum acceptable conductivity of soil solution or drainage water (dS m^{-1})			
	2	4	8	12
0.1	5.0	2.5	1.3	0.8
0.25	12.5	6.2	3.1	2.1
0.75	37.5	18.8	9.4	6.3
2.25	—	56.2	28.1	18.7
5.0	—	—	62.5	41.7

[a]Calculated from equation [10-5] (U.S. Salinity Laboratory).

STUDY QUESTIONS

1. Why are colloids dispersed in sodic soils but not in saline-sodic soils?
2. Why are crop yields decreased when grown on saline soils?
3. What are the preferred techniques to measure soil salinity and the amount of exchangeable sodium?
4. How would you predict if an irrigation water presents a salinity or sodium hazard to soils?
5. Briefly discuss the reclamation of saline, saline-sodic, and sodic soils.
6. What is a leaching requirement, and how might it be determined?
7. Why is "soft" water considered good-quality water for industrial use and household uses, such as bathing and washing clothes, but poor-quality water for irrigation or watering household plants? Would "hard" water be a good irrigation water?
8. How can you amend (decrease) the sodium hazard of an irrigation water?

11

Soil Organisms, Organic Matter, and Organic Soils

11.1 SOIL ORGANISMS

Soil organisms include the roots of higher plants, bacteria, actinomycetes, fungi, moles, badgers, crayfish, earthworms, protozoa, and numerous other organisms. Soil organisms include autotrophic and heterotrophic organisms. Autotrophic organisms utilize solar radiation or the bond energy of inorganic chemicals to drive their metabolisms, while heterotrophic organisms are dependent on the oxidation of reduced organic materials produced by other organisms for their metabolic energy. Soil organisms can be aerobic or anaerobic, symbiotic, parasitic, or free living. Aerobic organisms utilize elemental oxygen [O_2(gas)] as a terminal electron acceptor in their metabolisms and produce CO_2(gas) as a waste product. Anaerobic organisms use NO_3^-, Fe^{3+}, and a variety of other chemicals as terminal electron acceptors and produce a variety of waste products, such as methane or methyl or ethyl alcohol. Symbiotic organisms live in a mutually beneficial association with other species. Parasitic organisms live in an association with other species, but at the expense of the other species. Free-living organisms are independent of symbiotic or parasitic associations with other species.

Climate and the activity of organisms are the two dominant soil formation factors. Climate controls the intensity of soil formation, dictating the amount of heat energy and water available for soil formation. To a large extent, climate also controls the type of organisms found in and on the soil, as well as their activity. Organisms add organic matter to soils, mediate many important chemical reactions, mix the soil, increase the availability of nutrients, and in some cases, are important factors in plant and animal diseases. The presence of a variety of organisms and their associated activities are primarily what differentiates soil from the underlying regolith.

Five kingdoms of organisms are currently recognized.[1] These include organisms that have their genetic material not enclosed in a nucleus, but mixed through the cytoplasm of the cell (procaryotic), and organisms that have their genetic material enclosed in a nucleus (eucaryotic).

Monera. Procaryotic organisms, bacteria, cyanobacteria (blue-green algae), actinomycetes.

Protista. Unicellar eucaryotic organisms, protozoa, and slime molds.

Fungi. Unicellar and multicellar eucaryotic organisms, includes nonphotosynthetic organisms that obtain their energy from living organisms (mildews, rusts, and smuts) or that obtain their energy from dead organic materials (molds and mushrooms).

Plants. Unicellar and multicellar eucaryotic organisms; most obtain their energy from photosynthesis and are stationary; includes red, brown, and green algae, mosses and ferns, and seed-bearing plants.

Animals. Multicellar eucaryotic mobile organisms with central nervous systems; includes flat, round, and segmented worms, spiders, insects, mollusks, and mammals.

The soil contains numerous examples of each kingdom, many classified species, and probably many unknown species as well (Table 11-1).

11.1.1 Animals

Large animals that burrow in the soil, such as moles, badgers, gophers, and crayfish, mix the soil primarily through their burrowing action. The burrows help to aerate the soil, but affect water movement only when an excess of water occurs on the soil surface and the water can enter directly into the burrows at the surface. Burrows could facilitate the movement of water through impermeable layers, but most burrowing animals, with the exception of crayfish, build their burrows in well-drained soils. Because of their burrowing activities, large animals are destructive of crops, gardens, and lawns, and are considered to be undesirable by most farmers, horticulturists, and homeowners.

Smaller burrowing animals, especially earthworms, can have very beneficial effects on soil tilth and fertility. Earthworms are heterotrophic, utilizing the organic residues as an energy source. As such, earthworms constantly ingest organic residues and soil particles and pass them through their digestive tracts. The digestive processes initiate the decay of the organic residues and also increase the

---

[1]D. D. Ritchie and R. Carola, *Biology*, Addison-Wesley Publishing Co., Reading, Mass., 1979, p. 138.

TABLE 11-1   Approximate populations of selected organisms in a 1-gram soil sample

Kingdom	Example	Number of organisms/gram
Monera	Bacteria	10^8
Monera	Actinomycetes	10^7
Fungi	Fungi	10^5
Plants	Algae	10^4
Protista	Protozoa	10^4
Animals	Nematodes	10
Animals	Earthworms	0.001

solubilities of soil minerals. Earthworm casts, "excrement," have been shown to have a much higher nitrogen content as well as higher plant availability of many plant nutrients (Table 11-2). The higher nitrogen content comes from the incorporation of organic wastes into the casts within the digestive tract of the earthworm. The greater plant availability of nutrients is due primarily to the action of the digestive process within the earthworm, increasing the solubility of minerals.

Earthworms aerate and stir the soil, and as shown in Table 11-2, earthworm casts have lower bulk densities than the original soil and are much higher in organic

TABLE 11-2   Analysis of earthworm cast compared to the bulk soil

Property	Cast	Ekiti series (paralithic Ustropept)
Sand (%)	57.0	77.0
Silt (%)	30.3	12.7
Clay (%)	12.7	10.3
pH (1:1)	5.4	5.4
Bulk density	1.01 g/cm^3	1.33 g/cm^3
CEC	8.9 cmol (+)/kg	2.9 cmol (+)/kg
Exchangeable Ca	5.0 cmol (+)/kg	1.5 cmol (+)/kg
Exchangeable Mg	3.2 cmol (+)/kg	0.8 cmol (+)/kg
Exchangeable K	0.6 cmol (+)/kg	0.2 cmol (+)/kg
Exchangeable Na	0.10 cmol (+)/kg	0.06 cmol (+)/kg
Bray P_1	11.2 ppm	6.4 ppm
Total N (%)	0.36	0.09
Organic C (%)	3.05	1.43
Number of raindrops required to disrupt aggregate	884	50

Source: D. De Vleeschauwer and R. Lal, Properties of worm casts under secondary tropical forest regrowth, *Soil Sci.,* 132:175–181 (1981).

matter and fertility. In addition, earthworms improve structural stability through incorporation of humus. In one study (Table 11-2) it took 884 raindrops to disrupt aggregates in earthworm casts compared to 50 raindrops to disrupt unaltered soil aggregates. Earthworms prefer moist, well-aerated soils with pH values in the range 5.0 to 8.4. They prefer medium- to fine-textured soils high in organic matter with low salt contents but high in available calcium. Under ideal conditions earthworms can ingest as much as 33.6 mt of earth per hectare per year. This means that earthworms could injest the equivalent of an hectare-furrow-slice of soil in less than 100 years. Soil tillage tends to disrupt earthworm burrows and hence tends to lower the numbers of earthworms in a soil. Earthworms feed on most grasses and deciduous leaves but not on the waxy, resinous needles of conifers.

Ants and termites also play a role in soil formation. Termites are especially important in tropical and subtropical forests and grasslands. Termites, and to a lesser extent ants, mix subsoil materials with the topsoil during construction of their mounds. Both termites and ants are important as initiators of the decay process. Termites and the microorganisms in their intestinal tracts are very important in the breakdown of cellulose and hemicellulose in woody tissues. In tropical and subtropical areas, termites play a more important role in soils than do earthworms.[2]

Nematodes are small threadlike round worms whose size varies from microscopic to barely visible. Nematodes are omnivorous, predaceous, or parasitic. *Omnivorous* nematodes feed on dead plant and animal tissues, *predaceous* nematodes prey on algae, bacteria, fungi and other nematodes, while *parasitic* nematodes infest plant roots and can cause great economic damage to crops when present in large numbers. Nematodes are important disease organisms on soybeans, sugar beets, alfalfa and other crops.

11.1.2 Protozoa

Protozoa are heterotrophic organisms. This means that they are dependent on other organisms for a source of reduced organic carbon to drive their metabolisms. Most protozoa are predatory or parasitic organisms, in that they prey on other living organisms or they infect other organisms. In tropical climates, many important human diseases are the result of soil borne protozoa.

11.1.3 Fungi

Fungi are the dominant decay organisms in acid soils. Fungi live on dead organic materials or as disease organisms on living plant and animal tissues. Fungi are much more tolerant of low soil pH values than are bacteria, and because of their tolerance are very important in acid forest soils.

One group of fungi, the mycorrhizae fungi, form very important symbiotic

---

[2]R. Lal, *Tropical Ecology and Plant Edaphology*, John Wiley & Sons, New York, 1987.

relationships with plant roots. The hyphae of mycorrhizae fungi penetrate plant roots, ectomycorrhizae penetrate in between cells, and endomycorrhizae fungi penetrate into the cell interiors. Filaments extend outward from the plant root into the soil and dramatically increase the absorbing surface area of the root. The higher plant provides the fungi with energy from the products of its metabolism. Mycorrhizae increase the ability of the plant to extract water and nutrients, such as phosphorus, from the soil. Table 11-3 compares the nutrient content of corn in the presence and absence of mycorrhizae. The success or failure of introduced tree species is often dependent on the presence or absence of the correct mycorrhizae. Mycorrhizae infections make many tree species more competitive in nutrient poor soils. The benefit of mycorrhizae in well-fertilized soils is not as great, since the absorbing surface of the root is not as critical to nutrient uptake as in nutrient poor sites.

11.1.4 Higher Plants and Algae

Higher plants and algae are the primary producers in soils. Higher plants are photosynthetic organisms that utilize sunlight, carbon dioxide, and water to produce reduced organic carbon compounds. These compounds provide the energy for most of the biotic community in the soil. The type of plant, whether it is an annual or perennial grass or herb, a deciduous or conifer tree, has a major effect on the type of soil that forms. Compare the map (Figure 3-25) showing the distribution of major vegetation communities with the map (Figure 4-4) showing the distribution of soil orders in the United States. The type of plant community, coupled with climate, provides the most significant effect on the type of soil that forms.

The type and placement of the organic matter produced by the plant community also has a major effect on the soil horizons that form. Prairie species tend to produce a residue that is higher in basic cations than forest species. In addition,

TABLE 11-3 Nutrient content of corn with and without mycorrhizae infection

Element	Philo soil (Fluvaquentic Dystroept)	
	With mycorrhizae	*Without mycorrhizae*
P (%)	0.142	0.070
K (%)	2.40	2.16
Ca (%)	1.80	1.70
Mg (%)	0.54	0.58
Zn (ppm)	56.5	16.4
Cu (ppm)	8.2	5.5
Fe (ppm)	55	67
Mn (ppm)	73	62

Source: D. H. Lambert, D. E. Baker, and H. Cole, Jr., The role of mycorrhizae in the interactions of phosphorus with zinc, copper and other elements, *Soil Sci. Soc. Amer. J.* 43:976–980 (1979).

many prairie species are annual grasses that die each fall, contributing large amounts of organic matter not only to the soil surface, but also within the soil when their large fibrous root systems die. Trees, on the other hand, tend to survive for many years, contributing only their leaf or needle fall each year to the soil surface. Soils formed under prairie vegetation tend to have thick A horizons high in organic matter, while soils formed under forest tend to develop O horizons above thin A horizons which contain lower amounts of organic matter. The decay process tends to produce a more acidic leachate from forest leaves and needles than from grasses because of the lower base contents of the leaves and needles. Trees tend to dominate in wetter climates and grasses in dryer climates. Compare Figure 3-25, showing the distribution of vegetation in North America, with Figure 3-22, showing the amount of precipitation. The greater amounts of rainfall in humid forest regions tend to result in greater amounts of leaching, more weathering, and acid soils.

11.1.5 Bacteria and Actinomycetes

Bacteria and actinomycetes are the most important decay organisms in slightly acidic, neutral, and basic soils (Figure 11-1). In addition, bacteria and actinomycetes mediate many important soil reactions, such as the fixation of N_2(gas) and the oxidation of reduced sulfur compounds. These organisms can be classified in several ways: by their sources of energy and carbon, by their terminal electron acceptors, and by their ability and mechanisms to fix N_2(gas). Photoautotrophs are organisms that use light energy in the process of photosynthesis and obtain their carbon from CO_2(gas). Photoheterotrophs obtain their energy from light and their carbon from organic matter. Chemoautotrophs obtain their energy from the oxidation of reduced inorganic materials, such as sulfide, and their carbon from CO_2(gas). Chemoheterotrophs obtain their energy from the oxidation of organic matter and their carbon from organic matter. Most of the decay organisms are chemoheterotrophs.

Bacteria can be aerobic, that is, dependent on O_2(gas) as a terminal electron acceptor or anaerobic, that is, capable of using NO_3^-, Fe^{3+}, or a variety of other oxidized materials as terminal electron acceptors. Some bacteria are facultative anaerobes. They utilize O_2(gas) in well-aerated soils and other oxidized materials in reduced, poorly aerated soils. Bacteria and actinomycetes are responsible for mediating many important soil reactions, many of which are parts of the carbon, nitrogen, or sulfur cycles and are discussed in later chapters.

Bacteria and actinomycetes are also important nitrogen-fixing organisms. Nitrogen fixation consists of reducing N_2(gas) to NH_3 and then combining the ammonia into organic compounds. Organisms can fix nitrogen in symbiotic associations with higher plants and as free living organisms in the soil. Bacteria (*Rhizobium* and *Bradyrhizobium*) form symbiotic associations with plants of the legume family, and actinomycetes (*Frankia*) form symbiotic associations with several different plant families. Actinomycetes and some of the fungi form a variety of antibiotics in

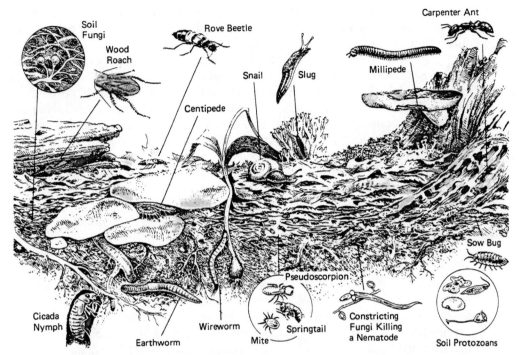

Figure 11-1 Examples of major groups of soil organisms. (*Source: Ecology of Field Biology*, Second Edition, by Robert Leo Smith. Copyright © 1966, 1974 by Robert Leo Smith. Reprinted by permission of Harper Collins Publishers Inc.

the soil. Antibiotics derived from actinomycetes include streptomycin, aureomycin, terramycin, and neomycin.

Aerobic bacteria thrive under the same conditions that are optimum for the growth of higher plants; nearly neutral pH values, moderate temperatures, and moist, well-aerated soils. Some species are very tolerant of extreme conditions. For example, *Thiobacillus ferroxidans* can tolerate soil pH values around 2. Some bacteria are thermophiles or psychrophiles, having optimum growth in very warm or very cold soils, respectively.

11.2 SOIL ORGANIC MATTER

Soil organic matter is an essential component of productive soils. Soil organic matter improves the physical and chemical properties of the soil and contains essential plant nutrients that are released upon its decomposition. Soil organic matter is a collection of plant, animal, and microbial residues in various stages of decomposition. Humus is the more or less stable fraction of soil organic matter. Humus is composed of a variety of substances, ranging from the decomposition products of plant and animal remains, to new microbial products, as well as condensation or polymerization products of both of these. The colloidal properties of humus, humic, and fulvic acids are discussed in Chapter 8.

Soil organic matter influences soil aggregation and thus tilth, permeability, soil aeration, and erodibility. Soil organic matter also increases the water-holding capacity and CEC of soils. Humic acid extracted from midwest soils has been shown to have CEC up to 394 cmol$_c$/kg of humic acid[3]; this is considerably higher than that found for inorganic colloids. Soil organic matter is the primary source of nitrogen in soils and is an important source of sulfur, phosphorus, and some micronutrients. As a general rule, within a given region soil productivity will be highly correlated with soil organic matter content. This does not mean that low organic matter soils are not productive—many sandy and desert soils are very productive under irrigation—but even in these areas, initial soil productivity, as well as the ability to sustain the level of productivity, are highly correlated with soil organic matter.

The amount of humus in a soil is a function of the soil formation factors. The effect of these factors on organic matter content are in the order climate > vegetation > topography = parent material > age. Climate, that is, rainfall and temperature, influence the amount and type of vegetation as well as the rate and efficiency of decomposition. The organic matter content of a soil increases with increasing precipitation up to the limit set by temperature. As a general rule, soil organic matter in the United States increases from the east side of the Rocky Mountains to the east with increasing precipitation (Figure 3-21), and from south to north with decreasing temperature. In soils, every 10°C (18°F) increase in mean annual temperature results in the organic matter content being reduced by $\approx$ 1/3 to 1/2, if all other factors are constant.

Vegetation (Figure 3-25) affects the type, amount, and placement of the organic residues that are decomposed to produce humus. Perennial species such as trees and shrubs add organic residues to soils primarily in the form of annual leaf fall. The accumulation of the litter is on the surface of the soil and lends itself to the formation of O horizons and thin A horizons. In addition, deciduous leaves and especially evergreen needles tend to have low contents of basic cations, such as Ca^{2+} and K^+. The decomposition products of these materials tend to be acidic and promote greater amounts of soil weathering. Annual species, such as grasses, tend to add organic residues not only to the surface, but due to the death and decay of the roots, almost equal amounts of residue are deposited annually in the root zone. The residue from annual species tends to have higher base contents than are found with the perennials. Both factors, the higher base content and placement of the residue, facilitate the formation of thicker, darker A horizons with higher organic matter contents.

Topography can have major effects on the organic matter content of soils.

[3]W. A. Gillam, A study on the chemical nature of humic acid, *Soil Sci.* 49:433–453 (1940).

Topography affects the amount of runoff and erosion. Water lost as runoff cannot be used by plants, and less organic matter is produced. In addition, erosion removes the portion of the soil that contains the highest humus levels. Soils that are subject to runoff and erosion tend to have thinner, lighter-colored A horizons. Soils that are in a landscape position such that they receive runoff water and eroded material tend to have both higher humus contents and thicker A horizons because of these additions. Soils that occur in these site positions are usually in aquic moisture regimes and have poor or very poor internal drainage. The poor drainage—actually the poor aeration caused by the poor drainage—creates anaerobic conditions. Anaerobic oxidation of organic residues is less efficient than aerobic oxidation, and consequently, poorly and very poorly drained soils have higher humus contents than that of the better drained members of a catena. Organic soils are the extreme example of this phenomenon. Organic soils form in shallow lakes, bogs, and swamps, where adequate moisture allows high organic matter production, but because of the poor aeration, the decomposition of the residues is very inefficient.

Parent materials and soil age also affect soil organic matter content. If the parent material is dominated by sand, the soil will tend to have less humus, and if the parent material is dominated by clay, the soil will tend to have more humus than the surrounding medium-textured soils. Sandy soils tend to be droughty and well aerated, and both conditions favor low humus contents. The droughty conditions reduce plant growth, while the well-aerated conditions favor rapid efficient decomposition of plant residues. Soils with high clay contents tend to have high humus contents. In fact, in any geographical region the humus content of soil tends to be highly correlated with clay content. Clay-textured soils tend to be wetter and more poorly drained than the associated medium-textured soils. Both properties favor higher humus contents. In addition, the high surface area of clay-textured soils stabilizes the organic residues, due to adsorption of the decay products onto clay surfaces. The adsorbed materials are more difficult to decompose than organic residues in the soil solution or those existing as discrete pieces of organic matter in the soil.

Maintaining soil organic matter. When virgin soils are brought under cultivation, the environment of the soil is modified. These changes usually result in a reduction in the organic matter content of the soil. Initially, there is a rapid decrease, but after a few years, a new equilibrium level of soil organic matter is established. It is the goal of agriculture to maintain the level of soil organic matter at an optimum level. The optimum level must consider both soil and economic constraints.

Soil organic matter can be maintained or in some cases increased by returning organic materials to the soil. These can be in the form of plant residues such as cornstalks, soybean stover, wheat straw, or animal manures. These materials provide the energy and carbon sources to drive the decay processes that produce humus. In Chapter 13 we discuss in detail the role of nitrogen in the production of humus. Organic residues in the form of municipal sewage sludge may be a suitable source of organic materials to be recycled to the soil. The extent of their use is often limited by the presence of trace metals, such as cadmium (see Chapter 17). A good soil management program must also control erosion. Erosion removes the surface soil, the soil that contains the highest humus contents and that in most cases is the most productive soil.

Crop residues are a major source of organic matter, exceeding farm manures and sewage sludges in total additions to the soil. For example, a 8.96-mt/ha (4-ton/acre) alfalfa crop produces approximately 7504 kg/ha of dry harvested hay and leaves approximate 5600 kg/ha of roots. In addition to the harvested grain, a 3359-kg/ha (50-bu/acre) wheat crop leaves 2463 kg/ha of roots and 5040 kg/ha of aboveground residue in the field, although much of the aboveground residue is often baled and removed from the field.

Farm manure is a valuable source not only of organic matter but also nutrients, such as nitrogen, phosphorus, and sulfur. On the average farm manure returns from 75 to 80% of the nitrogen, 80% of the phosphorus, 85 to 90% of the potassium, and 40 to 50% of the organic matter than has been consumed by the animals.[4] The actual content of farm manures varies from animal to animal, and large variations can occur due to handling and storage. Handling and storage especially affect the nitrogen content. Since the amount of humus produced is a function not only of the amount of organic matter returned to the soil, but also its nutrient content, handling can greatly influence humus production.

11.3 ORGANIC SOILS

Organic soils (Histosols) are soils composed almost entirely from organic materials. Histosols generally form in wet, poorly aerated sites, such as shallow lakes, ponds, swamps, and bogs and are the end product of natural eutrophication (Figure 11-2). Eutrophication is the process of shallow lakes and ponds becoming nutrient rich and subsequently filling with plants and plant remains. The plant remains accumulate in the bottom of the shallow lakes and ponds, due to poor aeration and the inefficiency of anaerobic decay.

Histosols have a wide distribution in wet areas but represent only a small percentage of land area compared to mineral soils. In the United States, Histosols are common in the Everglades of Florida, the Okefenokee swamp of Georgia, in northern Minnesota, and in other wet, poorly aerated areas, such as floodplains, deltas, and tidal marshes. Histosols are classified on the basis of the degree of decomposition of the organic materials, origin of plant materials, thickness of organic materials, temperature regimes, and the presence of other materials and horizons.

Histosols can be divided into three suborders, the Fibrists, Hemists, and

[4]D. W. Thorne and H. B. Peterson, *Irrigated Soils: Their Fertility and Management*, The Blakiston Company, New York, 1954, p. 245

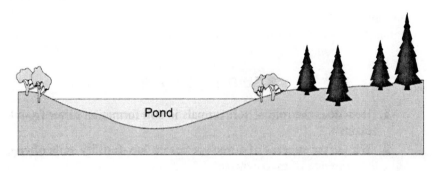

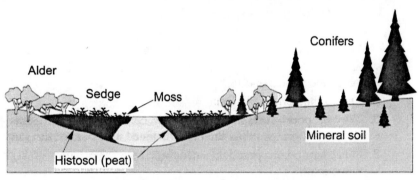

Figure 11-2 Development of a Histosol. Shallow pond fills with plant material that accumulates because of the poorly aerated conditions. Eventually, the pond will completely fill with organic materials.

Saprists. *Fibrists* are formed by the accumulation of relatively undecomposed *Sphagnum* moss in very wet conditions. *Hemists* are intermediate in degree of decomposition between Fibrists and Saprists. *Saprists* contain highly decomposed, highly humified, usually black colored organic materials. Saprists have higher bulk densities than those of other Histosols.

Histosols usually require drainage to make them suitable for agricultural production. After drainage, Histosols are often used for vegetable production. A high percentage of carrots, celery, lettuce, and onions in the United States are grown on Histosols. Histosols have high water holding capacities and very high CEC. Acid Histosols, because of their high CEC and buffering capacities, require large amounts of lime to alter their pH values. If fertilized with calcium, Histosols can produce crops at lower pH values than on minerals soils, because of less problems with aluminum toxicities. After drainage, the poor-aeration conditions that created the Histosols are improved and decomposition of the organic matter is rapid, often more than 1 cm/yr.

Histosols are subject to wind erosion when dry and after drainage can be destroyed by fire. Histosols provide little support for roads and buildings and generally have to be excavated and replaced with mineral material before roads and buildings can be constructed. Some Histosols, Fibrists, are excavated and used in greenhouse potting mixtures and to wrap plants for transport.

STUDY QUESTIONS

1. How does the role of soil animals in soil formation differ from that of higher plants?
2. Why is the success of a tree species on low-fertility soils often dependent on the presence of mycorrhizae?
3. Why do different soil horizons form under tree species than under grass species?
4. What are the sources of energy for chemoautotrophs and chemoheterotrophs?
5. How does anaerobic oxidation of organic residues differ from aerobic oxidation of the same residues?
6. Discuss the effect of climate on soil humus contents.
7. Why do poorly drained soils usually contain greater amounts of humus and thicker A horizons than the better-drained members of the same catena?
8. What agronomic practices encourage the maintenance of high levels of organic matter?

12

Soil–Plant Interactions

Soil provides plants and other organisms with many factors necessary for successful growth and reproduction. These include mechanical support, heat energy, water, nutrients, and oxygen for plant roots and other soil organisms. In addition, to these soil-supplied factors, light, heat energy, and oxygen are provided to the aboveground portions of the plant by the atmosphere. Negative factors, such as toxic levels of trace metals or organic compounds (herbicides), can also affect plant growth. Odum[1] expressed, in a combination of Liebig's (1840) law of the minimum and Shelford's (1913) law of tolerance, that "the presence and success of an organism or a group of organisms depends upon a complex set of conditions. Any condition which approaches or exceeds the limits of tolerance is said to be a limiting factor."

The concept of *limiting factors* states that the growth of a plant can be no greater than that allowed by the factor that most closely approaches a critical minimum or maximum. Figure 12-1 illustrates this concept using a water barrel with different-length staves. Each stave represents a different growth factor, such as nitrogen, zinc, phosphorus, or light. The length of each stave represents the amount of each factor available to the plant relative to the critical minimum for that factor. The amount of water the barrel can hold represents the amount of plant growth. To increase plant growth, the most limiting factor must be identified and increased. In this case, zinc must be supplied to the plant to increase growth, but growth will increase only up to the level set by the next most limiting growth factor, nitrogen.

[1]E. P. Odum, *Fundamentals of Ecology*, W. B. Saunders, Co., Philadelphia, 1959, p. 93.

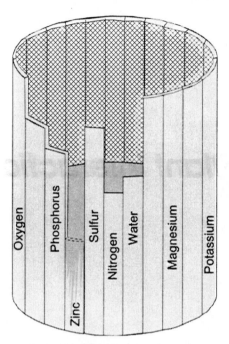

Figure 12-1   The concept of limiting factors.

The application of sulfur, calcium, or any other growth factor would not increase growth since they are not the limiting factors.

The water barrel example does not illustrate the potential interaction between growth factors. For example, high phosphorus levels can induce or exacerbate zinc deficiency. Nor does the example address the problem of excess levels of particular growth factors which can also limit growth. Figure 12-2 is a generalized growth response curve. The curve can be divided into three regions. The first region represents deficient levels of the nutrient, the second represents adequate levels, and the third shows toxic nutrient levels. In the deficient region, the nutrient is limiting plant growth, and fertilizing with the element will increase growth up to the level set by the next limiting factor. In the adequate region, application of the nutrient will not increase nor decrease plant growth. But once the upper limits of the adequate region are approached, continued applications of the nutrient can result in decreased plant growth. The toxicity may be due to the direct affect of the nutrient on a plant enzyme or due to the interaction of the nutrient with the uptake and utilization of other nutrients. As an example, excess nitrogen may predispose a plant to diseases or even result in inefficient water use. The actual size of each region of the curve depends on the nutrient. For many of the micronutrients there is a very small difference between deficient and toxic levels.

The role of the soil in supplying water, oxygen, mechanical support, and heat energy has been discussed in earlier chapters. This chapter and several to follow are

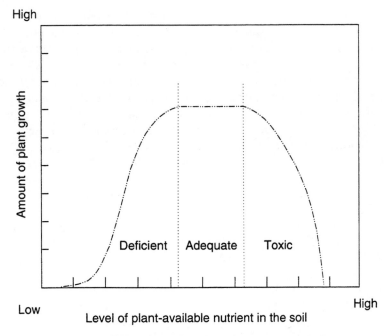

Figure 12-2 Theoretical growth response curve.

concerned with mineral nutrients and their availability to plants. Plant growth can also be limited by some potentially negative growth factors associated with soil pollution, as discussed in Chapter 17.

12.1 MINERAL NUTRITION

Plants and other organisms require certain essential elements. Three criteria must be satisfied for an element to be considered essential. First, a deficiency of the element must prevent the completion of the plant's life cycle (germination, growth, flowering, seed set). Second, the element cannot be replaced by a similar element. Third, the element must be involved directly in the metabolism of the plant.

Essential elements serve a variety of functions in the plant. Some are basically building blocks for plant growth. Some maintain electrical neutrality and others function as a part of specific enzyme systems. Some essential elements are obtained from the atmosphere or from water. The remaining essential elements are obtained from soil or from processes that take place in the soil. Plants obtain oxygen, hydrogen, and carbon from the atmosphere above the soil or from water, although the roots of most mesophytes are dependent on oxygen in the soil atmosphere. The plant obtains nitrogen, phosphorus, potassium, calcium, magnesium, and sulfur from the soil or from soil processes, such as nitrogen fixation that convert N_2(gas) into forms that are usable for higher plants. The aforementioned nine essential elements are known as the macronutrients, since they are used by the plant in fairly large amounts. The remaining seven essential nutrients are known as the micronutrients, since they are required in much smaller amounts (see Chapter 15). These essential micronutrients are iron, manganese, boron, molybdenum, chlorine, zinc, and copper. Cobalt, although not considered an essential element of higher plants, is a constituent of vitamin B_{12} and is essential for species of *Rhizobium*, a symbiotic nitrogen-fixation bacteria. Additional elements, such as sodium and silicon, have been shown to increase growth of certain species but have not been shown to be required for the plant to complete its life cycle. In addition, animals have been shown to need other elements, such as selenium and iodine that are not required by plants.

Several factors control the availability of essential nutrients to plants. These include the chemical form and concentration of the nutrient in the soil solution and the rate at which nutrients can be replenished upon plant uptake. For most nutrients the rate that the nutrient can be replenished in the soil solution after depletion by plant uptake is the limiting factor in nutrient availability.

For the essential nutrients that occur as cations, the available pool of nutrients includes the nutrients in the soil solution and those on the exchange complex. Exchangeable cations are considered available since they rapidly replace cations in the soil solution upon removal by plant uptake. It should be emphasized that the majority of exchangeable cations are absorbed by plants from the soil solution, not directly from the colloidal surfaces. Remember that the concentration of a particular type of cation in the soil solution is a function of amount of that cation on the exchange complex. When the exchangeable level of a nutrient that occurs as a cation is low, the availability of that nutrient is also low.

The replenishment, after depletion by plant uptake, of the essential nutrients that occur in the soil solution as anions or neutral molecules is dependent on a complex set of reactions (Table 12-1). These include the release of organically bound nutrients upon decay, the dissolution of soil minerals, or the diffusion of nutrients from nondepleted regions of the soil. These processes are also important for the cationic nutrients when the levels on the soil exchange complex adjacent to the root become exhausted.

12.1.1 Nutrient Uptake

The plant root absorbs nutrient elements from a dilute soil solution of highly variable composition. Often the process involves absorption or exclusion of nutrients against concentration gradients. For example, plants may contain lower concentrations of Na^+ or Mg^{2+} and higher concentrations of Cl^- or NO_3^- than are found in the soil solution and, at the same time, contain concentrations of K^+ comparable to the soil solution.

TABLE 12-1 Overview of the functions of essential macronutrients in plants

Carbon	Primary constituent of organic compounds, carbohydrates, proteins, lipids, and nucleic acids. Reduction and oxidation of carbon serves as the basis for photosynthesis and respiration. Reduced carbon compounds serve as energy sources for most organisms.
Oxygen	Constituent of most organic compounds. Role of free oxygen in plants is primarily as an electron acceptor. $$O_2 + 4e^- + 4H^+ \rightarrow 2H_2O$$ Oxygen is oxidized in photosynthesis and reduced in respiration.
Hydrogen	All organic compounds contain hydrogen. Hydrogen ions are ubiquitous and are important to ionic balance. Hydrogen ions are involved in a variety of biochemical and chemical reactions. Reduced forms of hydrogen, NADH, and NADPH are important reducing agents in oxidation-reduction reactions.
Nitrogen	Component of proteins, nucleic acids, important in protein and chlorophyll synthesis.
Phosphorus	Component of activated carbohydrates. Central role of phosphorus is in energy transfer.
Sulfur	Component of amino acids and proteins. Maintains protein structure through disulfide bonds. Involved in cell energetics.
Calcium	Involved in cell division. Important in cell walls and maintenance of cell membranes.
Magnesium	Component of chlorophyll and cofactor for many enzymatic reactions. Occurs primarily as free ion in cell solutions.
Potassium	Primary role is in osmotic and ionic regulation.

General rules for uptake.[2]   Nonionic hydrophilic substances are generally absorbed in an inverse proportion to their molecular size. Smaller molecules will be absorbed to a greater extent than larger molecules. This implies that uptake of these molecules is through pores in the hydrophilic portions of root membranes. Adsorption of these molecules is a passive process and follows concentration gradients, that is, movement from high to low concentrations.

Hydrophobic substances are absorbed in direct proportion to their lipid solubility. The greater the lipid solubility of the substance, the greater its absorption through the lipid portion of the root membranes. In general, as lipid solubility of a substance increases, its water solubility decreases. Absorption of these substances is also passive and follows concentration gradients.

Ionic substances may be absorbed passively as dictated by concentration gradients, or they may be converted to nonionic forms and be passively absorbed as nonionic hydrophilic substances. In addition, ionic species are often accumu-

[2]I. P. Ting, *Plant Physiology*, Addison-Wesley Publishing Co., Reading, Mass., 1982, p. 334.

lated or excluded against concentration gradients. This implies that root membranes are selective and suggests that absorption may sometimes be an energy-driven process.

Simple (passive) uptake. Simple uptake is driven by a concentration difference between the soil solution and the root interior. The amount of uptake of an ion by this mechanism is proportional both to the concentration gradient between the root and the soil solution and to the permeability of the root membranes:

Amount of uptake = membrane permeability × concentration gradient   [12-1]

Ion uptake by this mechanism has been shown to increase with increasing concentrations of the ion in soil solution up to a maximum and then not show further increases with increasing concentrations. At this point physical factors limit plant uptake even though solution concentrations increase. Figure 12-3 illustrates that ion uptake by plants reaches a maximum independent of nutrient concentrations in the soil solution. This is taken as evidence that ion uptake takes place at specific sites on root membranes called *carrier sites*. When these sites are saturated, further increases in the concentration of the ion in the soil solution will not result in increased uptake.

Active uptake. There is considerable evidence for active nutrient uptake mechanisms in plants. This includes the uptake of ions against concentration

Figure 12-3   Effect of soil solution concentration on ion uptake by plants.

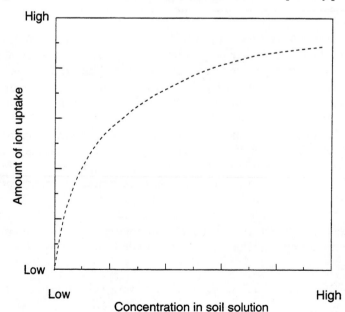

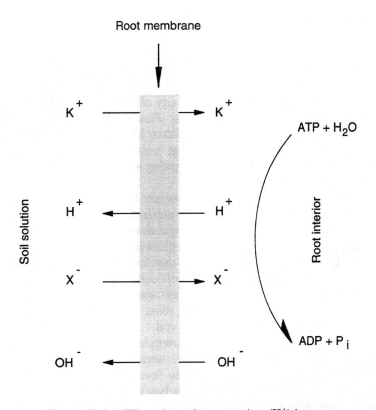

Figure 12-4   ATPase-dependent potassium (K^+) ion pump.

gradients, the exclusion of ions from the root, the decrease in ion uptake when plants are treated with respiration inhibitors, as well as the antagonism of one ionic species on the uptake of other ionic species. Active uptake occurs when the plant expends energy to move the ion out of the soil solution, across the root membrane into the plant. Figure 12-4 is one possible model of an energy-driven absorption process.

Energy for active uptake is provided by the hydrolysis of ATP to ADP. The energy results in K^+ ions being moved out of the soil solution, across the root membrane, and into the root. At the same time, H^+ ions are moved from the root interior across the membrane into the soil solution. This counter movement of K^+ and H^+ ions maintains electrical neutrality within the root and in the soil solution. Figure 12-4 also illustrates that anions (X^-) may be moved with the K^+ ions into the root. If this occurs, OH^- ions must be pumped out of the root into the soil to maintain electrical neutrality. Once inside the root, the K^+ ions can catalyze additional hydrolysis of ATP to ADP and stimulate additional ion pumping.

$$ATP + H_2O \overset{K^+}{\leftrightarrow} ADP + P_i \qquad\qquad [12\text{-}2]$$

12.1.2 The Plant Root

If a single measurement could be made that expressed the ability of a soil to support the growth of higher plants, that measurement would be the total volume of soil in contact with the plant's roots. The total volume of soil in contact with the roots is a function of rooting depth, the density of roots within a given region of the soil, and the total surface area of the roots. Plant roots seldom occupy more than 5% of the soil volume, even in the upper 100 to 150 cm of soil, where roots are abundant. The volume of soil occupied by plant roots decreases rapidly with soil depth, with the roots of many plant species occupying no more than 0.01 to 0.001% of the soil's volume at 0.5 m of depth.[3]

Plant roots perform many important functions. These include the absorption of nutrients and water, transport of nutrients and water to the stems, synthesis of many plant hormones and growth regulators, the storage of carbohydrate reserves, and the mechanical anchoring of the plant in the soil. The ability to absorb nutrients and water from soil is the function that has the greatest affect on the growth and yield of crops on different soils. If soil conditions promote a large rooting volume, the plant can exploit the nutrients and water stored in this volume of soil. If soil conditions inhibit root growth, the plant is restricted to the nutrients and water stored in a much smaller volume of soil.

Soil factors that affect root growth include soil-water content, soil temperature, soil physical properties, soil aeration, soil fertility, and soil chemical conditions (e.g., pH) that affect the level of toxic substances, such as soluble aluminum. Excess soil water, which is often encountered in humid regions in the spring, can inhibit root growth due to poor soil aeration. Later in the summer when the soil has dried, the plant is left with a restricted root system that is incapable of providing sufficient water and often nutrients to the plant. In arid climates, depth of rooting is often related to the depth of wetting of spring and summer rains. Root growth is often restricted below these depths by large (negative) matric potentials. Cool soil temperatures often restrict both root growth and nutrient uptake in the early spring. If the soil is slow to warm in the spring because of high soil water content, plant vigor can be affected and the incidence of various plant diseases increased.

Soil physical conditions also affect root growth. The "tilth" of the soil can affect root growth indirectly through effects on soil aeration and amount of plant-available water, and directly by mechanical impedance. Bulk densities greater than 1.5 have been shown to impede root growth severely. Pore size and distribution also affect root growth. Young roots of many species require a soil matrix with soil pores ≥ 0.3 mm in diameter for easy penetration and growth.[4]

[3]P. J. Gregory, Growth and functioning of plant roots, in *Russell's Soil Conditions and Plant Growth*, A. Wild (ed.), Longman Scientific and Technical Publishers, Essex, England, 1988.

[4]Ibid.

Plants rapidly deplete available water and nutrients in the vicinity of the root. Either the root must grow into undepleted regions or the water and nutrient must be replenished by some mechanism. Water can be replenished by rain or irrigation and nutrients by diffusion from undepleted zones or by application in the irrigation water. Diffusion is an important process only for short distances. Diffusion can move substantial quantities of water and nutrients only a few centimeters or inches in a reasonable length of time. Diffusion has the effect of increasing the volume of soil that a given root can exploit.

The ability of the plant to use water and nutrients at distances greater than a few centimeters from the root is dependent primarily on root growth into these regions. For example, wheat roots have been shown to have rates of extension of ≈ 10 mm/day, soybeans ≈ 20 mm/day, and corn roots in warm soils up to 60 mm/day.

STUDY QUESTIONS

1. What do soils provide to support the growth of higher plants?
2. Discuss the concepts "limits of tolerance" and "limiting factors."
3. Compare active and passive uptake of essential elements.
4. What soil and plant factors determine the availability of essential nutrients?
5. What soil textures would probably have nutrient deficiencies of plant essential cations?

13

Nitrogen and Sulfur

13.1 NITROGEN

Nitrogen, except for water, is the most limiting element worldwide in terms of the production of food and fiber. It is an integral part of plants and plant constituents, such as chlorophyll, proteins, and nucleic acids. Deficiencies of nitrogen result in visual symptoms that manifest themselves in yellowing of plant leaves and other aboveground plant parts. An adequate supply of plant-available nitrogen promotes rapid growth and dark, lush, green-colored leaves and stems. Nitrogen is unique in that while the soil often does not have an adequate supply of plant-available nitrogen, other forms of nitrogen are quite abundant. The atmosphere surrounding the earth contains approximately 79% nitrogen by volume, but in the form of N_2(gas), which is very stable and cannot be used by most organisms. Atmospheric nitrogen must be reduced by nitrogen-fixing organisms in soils to convert it to a plant-available nutrient. Nitrogen is also present, in large amounts in most surface soils, as a constituent of soil organic matter. Nitrogen associated with soil organic matter must be converted by decay processes from organic to inorganic forms before it can be absorbed by plants.

Nitrogen undergoes a host of transformations that make its chemistry interesting and unique. Because of the large number of transformations possible, nitrogen has often been called the elusive element. Nitrogen can be used by plants as either positively charged ammonium (NH_4^+) ions or negatively charged nitrate (NO_3^-) ions. Ammonium ions can also be adsorbed on cation-exchange sites, oxidized to NO_3^- by soil bacteria, and "fixed" by certain clay minerals, or when soil pH values are very high, volatilized from the soil as ammonia (NH_3) gas. In addition to being absorbed by plants and other organisms, nitrate can be lost from the soil by leaching into groundwater, or be converted by bacterial action into volatile nitrogen gases and lost from the soil. Because nitrogen occurs in so many forms, many of which are not available to higher plants, and because it can quickly be converted from one form to another, nitrogen is of great interest to soil and plant scientists.

13.1.1 The Nitrogen Cycle

Nitrogen fixation. As shown in Figure 13-1, the nitrogen cycle illustrates many of the unique features of this elusive element. Atmospheric nitrogen [N_2(gas)] can be added to soils in a process called nitrogen fixation. Nitrogen fixation can occur symbiotically, nonsymbiotically, or abiotically by natural or anthropogenic processes.

Symbiotic N Fixation by Legumes. As shown in Figure 13-1, *Brady-rhizobium* and *Rhizobium* are bacteria responsible for symbiotic fixation of nitrogen by various species of legumes. Symbiotic N fixation is a process whereby bacteria reduce N_2(gas) to NH_4^+ ions. This conversion takes place in abnormal root growth structures, called *root nodules*, of the host plants (Figure 13-2). Symbiotic fixation

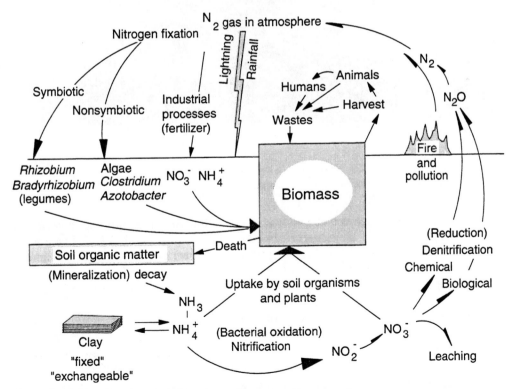

Figure 13-1 The nitrogen cycle.

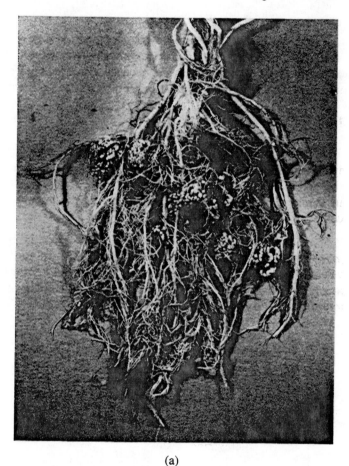

(a)

Figure 13-2   Nodules on a legume root (a) and expanded root nodule (b). (Courtesy of USDA Soil Conservation Service.)

of nitrogen by legumes can add as much as 600 kg/ha/yr of nitrogen, contributing most, if not all, of the nitrogen required for some plants (Table 13-1). Leguminous grain (soybean, peanut) and forage (alfalfa, clover) crops account for a significant acreage in the United States and result in the conversion of millions of tons of N_2(gas) to plant-usable nitrogen forms (Table 13-2). While symbiotic N fixation in leguminous grain crops accounts for over 2 million tons annually in the United States, it accounts for almost 10 million tons worldwide (Table 13-3).

Species of nitrogen-fixing bacteria are specific for an individual species of legumes. They invade the root hairs of plants, such as alfalfa, clover, peas, or peanuts and induce root nodules that become packed with modified forms of the *Bradyrhizobium* or *Rhizobium* cells called *bacteroides*. Bacteroides live in a mutually beneficial relationship with the plant, where they supply the plant's need for

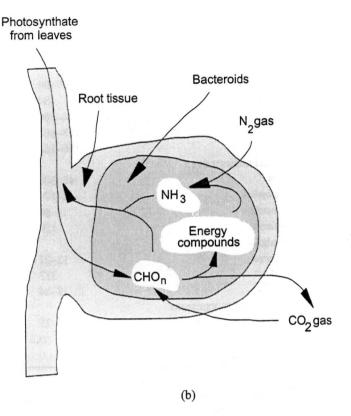

(b)

Figure 13-2   (cont.)

nitrogen and the plant supplies the bacteria with organic compounds for energy. The amount of nitrogen fixed by this mutually beneficial relationship depends on many factors, including the species of nitrogen-fixing bacteria present in the soil and the environmental conditions for plant and bacterial growth. However, the nitrogen needs of most leguminous plants are met from the atmosphere through this process.

Yields of plants that derive their nitrogen from symbiotic N fixation are usually not enhanced by additional nitrogen fertilizers. As shown in Table 13-4, the addition of over 400 kg/ha of fertilizer N did not result in increased nitrogen content in soybean grain or in increased yields. Some studies have shown that when nitrogen fertilizer is added, the amount fixed symbiotically is drastically reduced. The quantity of nitrogen fixed in symbiotic associations is a function of the host plant's nitrogen requirement.

Fixation of nitrogen by legumes is an increasingly important process for the worldwide production of protein. This is especially important as the cost of commercial nitrogen fertilizers increases and as concern for NO_3^- pollution of groundwaters and surface waters increases. In areas where legumes have been grown in the past, soils

TABLE 13-1 Relative rates of biological nitrogen fixation

Organism or system	N_2 fixed (kg ha^{-1} yr^{-1})
Legumes	
Soybeans	57–94
Cowpeas	84
Clover	104–160
Alfalfa	128–600
Lupins	150–169
Nodulated nonlegumes	
Alnus	40–300
Hippophae	2–179
Ceanothus	60
Coriaria	150
Plant–algal associations	
Gunnera	12–21
Azollas[a]	313
Lichens	39–84
Free-living microorganisms	
Blue-green algae	25
Azotobacter	0.3
Clostridium pasteurianum	0.1–0.5

Source: H. J. Evans and L. E. Barber, Biological nitrogen fixation for food and fiber production, *Science* 197:332–339 (1977).
[a]Very important in tropical rice production.

contain native strains of nitrogen-fixing bacteria, which can exist for many years. Where legumes have never been grown or where the strain of nitrogen-fixing bacteria present in the soil may not be compatible with the particular legume to be grown, *Bradyrhizobium* or *Rhizobium* can be introduced by direct inoculation of the soil or coating the seeds to be planted with a mixture of compatible strains.

Considerable effort has been expended in attempts to increase symbiotic N fixation by manipulation of either the bacteria or the host plant. These efforts have identified superior strains of nitrogen-fixing bacteria when the host plants are grown in controlled (i.e., greenhouse) conditions, but in general, the improved strains do not compete well with the native soil strains in the field. The future for further improvements looks promising. Biotechnology and genetic manipulation techniques could result in significant advances to improve the efficiency of symbiotic nitrogen fixation or extend this capacity to other species of plants. Such advances would help supply world soils with an efficient and economical supply of nitrogen.

Symbiotic Fixation by Nonlegumes. There are approximately 160 species of nonleguminous plants that can develop nodules that fix nitrogen symbiotically. The amount of nitrogen fixed by these species is generally less than that fixed by

TABLE 13-2   Estimates of area harvested and N_2 fixed by grain and forage legumes in the United States

Legume crop	Area harvested ($ha \times 10^3$)	N_2 fixed ($tons \times 10^3/yr$)
Grain legumes		
Soybean [*Glycine max* (L.) *Merr.*]	28,000	3,252
Dry edible beans (*Phaseolus* sp.)	716	36
Peanut (*Arachis hypogaea* L.)	613	29
Dry edible peas (*Pisum sativum* L.)	53	4
Others	33	2
	29,415	2,423
Forage legumes		
Alfalfa (*Medicago sativa* L.)	10,873	2,196
White clover (*Trifolium repens* L.)	5,431	755
Red clover (*Trifolium pratense* L.)	4,075	566
Sweet clover (*Melilotus* sp.)	1,128	157
Vetch (*Vicia* sp.)	760	65
Trefoil (*Lotus corniculatus L.*)	686	81
Crimson clover (*Trifolium incarnatum* L.)	328	46
	23,281	3,866
Total	52,696	6,289

Source: D. A. Phillips and T. M. DeJong, Dinitrogen fixation in leguminous crop plants, in *Nitrogen in Crop Production*, R. D. Hauck (ed.), American Society of Agronomy, Madision, Wis., 1984.

TABLE 13-3   Magnitude of N_2 fixation in grain legumes

Continent	Area harvested ($ha \times 10^3$)	N_2 fixed ($tons \times 10^6/yr$)
Asia	86,718	5.17
North America	29,965	2.42
Africa	18,269	0.96
South America	15,878	0.81
Europe	3,549	0.20
Oceania	269	0.02
World	154,648	9.58

Source: D. A. Phillips and T. M. DeJong, Dinitrogen fixation in leguminous crop plants, in *Nitrogen in Crop Production*, R. D. Hauck (ed.), American Society of Agronomy, Madision, Wis., 1984.

TABLE 13-4 Effect of fertilizer nitrogen on soybean yields and nitrogen content

Fertilizer N (kg/ha)	N in grain plus stover (kg/ha)	Yield (q/ha)
0	215	28
112	227	28
224	211	28
448	227	30

Source: L. F. Welch, *Nitrogen Use and Behavior in Crop Production,* Agricultural Experiment Station Bulletin No. 761, College of Agriculture, University of Illinois, Urbana, Ill., 1979.

legumes, but is still very significant. Actinomycetes can form symbiotic nitrogen-fixing associations with trees or woody shrubs, such as *Purshia, Comptonia,* and *Alnus.* These nonleguminous nitrogen-fixing associations are especially important for the nitrogen balance of forests, woodlands, and freshwater and marine habitats.

There is considerable interest and research effort under way to select or develop symbiotic nitrogen-fixing bacteria capable of living in association with the roots of cereal crops, such as corn and small grains. If these yet to be developed bacterial strains could supply adequate fixed nitrogen for optimum plant growth, then the need for nitrogen fertilizer would be lowered dramatically. Success in this endeavor would mark a significant milestone in food production.

Nonsymbiotic Nitrogen Fixation. In addition to symbiotic N fixation, nitrogen is fixed nonsymbiotically by a variety of microorganisms and associations. These organisms are not in association with specific plants, but grow freely in soils and incorporate atmospheric nitrogen into body tissue. When the organisms die and are decomposed, this nitrogen is released as inorganic forms that are available to plants. Nitrogen fixed by nonsymbiotic processes can account for as much as 25 kg/ha/yr, but more often is in the range 5 to 10 kg/ha/yr. These organisms include bacteria in soils, bacteria in rotting wood, free-living blue-green algae in terrestrial and marine environments, and associations of blue-green algae with fungi, ferns, mosses, liverworts, and higher plants. Certain photosynthetic bacteria and blue-green algae are able to fix their own carbon and nitrogen at the same time. These organisms are particularly adapted to flooded or wetland conditions. Much of the nitrogen available for rice production in the tropics is due to these organisms.

Abiotic Fixation. *Abiotic,* as the name implies, is the fixation of nitrogen in the absence of life or living organisms. For the most part, nitrogen, added to soil by abiotic fixation, comes from industrial sources, which through the expenditure of heat and pressure converts atmospheric nitrogen to plant-available forms. These, then, are added to the soil as fertilizer forms of nitrogen. Another source of abiotically fixed nitrogen is that added by rainfall. Nitrogen, thus added, accounts for approximately 2 to 10 kg/ha/yr in the United States (see Figure 17-14). Although this is not a significant amount for cultivated crops, it is vitally important to forest and other natural ecosystems having limited nitrogen sources. The source of much of this nitrogen originates as nitrogen oxide emissions from industry and automobiles. Some nitrogen oxides may also be released from soils in the process of denitrification. The nitrogen oxides are converted to nitric acid in the atmosphere and are returned to the soil as nitrates in the form of acid rain. N_2(gas) can also be oxidized in the strong electrical fields associated with lightning.

Mineralization and immobilization.   An important source of plant-available nitrogen in soils is due to the process of mineralization. Mineralization of nitrogen is defined as the transformation of nitrogen from organic forms to inorganic forms, NH_4^+ or NH_3. The process of mineralization is, in fact, the decay process carried out by heterotrophic soil organisms. The organisms utilize a wide variety of nitrogen-containing organic substances, such as amino acids, amino sugars, and nucleic acids found in soils. Mineralization of organic nitrogen is an important source of nitrogen usable by plants. The magnitude of available nitrogen released by mineralization is explored in the following example. A hectare-furrow-slice of soil is assumed to weigh 2,000,000 kg. If a soil contains 5% organic matter, it would contain approximately 100,000 kg/ha of organic matter. Of that, 5% might be nitrogen, or approximately 5000 kg of nitrogen per hfs of soil. An annual mineralization might be expected to be 2% under favorable conditions. This would result in the release of approximately 100 kg/hfs of nitrogen each year from the soil. Because mineralization is carried out by bacteria, the rate of decomposition, and thus the rate of nitrogen release, is dependent on climatic factors. Decomposition rates increase with increasing soil temperature, better soil aeration, and adequate soil water up to a maximum, above which further increases in temperature or soil water result in rapid reductions in the rate of decomposition.

There is, in soil, a continuous flux of organic nitrogen to mineral forms and mineral forms back to organic nitrogen. The reverse of mineralization, the transfer of inorganic forms (NH_4^+, NO_2^-, or NO_3^-) into organic forms, is defined as *immobilization*. Soil organisms have a need for nitrogen and thus absorb the inorganic compounds, transforming them into organic constituents in the form of cells, tissues, and soil biomass.

Nitrification.   Nitrification is the oxidation of ammonium, NH_4^+, to nitrate, NO_3^-. This process of nitrification is a rapid and almost universal process in soils where NH_4^+ is present and conditions for microbial activity are favorable. Nitrification is a two-step process. First is the oxidation of NH_4^+ ions to NO_2^-. This process is carried out by a special group of bacteria called *Nitrosomonas*. The second step of the process is the oxidation of NO_2^- to NO_3^-, by a separate but specific group of bacteria called *Nitrobacter*. Because the oxidation of NO_2^- is much more rapid in most soils than the oxidation of NH_4^+, only rarely are more than trace amounts of NO_2^- present. Because nitrification in most soils occurs readily, NH_4^+ ions do not normally accumulate in soils. The main factors that control nitrification rates in soils are (1) availability of NH_4^+ ions, (2) concentration of oxygen and carbon dioxide in the soil atmosphere, (3) pH, and (4) temperature.

Ammonium Availability. The concentration of NH_4^+ ions, subject to nitrification, is dependent on the rate of mineralization or the rate of addition of ammonium-containing fertilizers. If mineralization rates are slow or fertilizer additions minimal, nitrification will be limited. On the other hand, anhydrous ammonia fertilizer is sometimes applied in concentrated soil bands. The high concentration of NH_3 found in these bands of fertilizer can be toxic to *Nitrobacter*, thus limiting nitrification until concentrations decrease. Under the more normal conditions of warm temperatures, adequate water, and low NH_4^+ ion levels, NH_4^+ is oxidized to NO_3^- nearly as rapidly as it is formed.

Oxygen and Carbon Dioxide. The concentrations of O_2(gas) and CO_2(gas) in the soil will generally be adequate to support nitrification in soils that have good aeration and are not waterlogged or flooded.

Soil pH. Most observations indicate that nitrification proceeds very slowly when the pH of the soil is approximately 4.0 or less. Extremely acid soils have very slow nitrification rates.

Temperature. While the optimum temperature for nitrification may vary among different soils, nitrification proceeds more rapidly under conditions of warmer temperatures up to a maximum of $\approx 40°C$ (105°F). Nitrification rates are very slow in the range of 10°C (50°F), and this temperature is often used to regulate timing of fall-applied ammonium fertilizer. Waiting to apply anhydrous ammonia in the fall season until soil temperatures are below 10°C minimizes conversion of the added NH_4^+ to NO_3^- and hence nitrogen loss as NO_3^- due to leaching or denitrification. Minimum temperatures for nitrification would be close to biological zero (-5°C).

Nitrification Inhibitors. Control of the nitrification process is beneficial for nitrogen conservation, because NH_4^+ ions are not subject to loss by leaching or by denitrification as are NO_3^- ions. Several patented chemical nitrification inhibitors, such as N-serve (nitrapyrin or 2-chloro-6-trichloromethylpyridine), AM (2-amino-4-chloro-6-methylpyrimidine), ATC (4-amino-1,2,4-triazole), and Terrazole (5-ethoxy-3-trichloromethyl-1,2,4-thiadiazole), have been developed. Although these nitrification inhibitors are very effective in slowing nitrification, their effectiveness can be reduced by volatilization, adsorption, leaching, or microbial breakdown. The use of these chemicals can delay nitrification of fall-applied anhydrous ammonia and hence reduce the denitrification and/or leaching of nitrate by winter and early spring rains. From a practical standpoint, they are economically profitable to farmers only when leaching and denitrification are significant problems.

Denitrification. *Denitrification* is the collective name for biological processes by which oxidized forms of nitrogen are lost from the soil to the atmosphere. It is defined as the biological reduction of NO_3^- or NO_2^- to gaseous nitrogen either as molecular nitrogen (N_2) or as an oxide of nitrogen (mainly N_2O). The major requirements for denitrification are (1) the presence of suitable organisms, (2) an electron donor source such as organic carbon compounds, (3) restricted oxygen availability, and (4) the presence of nitrate or nitrogen oxides to serve as terminal electron acceptors. Denitrification is the major process leading to gaseous loss of N from soils.

Organisms. A majority of denitrifying organisms are facultative heterotrophs. Facultative heterotrophs oxidize organic materials and use O_2(gas) as a terminal electron acceptor in well-aerated soils. In poorly aerated soils they oxidize organic materials, but use other oxidized species, such as NO_3^-, as terminal electron acceptors. However, there have been approximately 23 genera of bacteria identified with the capacity to denitrify, including both autotrophs and heterotrophs. The most common are *Pseudomonas*, *Bacillus*, *Alcaligenes*, and *Paracoccus*. Most soils contain a variety of organisms capable of denitrification, so this is seldom the limiting step in denitrification.

Restricted Oxygen Availability. Denitrification occurs when denitrifying organisms cannot obtain enough O_2 to meet their metabolic requirements and use alternate oxidized materials, such as NO_3^- and NO_2^-. The most common situation occurs when well-aerated soils become temporarily waterlogged, due to high rainfall. Waterlogged soils are subject to very rapid losses of nitrogen if other conditions are favorable, with as much as 75% or more of applied nitrogen fertilizer being lost by denitrification. The effect of water content on denitrification results because the supply of oxygen to organisms in soils is strongly affected by the amount of water through which the oxygen must diffuse in moving from the atmosphere, where it is abundant, to the soil particle surfaces. Soil particle surfaces, especially colloidal surfaces, adsorb and concentrate both bacteria and organic materials and hence are microsites of biological activity. The rate of diffusion of oxygen through water is about 0.0001 of its diffusion in the atmosphere. Thus the thicker the water layers surrounding soil particles, the lower will be the oxygen content on the particle surface. Even if the soil is not completely flooded, there can be anaerobic "pockets" or microsites of saturated soil. If the soil is flooded, large areas of anaerobic conditions can develop very rapidly. While the exact mechanism of loss is not known, the pathway of denitrification is thought to be as follows:

valence state
of nitrogen   +5   +3   +2   +1   0
Compound NO_3^-(aq) $\leftrightarrow$ NO_2^-(aq) $\leftrightarrow$ NO(gas) $\leftrightarrow$ N_2O(gas) $\leftrightarrow$ N_2(gas) [13-1]

Organic Carbon Compounds. Most denitrifying bacteria are chemo-heterotrophs. This implies that they use organic carbon compounds as a source of carbon to build cells and as an energy source, rather than CO_2(gas) for a carbon source and sunlight for energy. Denitrification in soil is strongly dependent on the availability

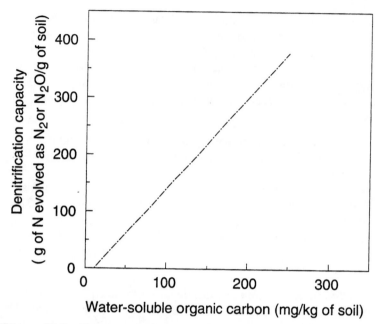

Figure 13-3 Correlation between water-soluble organic C and the denitrification capacity of the soil.

of organic compounds. For example, if waterlogging occurs without a readily oxidizable source or organic matter being present, little denitrification will occur. Because much of the carbon in soil organic matter is in complex forms, water-soluble or readily decomposable carbon provides a much better correlation with denitrification rates (Figure 13-3). Soils with high native organic matter contents, or to which plant residues or manures have been added, have greatly increased denitrification activity. Plant roots can also influence denitrification activity in localized zones, because they provide a readily available source of carbon and because root respiration can result in localized zones of reduced oxygen concentrations.

Presence of Nitrate. A prerequisite for denitrification in soil is the presence of inorganic N in the NO_3^- or NO_2^- form. While the kinetics or rates of denitrification are not easily measured under field conditions, almost any nitrate present is subject to conversion to N_2 and N_2O if the soil becomes saturated or inundated with water and temperature is favorable. The release of N_2O to the atmosphere is a matter of increasing concern to atmospheric scientists. N_2O reacts with ozone in the stratosphere, resulting in ozone destruction. Whereas ozone is harmful to plants and humans in the lower atmosphere, it is an important shield for ultraviolet radiation in the stratosphere. Increased ultraviolet radiation could potentially contribute to a phenomenon known as *global climate change* or *global warming*, which would have very serious effects on soils and crop production. It is not presently known whether or not this phenomenon is real, or to what extent N_2O emissions from soil contribute to ozone depletion.

Chemodenitrification. *Chemodenitrification* is defined as the non-biological reduction of NO_2^- to produce gaseous forms of nitrogen. In most soil systems NO_2^- does not accumulate because the oxidation of NO_2^- to NO_3^- is much faster than the oxidation of NH_4^+ to NO_2^-. When extremely high levels of NH_3 are applied to soils and where soil pH is alkaline, enhanced levels of nitrite sometimes occur. The buildup of NO_2^- in soils may be due, in part, to the toxicity of NH_3 to *Nitrobacter*, which is responsible for the oxidation of NO_2^- to NO_3^-. If high pH values or banding of anhydrous ammonia fertilizers exposes *Nitrobacter* to high levels of NH_3, NO_2^- can accumulate. Nitrite is relatively unstable and will undergo a series of reactions, leading to the formation of N gases. Certain constituents of soil organic matter have been suggested to react chemically with NO_2^- to form N gases. Nitrite can also be converted to N_2 by chemical reaction with NH_3 in the following reaction:

$$NH_3 + HNO_2 \leftrightarrow NH_4NO_2 \rightarrow 2H_2O + N_2(gas) \qquad [13\text{-}2]$$

As already discussed, the practical significance of chemodenitrification is thought to be relatively small, because of the low levels of NO_2^- in most soils.

Ammonia volatilization. Ammonia (NH_3) can also be lost directly from the soil surface by volatilization. This can be a serious problem with the addition of NH_4^+-containing fertilizers on calcareous soils. The loss of NH_3 at high pH values occurs because the equilibrium between NH_4^+ ions in the aqueous phase of soil solution and NH_3 gas is highly dependent on pH. As shown in Figure 13-4, at pH values less than 7, the equilibrium strongly favors the NH_4^+ form, while at pH 9 nearly 40% of the species would be NH_3. At pH values of 12 or greater, the equilibrium would be entirely in favor of NH_3, which can escape as a gas. While soils with a pH of 12 or more are very rare, soils with pH values of 8 or higher (calcareous) are common. Losses by NH_3 volatilization also increase with temperature, and they can be appreciable when neutral or alkaline soils containing NH_4^+ near the surface dry. Ammonia losses are greatest for sandy soils, with low cation-exchange capacities, since clay and organic matter in medium- and fine-textured soils tend to adsorb NH_4^+ as an exchangeable cation.

Surface application of animal manures also results in considerable loss of nitrogen as NH_3. These losses can be greatly reduced by incorporating the manure with a tillage implement or by injecting the manure in a vertical or horizontal band below the soil surface. Another potential serious loss of nitrogen by NH_3 volatilization can occur with the surface application of urea to grass lands or pasture. Hydrolysis of urea by the enzyme urease can result in the subsequent volatilization of NH_3, particularly if hot and dry climatic conditions exist following surface spreading. Hydrolysis and loss are also enhanced with pH values greater than 7.0. Overall losses of NH_3 in field situations are given in Table 13-5.

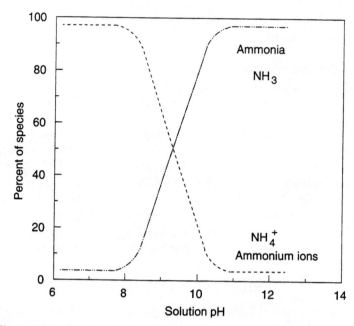

Figure 13-4 Fraction of ammonia and ammonium ion as a function of pH.

TABLE 13-5 Field measured ammonia losses

Fertilizer type and soil conditions	Added N evolved as ammonia gas (%)
Urea	
Silt loam soil, pH 6.3	19
Fine sandy loam soil, pH 5.6—5.8	9–40
Loamy sand, pH 7.7	22
Grass sod	20
Forest litter	3.5–25
Flooded rice soils	6–8
$(NH_4)_2SO_4$	
Silt loam soil, pH 6.3	4
Clay soil, pH 7.6	35
Surface of grass sod	50
Flooded rice soils	3–7
NH_4NO_3	
Loamy sand, pH 7.7	17

Source: F. J. Stevenson, *Cycles of Soil Carbon, Nitrogen, Phosphorus, Sulfur, Micronutrients.* John Wiley & Sons, New York, 1986.

$$H_2N–CO–NH_2 + H_2O \leftrightarrow 2NH_3(gas) + CO_2(gas) \qquad [13\text{-}3]$$
(urea)

Leaching, erosion and runoff. Nitrogen can also be lost from soils by leaching of NO_3^--nitrogen. Water containing NO_3^- finds its way to lakes and streams through tile drainage systems or though natural seepage flow. The amount of nitrogen lost in the United States each year has been estimated to be in excess of 2×10^9 kg. Leaching losses are greatest in the cultivated humid and subhumid regions, where there is sufficient rainfall to allow downward movement of water below the rooting zone and where nitrogen fertilizers are heavily used. The leaching of NO_3^- and possible contamination of groundwater is a topic of considerable research effort and public concern in both irrigated and nonirrigated agricultural systems. On a national basis, considerable N may be lost or at least moved by water and wind erosion. Current estimates of erosion losses are 4.5×10^9 kg of N annually.

Ammonia fixation. Soils can hold significant quantities of NH_4^+ in "fixed" or nonexchangeable, plant-unavailable forms. Expanding 2:1-type clay minerals, such as vermiculite and some smectites, can fix NH_4^+ when it substitutes for larger hydrated cations, such as, Ca^{2+}, Mg^{2+}, or Na^+ in interlayer positions. Because of its smaller diameter, NH_4^+ fits into hexagonal voids "holes" formed by the ring pattern of silicon tetrahedrons. This results in increased attraction between the two negatively charged layers and the NH_4^+ ions, so that the layers collapse and the NH_4^+ ions cannot move freely in or out of the interlayer (see Figure 8-4). High levels of potassium (K^+) can reduce the amount of NH_4^+ fixed, since K^+ has a hydrated radius similar to NH_4^+ and thus competes for fixation sites within the interlayer. Several studies have shown that fixed NH_4^+ represents less than 10% of the nitrogen in surface soils, especially in soils that contain higher organic matter contents. The percentage of soil nitrogen as fixed NH_4^+ increases with soil depth, because, while the total amount of N decreases quite rapidly with depth, the amount of fixed NH_4^+ decreases very little with depth. The amount of fixed NH_4^+ in an individual soil increases with increasing amounts of expanding clays, decreasing K^+ contents, increasing NH_4^+ additions, decreasing soil water, and increasing soil pH. The concentration of fixed NH_4^+ in the north central United States varies from less than 10 μg NH_4^+/g of soil to over 250 μg NH_4^+/g of soil.

13.1.2 Soil Nitrogen Balance

Plants remove large amounts of nitrogen from the soil (Table 13-6). Nitrogen, removed, must be replaced by nitrogen fixation, decay of crop residues and soil organic matter, or application of commercial fertilizers. The challenge to managers of the land is to ensure an adequate supply of plant-available nitrogen in the soil while minimizing losses that are economically painful and environmentally hazardous. Soil and crop management practices, which result in maximum economic yields (MEY), are often compatible with this challenge.

The amount of commercial fertilizers used depends on individual farm situations. If legumes are grown, they will supply most, if not all, of the nitrogen they need though nitrogen fixation. In most cases, crop residues return some nitrogen to the soil. If manures are available, they should be used as supplementary sources of nitrogen.

Soils tend to approach an equilibrium value for organic matter and soil nitrogen that depends on the cropping system employed, as well as the climate of the soil. Because nitrogen added in excess of that removed by crops tends to be lost from the soil, soil management practices should attempt to ensure availability of nitrogen at the proper times and in suitable amounts to minimize nitrogen loss (Table 13-6).

TABLE 13-6 Total amount of nitrogen removed in major crops[a]

Crop	Plant parts	Yield (kg/ha)	Total N (kg/ha)
Alfalfa	Total forage	18,000	510
Bermuda	Total forage	22,500	565
Brome grass	Total forage	7,000	175
Corn	Grain	10,000	150
	Stover	9,000	80
Cotton	Lint	1,690	
	Seed	2,530	105
	Stalks, leaves, burrs	5,000	70
Potatoes (Irish)	Tubers	6,000	170
	Vines	5,000	115
Rice	Grain	7,900	85
	Straw	10,000	40
Sorghum	Grain	9,000	135
	Stover	5,000	60
Soybeans	Grain	2,800	180
	Straw	5,400	75
Sugar beets	Roots	68,000	140
	Tops	36,000	145
Wheat	Grain	5,400	110
	Straw	6,000	45

Source: R. A. Olson and L. T. Kurtz, Crop nitrogen requirements, ultilization and fertilization, in *Nitrogen in Agricultural Soils*, F. J. Stevenson (ed.), American Society of Agronomy, Monograph No. 22, Madison, Wis., 1982.
[a]Substantial variation from these values can occur depending on soil N status and fertilization affected (i.e., total N of the end product continues to increase with added N beyond that required for maximum yield).

13.2 SULFUR

Sulfur is one of the essential nutrients needed for plant growth. Sulfur is needed in large amounts and is classified as a macronutrient. Analysis of many crops shows approximately equal quantities of sulfur and phosphorus in the plant material. Sulfur ranks in importance with nitrogen as a constituent of proteins. Also sulfur is an integral component of vitamins B_1 and H as well as several enzymes. Despite its importance, sulfur is often overlooked by scientists and growers. This is because in many areas crops have not responded to soil applications of sulfur fertilizers. Documented sulfur deficiencies, however, are being reported with increasing frequency, because of lower sulfur inputs from "high-analysis" fertilizers, less use of sulfur-containing pesticides, and reduced industrial emissions of sulfur oxides.

In soil systems, sulfur has many similarities to nitrogen. Both occur in a wide range of oxidation states and in various inorganic and organic compounds. The majority of both are found in soil organic matter in plant-unavailable forms. Both are subject to a variety of microbially mediated transformations and are mobile nutrients, readily lost from the soil. The sulfur content of soils ranges from 0.002% in very sandy soils to as much as 3.5% in soils developed in tidal areas, where sulfides have accumulated. The sulfur content of most soils is approximately 13% of the nitrogen content. Soil-forming factors that result in increased organic matter contents or nitrogen contents result in increased sulfur contents of soils.

13.2.1 The Sulfur Cycle

A generalized representation of the sulfur cycle is shown in Figure 13-5. Sulfur is added to soils as fertilizer, from atmospheric sulfur gases and from acid rain. Other important sources of sulfur additions to soil are plant and animal residues and soil amendments such as pesticides. Sulfur can also be added to soil when groundwater containing sulfur is used for irrigation. During the weathering process, sulfur is released from primary minerals. In plants, sulfur is converted to sulfur-containing amino acids, such as cystine and methionine, which are integral parts of plant and animal proteins. When plant and animal residues are returned to soils, microorganisms release, through the decay process, sulfur as sulfate (SO_4^{2-}). Sulfur is removed from soils by leaching, runoff, plant uptake, and in some cases, volatilization as a gas.

Chemistry of soil sulfur.   Like nitrogen, sulfur occurs in both organic and inorganic forms in soils. The ratio of organic to inorganic forms varies considerably, depending on conditions such as pH, drainage, organic matter content, parent material, and soil depth. In the humid and semihumid regions of the world, organic forms predominate and can account for 95 to 99% of the sulfur present.

Inorganic sulfur is vitally important, because this is the form of sulfur absorbed by plants. Most inorganic sulfur in soils is in the sulfate (SO_4^{2-}) form. Other forms of sulfur are the reduced species, thiosulfate ($S_2O_3^{2-}$), sulfide (SH^-), and

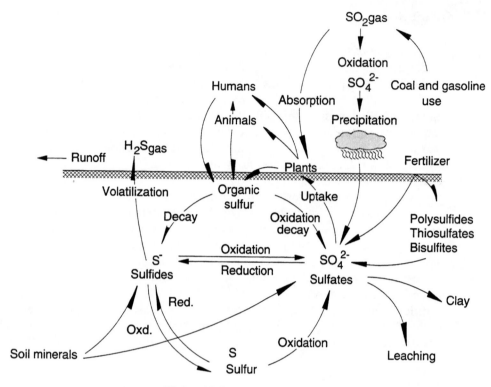

Figure 13-5   The sulfur cycle.

elemental sulfur (S^o). Organic sulfur consists of carbon-bonded sulfur, including three amino acids, cysteine, cystine, and methionine. Non-carbon-bonded organic sulfur is thought to occur as phenolic sulfates and sulfated polysaccharides. Much of the sulfur in soils occurs as complex organic forms whose structure has not been identified.

Atmospheric inputs of sulfur.   Increasing concern has been shown in recent years for sulfur in the atmosphere in the form of sulfur oxides and in the form of acid rain. Approximately two-thirds of the acidity in rainfall in the midwestern United States can be accounted for by hydrogen ions associated with sulfuric acid (H_2SO_4). The ratio is somewhat less in heavily populated urban regions, because of nitrogen emissions from automobiles. The source of part of the atmospheric sulfur is anthropogenic, including industrial emissions, with the largest single anthropogenic source being coal-burning power plants. Sulfur dioxide is also emitted by natural sources, such as volcanos and oceans. Approximately one-half of the total sulfur in the atmosphere is estimated to be from anthropogenic sources. The dramatic increase in sulfur emissions experienced over recent decades has slowed considerably because of increased concern over environmental effects and increased regulation of sulfur emissions.

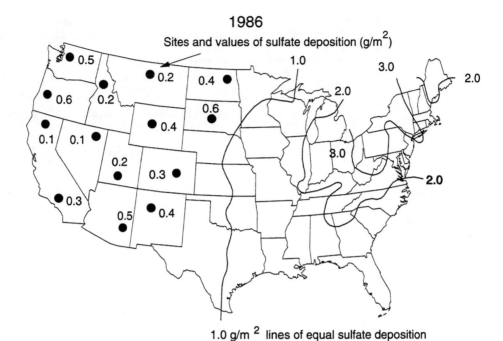

Figure 13-6   Annual (1986) sulfate ion deposition (g/m^2). Note: Deposition rates vary slightly from year to year. (Data from the NADP/NTN Annual Data Summary, *Precipitation Chemistry in the United States*, Natural Resource Ecology Laboratory, Colorado State University, Fort Collins, Colo.)

Atmospheric sulfur dioxide can be absorbed as a gas by soils directly from the atmosphere (dry deposition), or it may be oxidized in the air to sulfuric acid and deposited on the soil as acid rain (wet deposition). The amount of sulfur deposited in the United States as wet deposition is shown in Figure 13-6. The amount of sulfur deposited by wet deposition represents approximately 10 to 15% of the sulfur taken up by crops, such as corn, wheat, soybean, or cotton (Table 13-7).

Soil losses of sulfur.   Leaching of SO_4^{2-} is the major mechanism for loss of sulfur from soils. Some SO_4^{2-} can be retained by soils, as will be discussed later. The amount of SO_4^{2-} lost by leaching depends on the amount of rainfall, permeability of the soil, and the amount of SO_4^{2-} released by microorganisms. In humid regions, SO_4^{2-} seldom accumulates in soils, because it is relatively easily leached. In arid regions some soil horizons become enriched in gypsum (CaSO$_4$ · 2H$_2$O). Sulfur can also be lost by volatilization, if conditions are very reducing (flooded). The fact that sulfur does not accumulate in humid region soils is possibly one reason that the nitrogen/sulfur ratio of soils from many regions of the world is remarkably constant at about 10:1.3.

TABLE 13-7   Annual uptake of nitrogen and sulfur for high-yielding crops, average fertilization rates of nitrogen, and wet deposition of sulfur and nitrogen during the growing season in the United States (kg/ha)

		Nitrogen		Sulfur[a]	
Crop	Uptake	Fertilizer rate	Wet deposition	Sulfur uptake	Wet deposition
Corn	298	153	1.5	37	3.8
Wheat	153	67	1.3	23	3.0
Soybean	363	20	1.6	28	4.2
Cotton	201	91	1.3	34	2.7

Source: National Acid Precipitation Assessment Program VI, *Effects of Acidic Deposition.*
[a]Deposition values are weighted over several years for geographic regions where these crops are produced.

Sulfur mineralization and immobilization.   *Mineralization* of sulfur, like that of nitrogen, is defined as the transformation of organic forms to inorganic forms (SO_4^{2-}). This conversion of organic sulfur to SO_4^{2-} is a microbial process. *Immobilization*, by contrast, is the process where microbes incorporate SO_4^{2-} into microbial tissue as they grow. Mineralization of sulfur in soils is affected by the availability of organic matter, soil temperature, soil water, and soil pH. Soils, with higher organic matter contents, would be expected to experience higher mineralization rates. Mineralization of sulfur is increased as soil temperatures rise from 20°C to 40°C and is very slow at extremely cold or hot temperatures. The optimum water content of soils for sulfur mineralization is approximately 60% of field capacity. Mineralization of sulfur increases with increasing soil pH up to pH values of approximately 7.5.

Inorganic transformations.   The sulfur cycle (Figure 13-5) indicates several oxidation and reduction transformations for inorganic sulfur species. In many cases, these transformations are mediated by microorganisms. Sulfate can be reduced by bacteria to sulfides, resulting in formation of ferrous sulfide precipitates. This kind of reduction results in corrosion of steel pipes or cables buried in poorly drained soils. It also accounts for black stains sometimes observed in the B horizons of poorly drained soils. When conditions are extremely reducing, the conversion of SO_4^{2-} to hydrogen sulfide (HS^-) is carried out by specific bacteria of the genera *Desulfovibrio* or *Desulfotomaculum*. As was the case for NO_3^- in denitrification, SO_4^{2-} can serve as a terminal electron acceptor for microorganisms when anaerobic conditions exist in the soil, particularly in stagnant water basins and marine muds. Transformations by soil bacteria can also result in the formation of elemental S (Figure 13-5).

The reverse of the reducing reactions are transformations accomplished by sulfur-oxidizing bacteria under aerobic soil conditions. Many of the sulfur-oxidizing bacteria are obligate autotrophs, receiving their carbon from CO_2. Some of the sulfur-oxidizing species of bacteria are *Thiobacillus thiooxidans, T. ferrooxidans,*

T. thioparus, and *T. denitrificans*. Elemental sulfur is sometimes added to well-aerated soils to lower soil pH for certain plants, such as azaleas or blueberries. The production of sulfuric acid and subsequent lowering of soil pH in the presence of elemental sulfur is the direct result of oxidizing bacteria. The oxidation reaction can be described as follows:

$$S° + 1\tfrac{1}{2}O_2(gas) + H_2O \leftrightarrow H_2SO_4 \qquad\qquad [13\text{-}4]$$

Another important example of the effect of sulfur-oxidizing bacteria in nature is the development of so called acid-sulfate soils. These soils form from swamps or buried sediments that have accumulated FeS_2. As these wetlands are drained by natural processes or by people, opportunity for oxidation occurs and H_2SO_4 is produced. These soils, sometimes called "cat clays," can have extremely low pH values (less than 4.0). These soils occur in the delta regions, where the source of sulfide minerals in the mud was SO_4^{2-} from seawater (see Chapter 9). Because of the severe acidity that results from draining these soils, they are sometimes flooded and used for rice production. Environmental factors that favor sulfur oxidizers, such as temperature, water, and soil pH, are similar to those that favor nitrification.

Soil sorption of sulfate.   Many soils have some anion-exchange capacity, especially those with large amounts of kaolinitic clay or free iron and aluminum oxides. Anion-exchange sites serve as sites for SO_4^{2-} retention in soils. The sorption or retention of SO_4^{2-} in soils is important from a fertility standpoint as a source of sulfur for plant uptake. However, the amount of anion exchange in many soils is quite low, and in general, SO_4^{2-} is considered a mobile anion subject to considerable leaching. From an environmental viewpoint, SO_4^{2-} leaching is not a concern for groundwater pollution, since sulfur is not applied at rates as high as nitrogen.

S T U D Y   Q U E S T I O N S

1. What forms of nitrogen are available to higher plants?
2. Discuss the various forms of nitrogen fixation (i.e., symbiotic, nonsymbiotic, and abiotic).
3. Why does nitrification occur?
4. Under what circumstances does biological denitrification occur? What is the role of NO_3^- in anaerobic respiration?
5. What soil conditions would favor NH_3 volatilization from soils?
6. What are the primary functions of nitrogen and sulfur in plants?
7. Why do certain organisms oxidize reduced forms of sulfur, such as pyrite, to sulfuric acid?
8. What soil conditions favor SO_4^{2-} reduction? If NO_3^- is present in soils, will sulfate be reduced under these conditions? (See Figure 6-3.)

14

Phosphorus and Potassium

14.1 PHOSPHORUS

Phosphorus is an essential nutrient. This means by definition an element that is needed by plants to grow and to reproduce, that is, to complete their life cycles. Since P is needed in fairly large amounts, P is considered to be a macronutrient. Macronutrients are not more important to a plant's metabolism than micronutrients, but since they are needed in large amounts, deficiencies are more common with the macronutrients. Phosphorus is a component of many organic compounds in plants. It occurs as a component in activated carbohydrates, phospholipids, phosphoamino acids, and nucleic acids. Phosphorus is an important constituent of several highly reactive nucleotides that are important in energy transfer. These include adenosine triphosphate (ATP), adenosine diphosphate (ADP), and other phosphorylated compounds. The phosphorylation of compounds, such as carbohydrates, is important since it makes the compounds more reactive, that is, easier to metabolize. The major role of P in plants and other organisms is in the transfer of energy between compounds in the various metabolic cycles.

Phosphorus is a mobile element within the plant. When a deficiency occurs, P is translocated from the older to younger tissue, resulting in the older tissue showing deficiency symptoms first. Even though P plays a major role in energy transfer, phosphorus deficiency symptoms are often not well defined. Phosphorus-deficient plants may be stunted, or dwarfed and their fruits often mature more slowly. Minor P deficiencies, especially on young plants, often result in a dark purplish coloration due to anthocyanin production. Phosphorus deficiencies are often noted on young plants or new tissues when low soil temperatures interfere with the uptake of phosphorus. When the soil warms, normal phosphorus uptake returns and the plants grow out of the deficiency.

14.1.1 Forms of Phosphorus in the Soil Solution

Phosphorus occurs in the soil solution primarily as one of the ions of orthophosphoric acid (H_3PO_4). The following chemical equations illustrate the dissociation of orthophosphoric acid into its ions.

$$H_3PO_4 \leftrightarrow H_2PO_4^- + H^+, \qquad Ka_1 = \frac{(H_2PO_4^-)(H^+)}{(H_3PO_4)} \qquad [14\text{-}1]$$

$$= 7.6 \times 10^{-3}, \qquad pKa_1 = 2.12$$

$$H_2PO_4^- \leftrightarrow HPO_4^{2-} + H^+, \qquad Ka_2 = \frac{(HPO_4^{2-})(H^+)}{(H_2PO_4^-)} \qquad [14\text{-}2]$$

$$= 6.3 \times 10^{-8}, \qquad pKa_2 = 7.2$$

$$HPO_4^{2-} \leftrightarrow PO_4^{3-} + H^+, \qquad Ka_3 = \frac{(PO_4^{3-})(H^+)}{(HPO_4^{2-})} \qquad [14\text{-}3]$$

$$= 4.8 \times 10^{-13}, \qquad pKa_3 = 12.3$$

An equilibrium constant such as Ka_1 gives the relationship between the molar concentrations of H_3PO_4 and $H_2PO_4^-$ at equilibrium. Equations [14-1] through [14-3] can be used to establish the effect of soil pH on the form of orthophosphate that occurs in the soil solution. For example: What is the soil pH when the concentration of H_3PO_4 equals the concentration of $H_2PO_4^-$? When $(H_3PO_4) = (H_2PO_4^-)$ the ratio of $(H_2PO_4^-)/(H_3PO_4)$ in Ka_1 is equal to unity and the equilibrium constant Ka_1 is equal to the concentration of hydrogen ions (H^+) in the soil solution.

$$Ka_1 = (H^+) = 7.6 \times 10^{-3} \qquad [14\text{-}4]$$

In terms of pH, when $(H_3PO_4) = (H_2PO_4^-)$, soil pH $= -\log (7.6 \times 10^{-3}) = pKa_1 = 2.12$. When pH values are below 2.12, H_3PO_4 is the dominant species, and when pH values are above 2.12, $H_2PO_4^-$ is the dominant species.

The same reasoning can be applied to the second and third dissociation reactions of orthophosphoric acid and their equilibrium constants Ka_2 and Ka_3. When $(H_2PO_4^-) = (HPO_4^{2-})$, then (H^+) equals $Ka_2 = 6.3 \times 10^{-8}$ or pH equals 7.2. Hence $H_2PO_4^-$ is the dominant or stable species when pH is below 7.2, and HPO_4^{2-} is the dominant species above pH values of 7.2. Now consider the third dissociation of orthophosphoric acid (equation [14-3]). When $(HPO_4^{2-}) = (PO_4^{3-})$ the concentration of the hydrogen ion (H^+) equals Ka_3 and $(H^+) = 4.8 \times 10^{-13}$ or pH $= 12.3$. When pH values are below 12.3, HPO_4^{2-} is the dominant species, and when pH values are above 12.3, PO_4^{3-} is the dominant species.

Note: It can be generalized for a single-step dissociation that when (H^+) equals pKa (the numerical value of the equilibrium constant) the concentration of the species that is dissociating equals the concentration of the species that is being formed.

Figure 14-1 shows the distribution of orthophosphate species as a function of soil pH. Since normal soil pH values range from 4 to about 8.5, Figure 14-1 illustrates that $H_2PO_4^-$ and HPO_4^{2-} are the major forms of phosphorus in the soil solution. For soils with pH values in the range 4 to 7.2, $H_2PO_4^-$ is the major form of phosphorus in the soil solution. For soils with pH values in the range 7.2 to 8.5, HPO_4^{2-} is the major form of phosphorus in the soil solution. When soil pH is close to 7.2, both $H_2 PO_4^-$ and HPO_4^{2-} are present in approximately equal amounts.

Figure 14-1, which gives the distribution of phosphorus species as a function of pH, can also be used to partially explain the behavior of phosphorus fertilizer materials when added to the soil. *Example:* Three growers were discussing the effect of P fertilizers on soils. The growers farm in a region where soil pH values range from 6 to 8. The first grower stated that P fertilizers acidify soils, the second that P fertilizers raise soil pH, and the third that P fertilizers have no effect on soil pH. The growers were applying P fertilizers that contained either $Ca(H_2PO_4)_2$ or $CaHPO_4$ the most common forms of P in commercial fertilizers. Which of the growers is correct?

If $Ca(H_2PO_4)_2$(solid) is added to the soil, Ca^{2+} and $H_2PO_4^-$ ions will be produced upon dissolution of the fertilizer. If $H_2PO_4^-$ ions are added to a soil in the

Figure 14-1 Distribution of the dissociation products of orthophosphate acid as a function of soil pH.

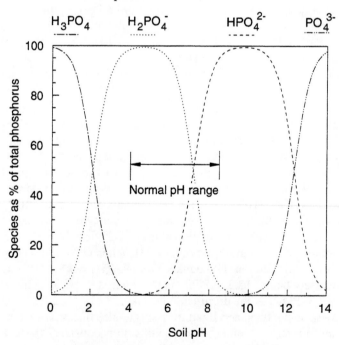

pH range 2.12 to 7.20, where $H_2PO_4^-$ is the stable form of orthophosphate, no change in soil pH will occur since $H_2PO_4^-$ will remain as $H_2PO_4^-$. If $H_2PO_4^-$ ions are added to the soil in the pH range 7.20 to 12.3, where HPO_4^{2-} is the stable form of orthophosphate, the $H_2PO_4^-$ ions will dissociate to form the stable form of orthophosphate and pH will decrease:

$$H_2PO_4^- = HPO_4^{2-} + H^+ \qquad \text{[14-5]}$$

If $CaHPO_4$(solid) is added to the soil, Ca^{2+} and HPO_4^{2-} ions will be produced upon dissolution of the fertilizer. If HPO_4^{2-} ions are added to the soil in the pH range 7.20 to 12.3, where HPO_4^{2-} is stable, no pH change will occur. If HPO_4^{2-} ions are added to soils with a pH range 2.12 to 7.2, where $H_2PO_4^-$ is the dominant form of orthophosphate, the HPO_4^{2-} ions will hydrolyze water to form the stable form of orthophosphate and soil pH will increase.

$$HPO_4^{2-} + H_2O = H_2PO_4^- + OH^- \qquad \text{[14-6]}$$

Since the effect of P fertilizers on soil pH is dependent on the form of P fertilizer added and on the pH range of the soil, it is possible that all three growers are correct. The orthophosphate ions $H_2PO_4^-$ and HPO_4^{2-} upon addition to a soil can increase, decrease, or have no effect on soil pH, depending on the ion added and the pH of the soil.

14.1.2 Solubility of Phosphorus Minerals and Phosphorus Fixation

The stable forms of phosphorus minerals in soils are dependent upon pH. In acid soils, iron and aluminum phosphates are very insoluble and are the stable forms of phosphorus. In neutral or basic soils, calcium phosphates are less soluble than the iron and aluminum phosphates and therefore are representative of the stable forms of phosphorus. When a mineral is present in a soil pH range that favors its precipitation, the mineral has a very low solubility and has little or no tendency to dissolve out of the soil. Under these conditions the mineral is considered to be stable. It is also the case, under these conditions, that the mineral would not be a good source of plant-available P because of its low solubility.

Figure 14-2, which is known as the Lindsay–Moreno diagram,[1] can be used to determine the effect of soil pH on the behavior of phosphorus minerals in the soil. The diagram provides insight into the solubility, hence stability of the various minerals, and also provides information about phosphorus-fixation mechanisms as well as the suitability of phosphorus minerals for fertilizers. The figure shows clearly that Fe and Al phosphates, such as strengite [$Fe(OH)_2H_2PO_4$] and variscite [$Al(OH)_2H_2PO_4$], are the least soluble and therefore the most stable phosphorus minerals in acid soils, while Ca phosphates, such as hydroxyapatite [$(Ca_{10}(PO_4)_6(OH)_2)$], are the least soluble and the most stable forms of phosphorus

[1]W. L. Lindsay and E. C. Moreno, Phospate phase equilibria in soils, *Soil Sci. Soc. Amer. Proc.* 24:177–182 (1960).

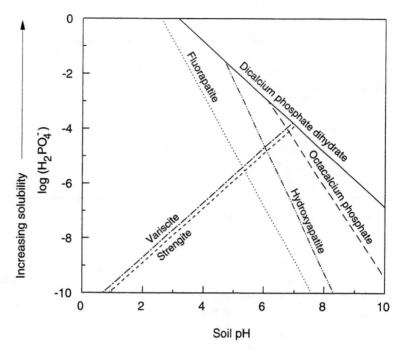

Figure 14-2 The solubility of representative phosphate minerals as a function of soil pH. [After W. L. Lindsay and E. C. Moreno, Phosphate phase equilibria in soils, *Soil Sci. Soc. Amer. Proc.* 24:177–182 (1960).]

in basic soils. If soluble forms of phosphorus such as dicalcium phosphate dihydrate are added to an acid soil as a P fertilizer, the materials will dissolve and a high percentage of the phosphorus will be precipitated as Fe and Al phosphates. If the same fertilizer materials are added to basic soils, the materials will dissolve, but a high percentage of the added material will be precipitated as less soluble Ca phosphates.

The precipitation of added P fertilizers, either as iron or aluminum phosphates in acid soils or calcium phosphates in basic soils, is one of the major problems associated with P availability. In strongly acid or basic soils a very high percentage of the added P fertilizer material is precipitated and made unavailable to plants. Figure 14-2 shows that the problem of P fixation is at a minimum for soil pH values where the iron and aluminum phosphates and calcium phosphate both have reasonably high solubilities, that is, in the pH range 6 to 7. One major benefit of liming acid soils is the increased solubility of iron and aluminum phosphates when the soil pH is increased from values around 4 to 5 up to the recommended pH values of 6 to 7. The effects of overliming can also be seen in Figure 14-2. If soil pH is raised above 7, the solubility of the calcium phosphates decreases rapidly, resulting in much lower P availability. The problem of P fixation is most severe in acid and basic soils, but even in neutral soils a substantial portion of the added fertilizer material is precipitated and therefore is unavailable to plants.

Figure 14-2 also helps to explain why untreated "raw" rock phosphate is a suitable source of P in acid soils but not in other soils. Rock phosphate contains primarily fluorapatites and hydroxyapatites. When rock phosphate is added to acid soils in which these minerals have a high solubility, it is a good fertilizer material. But when rock phosphate is added to neutral or basic soils, where hydroxyapatite and fluorapatite are much less soluble, it is not a suitable fertilizer material. Since most acid soils are limed to pH values around 6 to 7, untreated "raw" rock phosphate should not be considered to be a suitable P fertilizer material.

Adsorption of phosphorus. Phosphorus, in addition to forming precipitates with Fe^{3+} and Al^{3+} in acid soils and with Ca^{2+} in basic soils, is also strongly adsorbed by iron, aluminum, and calcium minerals. The reaction of phosphorus with iron oxides (Figure 14-3) is an example of chemisorption, in which there is an interaction of a specific funcional group of the adsorbing species with a hydroxyl or aquo site on the iron oxide surface. The sorption is very specific and represents a chemical bond being formed between the oxide surface and phosphorus. It has been determined experimentally that there is an active OH site every 22 to 24 $Å^2$ of an iron oxide-hydroxide mineral's surface. Since the base of the

Figure 14-3 Adsorption of phosphorus on iron oxide surfaces.

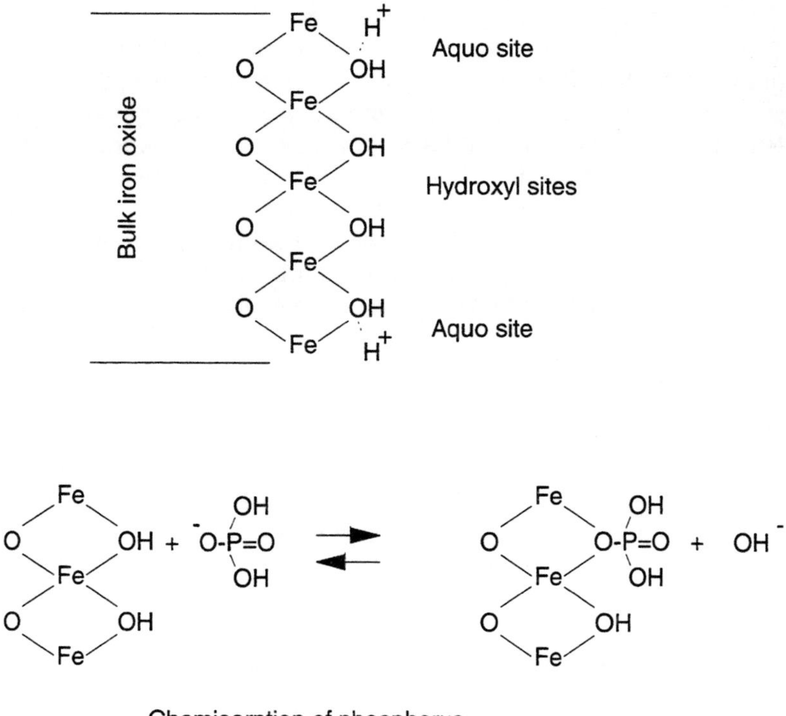

Chemisorption of phosphorus

phosphorus tetrahedron is about 21.6 Å^2, it appears that every hydroxyl is available for adsorption of phosphate ions without steric hindrance.

14.1.3 Management of Phosphorus

A major problem associated with phosphorus is the fixation of added fertilizer materials. In acid and basic soils fixation of added phosphorus can exceed 95%. The precipitation of added phosphorus can be minimized by liming acid soils to pH values in the range 6 to 7. Iron, aluminum, and calcium phosphates have their greatest mutual solubilities in this range. At higher pH values the solubilities of calcium phosphates are decreased, and at lower pH values, the solubilities of iron and aluminum phosphates are decreased.

The fixation of added phosphate fertilizers can also be affected by the method of application. Broadcasting phosphate fertilizer maximizes the contact between the soil and the fertilizer, while banding minimizes this contact and hence slows precipitation of the added phosphate materials.

Because of the strong fixation of phosphorus fertilizers, these materials are not mobile in the soil. Little phosphorus is lost due to leaching into the groundwater, but neither is it moved into the lower soil horizons, and these horizons are often very deficient in plant-available phosphorus. Because phosphorus is not mobile or subject to excess plant uptake, a grower may apply to the soil in one application all the phosphorus required by a crop rotation.

Mycorrhizae fungi have been shown to increase the uptake of nonmobile nutrients, such as P, especially by tree species. Conifers such as *Pinus strobus* are ectomycorrhizal. In experiments[2] where *P. strobus* was grown in soils both inoculated and uninoculated with mycorrhizae, P uptake was increased from 0.074 to 0.196% of dry weight upon inoculation. Mycorrhizal infection of tree roots greatly increases the absorption surface of root system. Trees tend to have coarse root systems with large interroot distances, which limits the availability of nonmobile nutrients, such as P. Mycorrhizal infection increases the effective surface area of the root system and increases uptake on nonmobile nutrients, especially in nutrient-poor soils.

14.2 POTASSIUM

Potassium is also an essential element. The role of potassium in the plant is primarily in osmotic and ionic regulation. It is abundant in plant vacuoles and is bound ionically to proteins and other negatively charged cell constituents. Potassium also functions as an activator for many plant enzymes. No organic forms of

---

[2]P. B. Tinker, Role of rhizosphere microorganisms in phosphorus uptake by plant, in *The Role of Phosphorus in Agriculture*, F. E. Khasawneh (ed.), American Society of Agronomy, Madison, Wis., 1980.

potassium are known to exist in the plant. Potassium is mobile within the plant, hence deficiency symptoms occur first on older tissue.

Potassium is an alkali metal, meaning that it forms +1 ions upon losing an outer $4s$ electron. Potassium occurs in the soil solution and in soil minerals as the K^+ ion and is not subject to oxidation or reduction in the soil. Potassium has a crystal radius of 0.133 nm and a hydrated radius of $\approx$ 0.331 nm. The lithosphere contains approximately 2.6% potassium. Salic rocks, which are rich in aluminosilicate minerals, contain high amounts of potassium, while femic rocks, which are rich in ferromagnesium minerals, are low in potassium. The potassium content of sedimentary rocks varies with the texture and composition of the sediments that compose the rock. Shales have high potassium contents ($\approx$ 27 g kg^{-1}), while sandstones have lower potassium contents ($\approx$ 11 g kg^{-1}) and limestones contain even less potassium ($\approx$ 2.7 g kg^{-1}).

Mineral soils contain from 0.04 to $\geq$ 3% potassium, with an average value of 0.83%. The difference in potassium content between rock materials and surface soils reflects the loss of K^+ during soil formation. Most of the potassium in soil (90 to 98%) occurs in the crystal lattices of feldspars, such as orthoclase and microcline, and in the interlayers of micaceous minerals, such as muscovite and biotite. The potassium balances the negative charge of the structural units of these minerals. Potassium in the interlayers of micas and within the three-dimensional tetrahedral structure of feldspars is tightly held and is released into the soil solution only upon weathering of the mineral. Potassium (1 to 10%) also occurs "fixed" in interlayer positions of vermiculites and possibly some highly charged smectites. One to 2% of the total potassium in the soil occurs as K^+ ions on the exchange complex and in the soil solution.

14.2.1 Potassium Status of Soils

The potassium status of mineral soils depends on the type of parent material, the degree of weathering, the amount of potassium fertilizer added, crop removal of potassium, potassium fixation, the amount of erosion, and the extent of leaching losses of potassium. Figure 14-4 shows the factors that control the potassium status of soil arranged in a potassium cycle. The actual amount of plant available potassium in the soil depends on the degree of transport between the various parts of the cycle.

Parent materials and weathering. The most important sources of potassium in soil are the micaceous minerals and their weathering products.[3] The degree of weathering of the minerals is as important as the types of minerals that are present. The degree or extent of weathering is a function of rainfall, temperature,

[3]P. M. Bertsch and G. W. Thomas, Potassium status of temperate region soils, in *Potassium in Agriculture*, R. D. Munson (ed.), American Society of Agronomy, Madison, Wis., 1985.

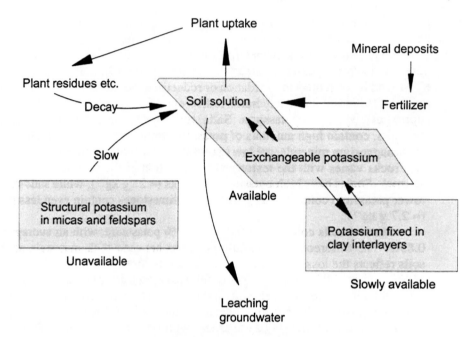

Figure 14-4   The potassium cycle.

and time. Weathering partially opens the micaceous interlayers, allowing K^+ to be replaced by hydrated cations that further expand the interlayer and prevent its collapse upon drying. Weathering also alters the mineralogy as shown in Figure 14-5.

Even though the potassium content is lower in the weathering products of micas, it is held with lower energies and is released into the soil solution at much faster rates than from the original minerals. In general, the highly weathered soils of the northeastern and southeastern United States have the lowest total potassium contents, soils in the corn belt, the north central United States, have higher potassium contents, and soils in the western United States have the highest total potassium contents. Sandy soils in all regions tend to have low potassium contents.

Figure 14-5   Weathering sequence of micaceous minerals.

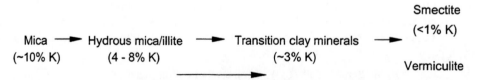

Fixation of potassium. The fixation of potassium is due primarily to the small hydrated size of K^+ (0.331 nm) compared to other exchangeable cations, such as Ca^{2+} (0.412 nm). The small size of the hydrated potassium ion allows potassium to fit into the "hole" created by the ring pattern of the silicon tetrahedrons of a tetrahedral sheet (see Figure 3-10). When the unit-layer charge of the mineral is $\geq$ 1, the combination of the closeness of approach of the small hydrated K^+ to the source of charge, isomorphous substitution in the tetrahedral sheet, and the high unit-layer charge results in the K^+ being more tightly bound than larger exchangeable cations. The localization of K^+ in the "holes" and the stronger bonding dramatically lowers the probability of exchange, and the K^+ ion is said to be fixed. Even though the K^+ is considered to be "fixed," there is a small but finite probability of exchange occurring, that is, K^+ being released into the soil solution. Potassium fixation by this mechanism is due primarily to vermiculitic minerals, although some highly charged smectites, unit-layer charges $\approx$ .8 to $\approx$ .9, may also "fix" potassium.

Potassium fixation by this mechanism has been shown to have a negative influence on plant uptake of potassium, particularly in soils that are very depleted in potassium. Some researchers suggest that fixation by this mechanism may be beneficial in soils with more reasonable potassium levels, as the fixation lowers the luxury consumption of potassium and minimizes the losses of potassium by leaching. In these soils the slow release of "fixed" potassium would serve as a buffer to the already adequate plant-available pool of potassium.

Potassium fixation, or at least lower levels of K^+ ions in the soil solution, has been shown to result from liming. The lower levels of K^+ in the soil solution can be attributed to three possible mechanisms. First, liming of acid soils increases the pH-dependent charge of the colloids. Liming may also result in removal of interlayer aluminum materials, exposing additional exchange sites. Both processes result in greater amounts of exchangeable K^+, which lowers the solution levels. Second, the higher pH values result in lower concentrations of Al^{3+} in the soil solution, resulting in more exchangeable K^+ and lower solution levels. Third, there is less displacement of K^+ by H^+ from edge and interlayer positions of potassium-containing primary minerals, such as micas, in basic soils as compared to acid soils. It has also been suggested that potassium fixation in acid soils can be caused by precipitation with high concentrations of soluble aluminum, sulfate, and/or phosphate.

Substantial losses of potassium due to leaching and erosion have also been demonstrated. Losses of 0.6 to 1.5 kg/ha/yr have been demonstrated on fallow fine-textured Illinois soils[4] and amounts as high as 12.8 kg/ha/yr on coarser-textured South Carolina Coastal Plain Soils.[5] Both of these experiments were performed on fallow soils, and as such, they probably represent maximum amounts of leaching, since crop production would lower potassium levels in the soil solution. Erosion losses of potassium can be very high. Losses as high as 150 kg/ha/yr have been

[4]R. S. Stauffer, Runoff, percolate, and leaching losses in some Illinois soils, *J. Am. Soc. Agron.* 34:830–835 (1942).

[5]Allison et al., *USDA Tech. Bull.* 1199:1–62 (1959).

reported.[6] The extent of erosional losses of potassium will depend upon the potassium levels of the surface soil and the amount of erosion.

14.2.2 Management of Potassium

Plant-available potassium is generally considered to be K^+ ions in the soil solution and on the exchange complex. When plants absorb K^+ from the soil solution it is replaced, that is, buffered, by the equilibrium between exchangeable K^+ and K^+ in the soil solution. Since plants rapidly deplete the concentration of K^+ in the vicinity of the plant root, plant uptake of potassium has been shown to be limited by the rate of K^+ diffusion from nondepleted zones into depleted zones.[7]

Figure 14-6 Luxury consumption of potassium. Crop yield levels off but potassium uptake increases with increasing soil potassium levels, resulting in excess crop uptake of potassium.

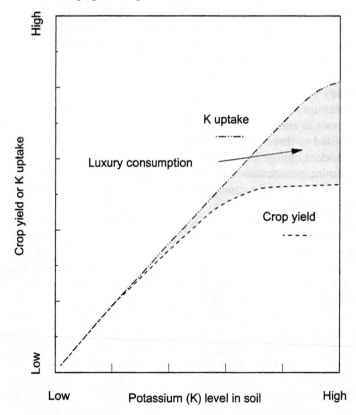

[6]Lipman and Conybeare, New Jersey Agric. Exp. Sta. Bull. No. 607, 1936.

[7]S. A. Barber, Potassium availability at the soil–root interface and factors influencing potassium availability, in *Potassium in Agriculture*, R. D. Munson (ed.), American Society of Agronomy, Madison, Wis., 1985.

Soil tests for potassium usually involve extraction of the soil with a reagent, such as 1 N NH$_4$-acetate, that extracts exchangeable potassium. Since the level of K$^+$ in the soil solution is a function of the percentage potassium saturation of the exchange complex, the amount of potassium fertilizer needed is a function of both the crop to be grown and the cation-exchange capacity of the soil. Soils with high cation-exchange capacities will require greater amounts of fertilizer to achieve a given level of K$^+$ in the soil solution than will soils with lower cation-exchange capacities.

The amount of potassium needed by crops varies to a greater extent than the amount of phosphorus and many other plant nutrients. For example, a 8779-kg/ha (140-bu/acre) corn crop removes 18 kg of K$_2$O equivalent when harvested as a grain crop and 90 kg of K$_2$O equivalent when harvested as silage. A 13.4-mt/ha crop of alfalfa removes 136 kg of K$_2$O equivalent from the soil. Plants, particularly forage species, often remove much greater amounts of K$^+$ from the soil than is needed for metabolic purposes. This excess removal (luxury consumption) does not increase crop yields or crop quality (Figure 14-6). Normally, annual applications of potassium are recommended to minimize the luxury consumption that might occur if all the potassium fertilizer required over the period of a rotation was applied in one application.

S T U D Y   Q U E S T I O N S

1. What are the roles of phosphorus and potassium in higher plants?
2. What forms of phosphorus are found in the soil solution?
3. Explain how phosphorus fertilizer can in one case increase soil pH and in another case decrease pH.
4. Why is phosphorus availability to higher plants greatest in the pH range 6 to 6.5?
5. Discuss the fate of phosphorus fertilizer added to acid and basic soils.
6. Why is K$^+$ fixed by certain types of clay minerals, but not Na$^+$?
7. What is luxury consumption?
8. Discuss the statement: "Anions are more subject to leaching than cations, because cations are attracted to the negative charge of soil colloids."
9. What types of soil conditions produce K$^+$ deficiency?

15

Micronutrients

Micronutrients are essential elements that are needed in very small quantities compared to the macronutrients. As essential elements, micronutrients are needed for the plant to grow and complete its life cycle. The term *micro* implies that only small amounts of these elements are needed, not that they are less important to the plant. A deficiency of a micronutrient will affect plant growth as much as a deficiency of any of the macronutrients. Micronutrients include the following elements:

iron	Fe
copper	Cu
boron	B
molybdenum	Mo
chlorine	Cl
zinc	Zn
manganese	Mn

Several other elements have been shown to increase the growth of certain species of plants, but their essentiality has not been demonstrated. That is, although they increase plant growth, the plant can apparently complete its life cycle when these elements are not present. These include:

cobalt	Co
silicon	Si

vanadium	V
sodium	Na

In addition, certain elements are required by animals that have not been shown to be essential for plants. For example:

iodine	I
fluorine	F
sodium	Na
selenium	Se

One of the unique aspects of micronutrients is the small amounts needed to satisfy plant needs. Table 15-1 gives the average content of macro- and micronutrients in plants on a dry weight basis. Table 15-1 shows that the content of micronutrients in plants range from 0.1 to 0.0001 that of sulfur, the macronutrient needed in the smallest amount. Table 15-2 gives the range of micronutrients found in soils. For some micronutrients, such as iron, the content in soils is not usually the limiting factor in plant availability. Rather, soil conditions, such as pH, that affect solubility of iron minerals limit the amount available for plant uptake. For other micronutrients, such as boron, there is little interaction with the soil and plant availability is controlled by the level in the soil.

There are several general conditions that result in micronutrient deficiencies in soils. These include:

1. Low level of a micronutrient in the soil, a condition either inherited from the parent material, created by the loss of micronutrients during soil formation, or created by long-term crop removal. It is obvious that with many years of crop removal, soil levels will be reduced and may become limiting.

TABLE 15-1   Content of essential nutrients in an average plant

Macronutrients	% dry wt	Micronutrients	% dry wt
Carbon	45	Chlorine	0.01
Oxygen	45	Iron	0.01
Hydrogen	6	Manganese	0.005
Nitrogen	1.5	Boron	0.002
Potassium	1.0	Zinc	0.002
Calcium	0.5	Copper	0.0006
Magnesium	0.2	Molybdenum	0.00001
Phosphorus	0.2		
Sulfur	0.1		

Source: D. W. Rains, in J. Bonner, *Plant Biochemistry*, Academic Press, New York, 1976.

TABLE 15-2 Range of micronutrients in soils

Micronutrient	μg/g soil
Chlorine	20–900
Iron	7000–550,000
Manganese	20–3000
Boron	2–100
Zinc	10–300
Copper	2–100
Molybdenum	0.2–5
Cobalt	1–40

Source: W. L. Lindsay, *Chemical Equilibria in Soils*, Wiley-Interscience, New York, 1979.

2. Presence of soil conditions that affect the plant availability of the micronutrient, such as acidity or alkalinity.

3. High crop yields; these result in greater removal of micronutrients from soils, and levels of micronutrients that were adequate with lower crop yields may not be adequate for higher yields. For example, a 9300-kg/ha corn crop requires approximately 50% more micronutrients than does a 6200-kg/ha corn crop.

4. Modern high-analysis fertilizers contain much lower levels of micronutrients as impurities than did many older fertilizer materials. In the past, normal fertilization practices with macronutrients would result in the inadvertent addition of micronutrients as impurities.

In addition to these factors, there has been a general increase in the awareness of and ability to test for as well as identify micronutrient deficiency symptoms. Often in the past, many plant conditions were not recognized as micronutrient problems.

15.1 IRON

Iron is a transition metal. Like many of the transition metals, iron can occur in multiple oxidation states in the soil environment. In soils, iron occurs primarily in the ferrous (iron II) and ferric (iron III) states. In plants, iron is essential for the synthesis of chlorophyll and is involved primarily in oxidation-reduction reactions. Iron is an integral part of the heme molecule, an important constituent of iron porphyrins. Iron also occurs as ferredoxin, one of the most electronegative compounds in plants. Ferredoxin plays an important role in electron transport in photosynthesis and in the reduction of N_2(gas) in nitrogen fixation. Iron is not readily transported in plants since it is a component of organic constituents; hence deficiency symptoms occur first in young tissues. Iron-deficient leaves become chlorotic in the interveinal regions while the veins remain green. Severe iron deficiency sometimes produces chlorotic symptoms that are almost white in color.

Solubility of iron minerals.   There is usually an adequate supply of total iron in most soils to supply the needs of higher plants. The exception would be acidic highly weathered soils developed from iron-poor rocks. Most iron-deficiency symptoms result from the effect of soil pH on solubility of iron minerals. Iron occurs in a variety of crystalline and amorphous ferric and ferrous minerals. The ferric compound, $Fe(OH)_3$(amorphous), and the ferrous compound, $FeCO_3$(siderite) can be used to illustrate the effect of soil pH on iron solubility.

For the ferric compound:

$$Fe(OH)_3(\text{amorphous}) + 3H^+ \leftrightarrow Fe^{3+} + 3H_2O, \qquad K_{sp} = 10^{4.83} \qquad [15\text{-}1]$$

$$K_{sp} = \frac{(Fe^{3+})}{(H^+)^3} \qquad [15\text{-}2]$$

By convention, the solid phase, [$Fe(OH)_3$(amorphous)] and H_2O are assigned values of unity and hence disappear from the solubility constant expression. The solubility constant expression can be solved to give the concentration of iron (Fe^{3+}) as a function of soil pH.

$$(Fe^{3+}) = K_{sp}(H^+)^3 \qquad [15\text{-}3]$$

Equation [15-3] can be solved for different pH values to illustrate the effect of soil pH on $Fe(OH)_3$(amorphous) solubility (Figure 15-1).

pH	(H^+) hydrogen concentration in the soil solution	(Fe^{3+}) iron concentration in the soil solution
4	10^{-4}	$10^{-7.17}$
5	10^{-5}	$10^{-10.17}$
6	10^{-6}	$10^{-13.17}$
7	10^{-7}	$10^{-18.17}$
8	10^{-8}	$10^{-21.17}$

$Fe(OH)_3$ is clearly much more soluble under acidic soil conditions. For every unit increase in soil pH the solubility decreases 1,000-fold. For soil pH values of 5 and above, $Fe(OH)_3$ is so insoluble that it cannot supply plant needs.

For the ferrous mineral, $FeCO_3$(siderite):

$$FeCO_3(\text{siderite}) + 2H^+ \leftrightarrow Fe^{2+} + CO_2(\text{gas}) + H_2O, \qquad K_{sp} = 10^{7.32} \qquad [15\text{-}4]$$

$$K_{sp} = \frac{(Fe^{2+})PCO_2(\text{gas})}{(H^+)^2} \qquad [15\text{-}5]$$

Siderite's solubility constant expression can be solved for the amount of (Fe^{2+}) in equilibrium with the mineral at different pH values.

$$(Fe^{2+}) = \frac{K_{sp}(H^+)^2}{PCO_2(gas)}$$
[15-6]

The amount of carbon dioxide gas can be set to be equal to the partial pressure in the atmosphere [$PCO_2(gas) = 0.0003$ atm] and the concentration of the Fe^{2+} ion calculated at different soil pH values.

pH	(H⁺) hydrogen concentration in the soil solution	(Fe²⁺) iron concentration in the soil solution
4	10^{-4}	$10^{-4.15}$
5	10^{-5}	$10^{-6.15}$
6	10^{-6}	$10^{-8.15}$
7	10^{-7}	$10^{-10.15}$
8	10^{-8}	$10^{-12.15}$

Figure 15-1 illustrates that the ferrous mineral is much more soluble than the ferric mineral. For example, at a pH of 5 the ferrous mineral is 10,000 times more soluble than the ferric mineral. Because of this great difference in solubility of ferrous and ferric minerals, plants use Fe^{2+}. Figure 15-1 also illustrates that soil pH has a great effect on the concentration of Fe^{2+} in the soil solution. Each unit increase in pH results in a 100-fold decrease in the solubility of siderite. At a pH of 7, if a soil is in equilibrium with siderite, there is less than 10^{-10} M Fe^{2+} in the soil solution.

Figure 15-1   Solubility of ferrous and ferric minerals.

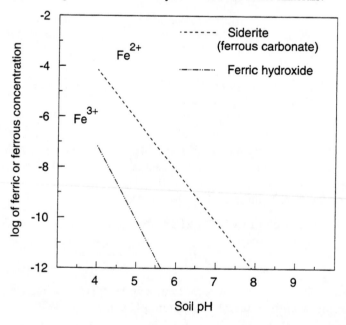

Treatment of pH-induced iron chlorosis is complicated by the fact that most of the iron added to soil to treat the deficiency is precipitated by the high pH that initially caused the problem. To correct such a deficiency it is necessary to add iron in a form (chelate) that prevents or minimizes the interaction of iron with soil or to apply iron directly to the plant, as in a dormant or chelated micronutrient spray.

Oxidation and reduction of iron. *Redox* is the term used to describe reactions that involve either the oxidation (loss of electrons) or reduction (gain of electrons) of constituents. As are many of the transition metals, iron is subject to oxidation and reduction in the soil environment. Oxidation of ferrous minerals and ions occurs when they are exposed to O_2(gas). Reduction of ferric ions and minerals occurs when these species are used as terminal electron acceptors by anaerobic organisms in poorly aerated soils. The redox relationship between $FeCO_3$(siderite) and $Fe(OH)_3$(amorphous) is given by the equation

$$Fe(OH)_3(amorphous) + H^+ + CO_2(gas) + e \leftrightarrow FeCO_3(siderite) + 2H_2O, \quad [15\text{-}7]$$
$$E° = 0.513 \text{ V}$$

where $E°$ is the standard redox potential for the reaction. The redox relationship illustrated in equation [15-7] cannot be solved by using an equilibrium constant, since the transfer of electrons is involved. While electrons are written as reactants in the equilibrium expression, they are not found as free species in the soil and are not included in the equilibrium constant expression. Rather, they are passed by the microbes metabolic processes from the species that is being oxidized, soil organic matter, to the species that is being reduced, $Fe(OH)_3$(amorphous).

The Nernst equation is used to describe the relationship between the soil's actual redox potential (Eh), the standard redox potential, and the concentrations of the reactants and products of the reaction.

$$Eh = E° - \frac{0.059}{n} \log \frac{(\text{reduced})}{(\text{oxidized})} \qquad [15\text{-}8]$$

where n is the number of electrons involved in the reduction and (reduced) and (oxidized) are, respectively, the products of the concentrations of the species on the reduced and oxidized side of the equation.

For the reduction of $Fe(OH)_3$(amorphous) to $FeCO_3$(siderite) the Nernst equation becomes

$$Eh = 0.513 - \frac{0.059}{1} \log \frac{1}{(H^+)PCO_2(gas)} \qquad [15\text{-}9]$$

By substituting in different values for (H^+) into equation [15-9] and using the partial pressure of carbon dioxide found in the atmosphere, Figure 15-2 can be constructed, which shows the stability regions of siderite and amorphous ferric hydroxide. When Fe^{2+} occurs in soils with Eh and pH values that are above the stability line, then it can be oxidized. When Fe^{3+} occurs in soils with Eh and pH values that lie below

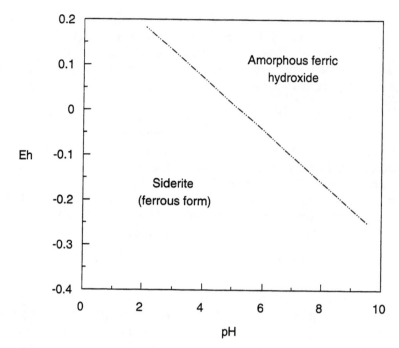

Figure 15-2 Eh–pH stability diagram for siderite and amorphous ferric hydroxide.

the line, then reduction can occur. Similar redox diagrams can be constructed for nitrogen, manganese, sulfur, and other species that are subject to oxidation and reduction in soil environments. The diagrams can be used to predict when a species may be oxidized or reduced. Reduction or oxidation can occur outside these boundaries, but only when mediated by an organism and at an expense of metabolic energy. Photosynthesis is a classic example of this situation.

Several generalizations concerning soil redox processes can be made from Figure 15-2. First, oxidation increases both the redox potential (Eh) and the H^+ concentration of the soil, while reduction decreases both the redox potential and the H^+ concentration of a soil. For example, consider the reduction of $Fe(OH)_3$(amorphous) to $Fe(OH)_2$(amorphous).

$$Fe(OH)_3(\text{amorphous}) + e \leftrightarrow Fe(OH)_2(\text{amorphous}) + OH^- \qquad [15\text{-}10]$$

Reduction reduces the valence of iron from 3+ to 2+ and results in the release of a hydroxyl into the soil solution with the subsequent increase in pH, that is, decrease in hydrogen ions. Oxidation increases the valence of iron from 2+ to 3+ and results in the removal of a hydroxyl from solution with an ensuing decrease in pH, that is, increase in hydrogen ions.

A second generalization that can be made about redox processes in soils is that the reduction of oxidized species, such as Fe^{3+} or NO_3^-, is almost always coupled with the oxidation of organic residues in soils by anaerobic organisms. In the absence of O_2(gas), organisms use these materials to serve as terminal electron acceptors. Oxidation, on the other hand, is generally the result of some geologic or anthropogenic process placing a reduced material into an oxidizing environment, such as a well-aerated soil. The oxidation agent is usually O_2(gas), but the reactions are often mediated by chemoautotrophic bacteria. In the absence of the bacteria, the oxidations can still occur but usually at very slow rates.

15.2 MANGANESE

Manganese is another essential nutrient for plants. Manganese is essential for the Hill reaction of photosynthesis, in which water is split into hydrogen ions and oxygen gas. Manganese also functions as a enzymatic cofactor in many reactions, although its essentiality in these cases is not clear, as other divalent cations appear to substitute for Mn. Manganese deficiency occurs primarily in basic soils. Manganese deficiency symptoms appear as spots and flecks and some interveinal chlorosis. Symptoms first appear on young leaves, as Mn is not translocated in the plant, and later on older leaves. Manganese is also toxic to plants, particularly in acid soils, where the solubility of manganese minerals is much higher than in basic soils.

Manganese occurs in the soil solution as the Mn^{2+} ion because the other possible oxidation states, MnO_4^-, Mn^{3+}, and Mn^{4+}, are not stable in the redox environment of the soil. The concentration of Mn^{2+} in basic soils could be controlled by the dissolution of manganese minerals, such as pyrolusite (β–MnO_2).

$$\beta\text{–}MnO_2(\text{pyrolusite}) + 4H^+ + 2e \leftrightarrow Mn^{2+} + 2H_2O, \qquad E^\circ = 1.240 \quad [15\text{-}11]$$

If equation [15-11] is solved, very high Mn^{2+} concentrations would be found to be in equilibrium with pyrolusite in acid soils. This suggests that either the soil is in equilibrium with minerals that are much less soluble than pyrolusite or that the concentration of Mn^{2+} in the soil solutions of acid soils is controlled by cation-exchange and/or complexation reactions with soil humic substances.

Manganese deficiency is often the result of high pH values lowering the solubility of manganese solid phases and, less often, the result of low total manganese levels in the soil. Like iron, manganese deficiencies are usually treated with complexed or chelated forms of the nutrient, as treatment with inorganic manganese is rendered quickly ineffective due to precipitation of the added manganese. Manganese toxicity is treated, as is aluminum toxicity, by liming the soil. Liming raises the soil pH and dramatically lowers the concentration of Mn^{2+} in the soil solution.

15.3 ZINC

Although zinc is a transition metal, it occurs in soils only in the Zn(II) oxidation state. The zinc content of the lithosphere is approximately 80 μg/g, with a common range in soils of 10 to 300 µg/g. The dominant form of zinc in the soil solution below pH values of 7.69 is the zinc ion, Zn^{2+}.

$$Zn^{2+} + H_2O \leftrightarrow ZnOH^+ + H^+, \qquad K = 10^{-7.69} \qquad [15\text{-}12]$$

The dominant form of zinc in the soil solution above a pH of 7.69, where H^+ concentration equals the numerical value of the equilibrium constant (equation [15-11]), is the $ZnOH^+$ ion.

    Zinc plays an essential role in oxidation-reduction reactions in plants. Zinc serves as a cofactor for enzyme systems. For example, zinc controls the synthesis of indoleacetic acid. Zinc deficiency causes mottling and necrosis of the leaves, as well as a condition known as "little leaf disease." Zinc deficiencies occur primarily in high-pH soils, where zinc appears to be precipitated as an insoluble mineral. Zinc deficiencies have also been noted on soils with high phosphate levels. For example, when soils that have been used to raise turkeys are put back into production, the high phosphate levels caused by manure applications often results either in zinc precipitation in the soil or interference with the uptake and utilization of zinc by the plant. Since most zinc deficiencies occur on high-pH soils, soil treatment with inorganic forms of zinc are usually not effective. Deficiencies are best treated with organically complexed or chelated forms of fertilizer (see Figure 15-3). Zinc deficiencies are often associated with operations that remove the higher-organic-matter topsoil, such as land leveling operations in southern Mississippi alluvium.

15.4 COPPER

Copper occurs primarily in organic complexes in the plant, whereas zinc and manganese occur primarily as inorganic species. A deficiency of Cu interferes with protein synthesis and causes a buildup of soluble nitrogen compounds in the plant. Copper is an essential part of several enzymes that are important in electron transport. Copper deficiencies are not common, but when they occur they are often associated with the death of the growing points in diseases, such as "dieback of citrus."

    Copper occurs primarily as the Cu^{2+} ion in soil with pH values below 7.7 and as the $CuOH^+$ species above 7.7.

$$Cu^{2+} + H_2O \leftrightarrow CuOH^+ + H^+, \qquad K = 10^{-7.70} \qquad [15\text{-}13]$$

    Copper is unique in that a substantial portion of the copper in a soil is bound to soil organic matter. The level of copper in the soil solution is probably controlled more by the release of copper from the organic forms than it is by the dissolution of copper minerals. Copper deficiencies are most common in sandy low-organic-matter soils.

15.5   MOLYBDENUM

Molybdenum is associated with nitrogen metabolism in plants. It plays a specific role in nitrogen fixation and functions as a cofactor of the enzyme responsible for nitrate reduction, nitrate reductase. Molybdenum occurs in soils as the molybdate anion, MoO_4^-. The chemistry of molybdate in soils parallels phosphate, in that MoO_4^- is adsorbed by iron oxides and the edges of clay minerals in acid soils. The mechanism appears to be very similar to phosphate, with the MoO_4^- ion replacing a hydroxyl that is bound to a surface iron or aluminum mineral. Adsorption of molybdate is at a maximum in the pH range of 3 to 5, with adsorption decreasing with increasing soil pH. Molybdenum is not as strongly adsorbed as phosphate. Since the plant requires much less molybdenum than phosphorus and since its reaction with the soil is not as strong, deficiencies of molybdenum are not common. Most deficiencies are associated with extremely low molybdenum contents of the parent materials. Unlike most micronutrients, molybdenum is more available in high-pH soils. Molybdenum is required in such small quantities by plants that coating seed with amounts of Mo that are equivalent to 40 g of Mo per hectare will correct soil deficiencies.

15.6   BORON

Deficiencies of the essential micronutrient boron have been reported for a variety of crop species in 43 states. Boron is associated with carbohydrate and nucleic acid metabolism. Boron may also form complexes with sugars and facilitate their transport within the plant as borate complexes. Boron appears to play an important role with calcium in cell-wall metabolism. Optimal plant growth appears linked to a constant calcium-to-boron ratio.

Boron deficiencies occur most commonly in high-rainfall regions. Soils formed from marine sediments tend to have higher boron contents than do soils formed from igneous rocks. The original source of B in soils is, primarily, the mineral tourmaline. Tourmaline contains 3 to 4% boron. Boron can also substitute for silica in tetrahedral sheets and be very slowly released as the minerals weather. A substantial portion of boron is associated with soil organic matter. The boron was incorporated into the organic matter during the process of plant growth and is released only upon decay. In humid-region soils, boron content of the soil is positively correlated with the humus content of the soil.

Boron occurs in dilute solutions as the weak Lewis acid, boric acid, $B(OH)_3$. Equation [15-13] illustrates how boric acid behaves as an acid.

$$B(OH)_3 + H_2O \leftrightarrow B(OH)_4^- + H^+, \qquad K = 10^{-9.24} \qquad [15\text{-}14]$$

$$K = \frac{(B(OH)_4^-)(H^+)}{(B(OH)_3)}$$

When the hydrogen ion concentration equals the value of the equilibrium constant, that is, at a pH of 9.24, the concentrations of $B(OH)_3$ and $B(OH)_4^-$ are equal. When pH values are below 9.24, the dominant form of boron in the soil solution is the neutral molecule, $B(OH)_3$. For most soils, the form of boron in the soil solution is the neutral molecule, $B(OH)_3$. This explains why boron is absorbed by plants as the neutral molecule and why its absorption tends to be passively associated with the transpiration stream. It also helps to explain why boron deficiency tends to occur in humid regions, where leaching of the neutral species would be the greatest.

The level of boron in the soil solution and its availability to plants are controlled by the equilibrium between boron in the soil solution and boron adsorbed by clay minerals, hydrous oxides and soil organic matter. Boron appears to be adsorbed primarily by the edges of clay minerals, not by the planar external or internal surfaces. Boron displaces OH– and H_2O groups from the clay edges and forms weak covalent bonds with the structural cations by approximately the same mechanism as those of phosphate, molybdate, and sulfate. The strength of bonding of boron to mineral surfaces is much weaker than is found with phosphate.

Like many of the micronutrients, boron can be deficient, in adequate supply, or toxic to plants over a very narrow range of soil concentrations. Plant-available boron has been determined using hot-water extracts of the soil. In general, hot-water-extractable levels of boron that are greater than 0.046 mol B per megagram (Mg; 10^6 g) are considered sufficient for most crops, although there is considerable variation for specific crops. For example, 0.046 to 0.093 mol B/Mg is considered normal for red beets, whereas 0.0099 mol B/Mg is normal for corn, wheat, and oats.

Potential boron toxicities are predicted using cold-water-saturated soil extracts. Sensitive crops are affected when boron levels in saturated extracts are in the range 0.046 to 0.093 mol B/m^3 of soil. Tolerant crops are not affected until boron levels in saturated extracts exceed 0.46 mol B/m^3 of soil.[1] Maximum permissible concentrations of boron in irrigation waters has been established as 0.028 to 0.093 mol B/m^3 of irrigation water for sensitive crops and 0.19 to 0.37 mol B/m^3 for tolerant crops.

15.7 CHLORINE

Chlorine occurs in the soil solution as the Cl$^-$ ion. Chlorine does not interact with soil constituents as do many of the other micronutrients. No Cl$^-$ deficiencies have been demonstrated in crops in the field. In coastal areas, sufficient Cl$^-$ is supplied by salt spray and dissolved in precipitation. In arid regions, leaching has not been sufficient to lower chloride concentrations to deficient levels. Chlorine has been

[1]R. Keren and F. T. Binghan, Boron in water, soils and plants, *Adv. Soil Sci.* 1:229–276 (1985).

shown to be necessary for the Hill reaction in photosynthesis, in which water is split into hydrogen ions and oxygen gas.

15.8   COBALT

Cobalt has not been demonstrated to be an essential element for higher plants, but it is an essential element for nitrogen fixation by *Rhizobium* and *Bradyrhizobium* species in symbiotic relationships with legumes. Cobalt is a component of cobalamines, such as vitamin B_{12}. Many algae show deficiency symptoms if grown in the absence of cobalt.

Cobalt occurs in the soil as the Co^{2+} ion and forms $CoOH^+$ at higher pH values. Cobalt availability decreases as pH is raised in a manner similar to the other micronutrients that are transition metals. Cobalt is often deficient in soils with fluctuating water levels. Fluctuating water tables result in the formation of Fe-Mn concretions, and Co^{2+} is adsorbed to the surface of the concretions and becomes incorporated into them as they grow.

15.9   MANAGEMENT OF MICRONUTRIENTS

Two general situations exist for micronutrient deficiencies. The first is a deficiency of a micronutrient that is not precipitated or otherwise converted to an unavailable form due to soil conditions. Deficiencies of boron or molybdenum would be examples of this case. The second situation results when the deficiency is caused by soil conditions. In this case there may be sufficient micronutrient present to meet plant needs, but soil conditions, such as high pH values, result in the micronutrient being unavailable to the plant. Deficiencies of iron, manganese, and other transition metals are often examples of this situation.

Deficiencies that fall in the first case can be treated by application of a soluble form of the micronutrient to the soil. For example, a deficiency of chlorine could be treated with KCl and a deficiency of boron with an application of boric acid [$B(OH)_3$]. Conversely, toxicities of these elements are difficult to correct since altering soil conditions, such as raising soil pH by liming, would not decrease solubilities or lower plant availabilities of these elements.

Deficiencies that fall in the second case, where soil conditions cause the problem, can be treated in several ways. The first option is to modify the soil. For example, if high pH values are causing iron deficiency of azaleas, aluminum sulfate, sulfur or sulfuric acid could be added to lower the soil pH. This option is usually restricted to small areas of soil because of the expense involved (see Chapter 9). The second option is to apply the nutrient directly to the plant as a foliar spray, avoiding contact with the soil. Nutrients applied as foliar sprays are absorbed through plant leaves. Foliar-applied micronutrients are more readily available to plants than are soil-applied micronutrients, but foliar applications do not provide

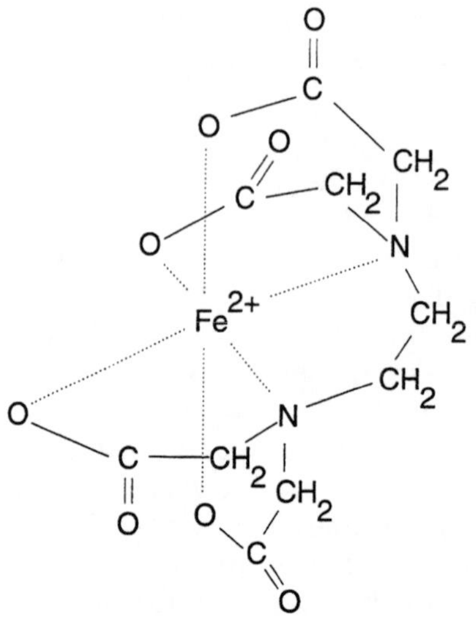

Figure 15-3   Structure of iron(II)–EDTA chelate. EDTA and other chelating agents are used to provide other micronutrients, such as Zn^{2+} or Mn^{2+}.

continuous nutrition and may have to be repeated several times during the growing season. Solutions of iron sulfate have been used as dormant sprays, with limited success, to treat iron deficiency of orchards. The spray is applied while the trees are dormant and the nutrient is absorbed through the bark and buds. Galvanized, zinc-coated nails have been driven into the trunks of zinc-deficient trees with some success in treating the deficiency. The most successful material for foliar application is to apply the micronutrient as a chelate. Iron deficiency of mature pear and apple trees has been successfully treated using Fe-EDTA. The third option is to apply the micronutrient in a chelated form directly to the soil. Fe-EDDHA has been the most successful iron chelate for treatment of iron deficiency of calcareous soils.

Fe-EDTA (iron ethylenediaminetetracetic acid) and Fe-EDDHA (iron ethylenediamine di-o-hydroxyphenylacetate) are both examples of iron chelates (Figure 15-3). A chelate is an organic molecule that forms a soluble complex with the micronutrient. The complex minimizes the interaction of the micronutrient with the soil (e. g., precipitation of iron at high soil pH values) but the chelated form of the micronutrient can still be absorbed and used by the plant. Sequestered or complexed micronutrients are alternative terms used to refer to chelated micronutrients.

STUDY   QUESTIONS

1. What general soil conditions result in micronutrient deficiencies?
2. Why do plants use ferrous forms of iron in well-aerated soils?

3. Iron deficiency has been diagnosed on two soils, an acid-sandy soil and a calcareous soil. How would you treat the deficiencies?

4. What is a chelate? Why are chelated forms of micronutrients often used to treat deficiencies?

5. The behavior and chemistry of iron, copper, cobalt, zinc, and manganese in soils are very similar. Why?

6. Why is boron passively absorbed with a plant's transpiration stream?

7. Why are hot-water extracts used to diagnose boron deficiencies and cold-water-saturated extracts used to diagnose boron toxicities?

16

Control of Soil Fertility and pH

16.1 FERTILIZERS

Any material that is applied to soils to supply nutrients essential for plant growth can be considered a fertilizer. The term *fertilizer*, therefore includes natural materials, such as organic composts, green manures, and cover crops, and commercial fertilizers or materials containing one or more of the primary plant nutrients: nitrogen, potassium, or phosphorus. Fertilizers have been in the forefront of the effort to keep food production at a level that will sustain the ever-increasing population of this planet. It is the use of fertilizers coupled with plant varieties that can utilize the increased soil fertility that has made this possible. Fertile soils have a well-balanced supply of all essential elements in available forms. Fertile soils are not necessarily productive soils unless they also have good tilth and are not restricted by problems such as aridity, coldness, or shallowness. A productive soil has good soil fertility and other the plant growth factors are not limiting. The major soil nutrients and their interactions in soils have been discussed in other chapters. In this chapter we focus on the kinds of materials that constitute commercial fertilizers and can provide the major or microelements necessary for increased plant growth. Table 16-1 lists common commercial fertilizer materials for the three major elements, N, P, and K, that are normally added to soils. Also listed in Table 16-1 are multinutrient fertilizers, called *mixed fertilizers*.

TABLE 16-1 Nutrient Content of Principal Fertilizer Materials (percent)

Material	N	P	K	Ca	S
Nitrogen					
Anhydrous ammonia	82				
Aqua ammonia	16–25				
Ammonium nitrate	33.5				
Ammonium nitrate-lime	20.5			7	
Ammonium sulfate	21				24
Ammonium sulfate-nitrate	26				15
Calcium cyanide	21			39	
Calcium nitrate	15			19	
Nitrogen solutions	21–49				
Sodium nitrate	16				
Urea	46				
Phosphate					
Basic slag		4–5		29	
Normal superphosphate		8–9		20	12
Concentrated superphosphate		18–22		14	1
Phosphoric acid		23–26			
Superphosphoric acid		30–33			
Potash					
Potassium chloride			50–51		
Potassium sulfate			42		18
Potassium-magnesium sulfate			18		23
Multinutrient materials					
Diammonium phosphate	16–21	20–23			
Monoammonium phosphate	10–11	21–24			
Nitric phosphates	14–22	4–10		8–10	0–4
Potassium nitrate	13		37		

Source: R. D. Young and F. J. Johnson, Fertilizer products, in *The Fertilizer Handbook*, W. C. White and D. N. Collins (eds.), The Fertilizer Institute, Washington, D. C., 1982.

16.1.1 Nitrogen Fertilizers

Ammonia (NH₃). Anhydrous ammonia is the most concentrated form of the nitrogen fertilizer materials. Anhydrous ammonia contains 82% N (Table 16-1). It is the most commonly used source of nitrogen fertilizer in the United States, in large part because it is the least expensive source of commercial fertilizer nitrogen on a per unit nitrogen basis. Ammonia is a gas and must be applied beneath the soil surface to prevent loss. As the gas is released below the soil surface, from an anhydrous ammonia applicator, it rapidly reacts with water in the soil, preventing loss. Approximately 40% of the total nitrogen fertilizer used in the United States is applied as anhydrous ammonia. The term *anhydrous ammonia* is used to describe ammonia gas under sufficient pressure to convert it into a liquid. The fact that ammonia is stored under high pressure and must have specialized farm equipment for application limits its use in developing countries or where soil conditions or climate do not readily allow the application of ammonia.

Ammonia gas may be dissolved in water, yielding ammonium hydroxide, NH_4OH, prior to application. In this form, called *aqua ammonia*, it contains 16 to 25% N. While aqua ammonia must be incorporated in the soil to prevent ammonia loss, it does not require the high pressure and sophisticated equipment used to apply anhydrous ammonia. Other important sources of nitrogen are nitrogen solutions, which are usually a combination of ammonium nitrate, urea, and water. Nitrogen solutions are gaining popularity as sources of nitrogen in U.S. agriculture.

Ammonium nitrate (NH_4NO_3).   Ammonium nitrate contains about 33.5% nitrogen (Table 16-1) and was a popular source of high-analysis nitrogen fertilizer until the mid-1970s, when it was replaced by nitrogen solutions. Ammonium nitrate is a solid fertilizer that can be bagged and shipped as a solid and also spread on the soil surface with commercial spreaders. Ammonium nitrate does not require incorporation into the soil to prevent hydrolysis and loss as found with urea. Ammonium nitrate is unique in that it contains both ammonium and nitrate ions. As was discussed in Chapter 13, ammonium ions under most soil situations are rapidly oxidized to nitrate ions, and thus most nitrogen removed from soil by plants is in the nitrate form. Ammonium nitrate must be handled and stored properly since it is an explosive. The use of ammonium nitrate as a solid nitrogen fertilizer is gradually declining in the United States.

Urea [$CO(NH_2)_2$].   Urea is the most concentrated of the solid forms of nitrogen fertilizers. While use of urea in the United States is only slightly greater than that of ammonium nitrate and considerably lower than nitrogen solutions or anhydrous ammonia, it has gained popularity in the international market, particularly in developing countries, because of the ease of storage, shipping, and spreading of this N source. Urea is hydrolyzed quite rapidly in most soils by the enzyme urease to produce ammonium carbonate [$(NH_4)_2CO_3$] and eventually ammonia [$NH_3(gas)$]. The ammonia released by the hydrolysis reacts in the soil the same as ammonia from anhydrous ammonia.

$$CO(NH_2)_2 + 2H_2O \xrightarrow{\text{urease}} (NH_4)_2CO_3 \leftrightarrow 2NH_3(gas) + CO_2(gas) + H_2O \quad [16\text{-}1]$$

Ammonium sulfate [$(NH_4)_2SO_4$].   Ammonium sulfate is a unique fertilizer in that it supplies not only nitrogen but also sulfur to the soil. While the use of ammonium sulfate as a nitrogen source in the United States is relatively small, some usage continues because it is a by-product of the steel and chemical industries. Ammonium sulfate is also important because it is used to manufacture mixed fertilizers. Ammonium sulfate is used extensively in the production of rice.

Production of nitrogen fertilizers. The building block of most of the nitrogen fertilizers on the market is ammonia gas, NH_3. While the air contains 79% nitrogen, this nitrogen cannot be absorbed by plants or stored in soils. It must be reduced to ammonia, which is either applied directly as ammonia gas, or converted to one of the other nitrogen fertilizers (Figure 16-1). Production of ammonia gas is carried out by the reaction of nitrogen gas from the atmosphere with methane. This reaction requires extreme temperatures of up to 1480°C, depending on the particular process used, and extreme pressures, commonly up to 34.5 MPa (5000 lb/in.^{2}) or as high as 103.5 MPa, again depending on the compression method used for synthesis. Basically, production of ammonia involves synthesizing hydrogen gas and adding stoichiometric amounts of nitrogen. The starting materials in the synthesis are usually natural gas, steam, and air. Hydrogen for the production of ammonia can also be obtained from water, oil, or coal, while in all processes of ammonia formation, the nitrogen comes from the air. Ammonia gas is less dense than air, but can be compressed and cooled to form a liquid about 60% as dense as water. As already mentioned, it readily absorbs water, resulting in liquid solutions of ammonia. Because anhydrous ammonia has a very high vapor pressure, it is stored and

Figure 16-1   Production of N fertilizers.

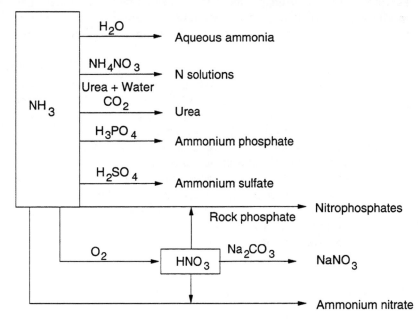

transported in pressurized containers. Gaseous ammonia is dangerous in that it is extremely irritating to the eyes and the respiratory system, and a sudden release of ammonia from a broken valve or line can cause blindness or be fatal.

Slow-release N fertilizers. The efficiency of a fertilizer is increased when it is manufactured in such a way that the rate of release, or dissolution of the nitrogen fertilizer, is controlled. This can be done by developing compounds with limited water solubility or by modifying water-soluble materials to delay the release of their nitrogen into soil solution. Manufactured slow-release nitrogen fertilizers are basically of three types: (1) water-soluble materials, containing nitrogen, in which dissolution is controlled by a physical means such as a coating; (2) materials that have low water solubilities and that contain plant-available forms of nitrogen, such as nitrogen complexed by metals; and (3) materials of low water solubility that release plant-available nitrogen upon chemical or microbial decomposition.

Many materials have been tested as coatings for solid nitrogen fertilizers, but the most important are waxes, polymers, and sulfur. One of the most common slow-release coated nitrogen fertilizers is sulfur-coated urea. In the manufacturing process, molten sulfur is sprayed on a falling curtain of preheated urea particles in a rotating drum. A sealant is then applied to close the pores in the sulfur coating, followed by conditioner to improve the handling properties of the fertilizer. The controlled release of nitrogen from sulfur-coated urea occurs because the fertilizer particles are imperfectly coated. Small holes in the coating eventually allow the entrance of water and the release of the nitrogen. Osmocote is the trade name of a polymer-coated fertilizer produced in the United States. Nitrogen release from this fertilizer is controlled by the thickness and composition of the polymer coat. Polymer-coated materials are used mostly on lawns and in professional turf management or horticulture.

The second class of slow-release fertilizers are the uncoated inorganic materials. These materials are generally compounds that include a divalent metal ion and a phosphate complex of ammonium. They are slightly water soluble and may contain potassium as well as nitrogen. An example is magnesium ammonium phosphate, which has been used worldwide for more than a century and is still used to a limited extent in the United States. Some of these products are said to release nutrients into the soil solution for one or two years.

The third group of slow-release fertilizers is composed of inorganic compounds that release ammonia upon decomposition. An example of these types of compounds is urea-aldehyde, which is produced by reacting urea with a variety of aldehydes, to form a product that is sparingly soluble in water. Urea-aldehyde condensation products of this nature have nitrogen contents of approximately 30% and decompose slowly in soils by chemical and/or biological reaction. Ideally, a slow-release fertilizer results in the release of nitrogen for plant uptake at the same time and rate as plants absorb it, eliminating losses by leaching or denitrification.

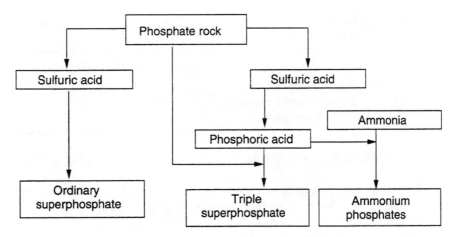

Figure 16-2 Production of phosphate fertilizers.

16.1.2 Phosphorus Fertilizers

Phosphate rock is the starting point for most phosphorus fertilizers. The principal phosphate mineral in the rock is francolite, a carbonate fluorapatite. Francolite can be represented chemically as $Ca_{10}(PO_4)_6F_2 \cdot xCaCO_3$, where x designates impurities such as Fe, Al, and Mg. In the processing of phosphate rock to make fertilizer, the fluorine is removed and used, although there have been some problems reported with floride contaminated soils near processing plants. Phosphate rock is treated with sulfuric and/or phosphoric acid to produce many phosphorus-containing fertilizers, including normal superphosphate, concentrated superphosphate, or, with the input of ammonia, ammonium phosphates (Figure 16-2).

By far the most popular phosphate-containing fertilizers in the United States are mixed fertilizers, including diammonium phosphate (DAP) and monoammonium phosphate (MAP). Together, these two account for more than half of the phosphorus used. Technology allowing ammonium granulation of DAP has resulted in this material being the most popular form of phosphorus fertilizer. Normal superphosphate (see Table 16-1) contains 8 to 9% phosphorus and was once a very popular source of phosphorus. Today, however, normal superphosphate accounts for a very small percentage of phosphorus fertilizers used. Concentrated superphosphate (CSP) is the most common nonnitrogen-containing phosphorus fertilizer in the United States, but it accounts for only about 10% of the total use. A comparison of the production of these forms is given in Table 16-2.

World reserves of phosphate rocks approach 50 billion metric tons of known reserves that can, with technology, be economically recovered, while the production rate is currently on the order of 150 million metric tons annually. Thus, adequate reserves have been identified, and more are likely to be found as mining continues. In the United States, reserves are the largest in the western states, but current production is highest in Florida.

TABLE 16-2 Production of Phosphorus Fertilizers in the United States, 1981

Fertilizer	Million metric tons of P
Ammonium phosphates	2.4
Concentrated superphosphate	0.7
Normal superphosphate	0.13

16.1.3 Potassium Fertilizers

Potassium follows nitrogen in terms of the amount of fertilizer used. This fertilizer is sometimes referred to as "potash," a carryover term that comes from the production of potassium carbonate by leaching of hardwood ashes during colonial times. However, the present source of potassium in fertilizers is, primarily, potassium chloride (Table 16-3), which accounts for over 95% of the potassium fertilizers used in the United States. It is the most easily obtained and the most economical source of potassium fertilizer. While state and federal laws require potassium to be expressed in terms of equivalent K_2O content, neither the natural source of potassium minerals nor the fertilizer itself contain potassium oxide.

Potassium fertilizers are obtained from mineral ores mined from underground deposits using the room and pillar method, or the solution method. In the solution method, wells are drilled into the ore deposits, and hot sodium chloride brine is pumped into the wells, dissolving the potassium-rich ores, which are then pumped to the surface. As the recovered brine cools, potassium chloride precipitates as a crystalline material that is easily shipped and easily mixed with other fertilizers. The original ore deposits are primarily mixtures of sylvite (KCl) containing approximately 52% K, or sylvinite (KCl and NaCl) mixtures with variable potassium composition.

Potassium chloride. This material, called muriate of potash, is the most common dry fertilizer material and one of the more concentrated potassium sources

TABLE 16-3 Important Potassium Minerals

Mineral	Composition	% K
Sylvite	KCl	52.5
Sylvinite	KCl · NaCl mixture	Variable
Carnallite	KCl · MgCl · 6H$_2$O	14.1
Kainite	KCl · MgSO$_4$ · 3H$_2$O	15.7
Langbeinite	K$_2$SO$_4$ · 2MgSO$_4$	18.8
Nitre	KNO$_3$	38.5
Polyhalite	K$_2$SO$_4$ · MgSO$_4$ · 2CaSO$_4$ · 2H$_2$O	12.9

(52%). It is an excellent source of potassium for most crops. Although KCl is somewhat water-soluble, clear liquid fertilizers are limited, because of solubility, to approximately 8% potassium, while liquid suspensions may contain as high as 25% potassium.

Potassium sulfate. Potassium chloride can react with sulfuric acid to form K_2SO_4. This fertilizer material is more expensive to produce, but can be used where it is desirable to limit chloride uptake by plants. Examples are the potato chip industry, where excess chloride results in potato chips with less crispness, or the tobacco industry, where very large amounts of potassium fertilizers are essential, but excess chloride results in less desirable burning properties.

Potassium magnesium sulfate. Where soils require a magnesium source, potassium magnesium sulfate can provide an excellent source of water-soluble K (18%) and Mg (11%), as well as supplying S (22%), if needed.

Potassium nitrate. KCl can also react with nitric acid to produce KNO_3, which is an adequate potassium source (37%) as well as a limited source of nitrogen (13%). Potassium nitrate is also used in liquid fertilizers to some extent.

16.1.4 Mixed Fertilizers

Most fertilizer materials supplied to growers contain a combination of two or all three of the fertilizer elements (N, P, K) as high-analysis fertilizers. Mixed fertilizers comprise approximately one-half of the total amount of fertilizers consumed in the United States.

Basis for guaranteed fertilizer analysis. Every substance sold commercially as a fertilizer material must carry an assurance of the content of major fertilizer nutrients: nitrogen, phosphorus, and potassium. This fertilizer guarantee is not regulated by the federal government, but the fertilizer industry has supported state controls, which exist in every state except Alaska and Hawaii. An example of the type of information included in the guarantee on a bottle of fertilizer for house plants, a bag of fertilizer for the lawn or garden, or bulk or bagged fertilizer sold for large-scale agricultural production is shown in Figure 16-3. The analysis shown in this example is the guaranteed analysis, stated in whole numbers, as a minimum percentage by weight of total N, available phosphate (percentage by weight as P_2O_5 equivalent), and soluble potash (percentage by weight as K_2O equivalent). In all likelihood the bag does not contain any elemental N, P_2O_5, or K_2O, but the amounts of nitrogen, phosphate, and potassium in the fertilizer are expressed in terms of these forms. In the official method adapted by many states the first number refers to the nitrogen content expressed as elemental nitrogen. The second number refers to "available" P, which usually means P that is extracted with water or with neutral ammonium citrate heated to 60°C. Any citrate-insoluble P is not included in the

Figure 16-3   Example of information provided by a fertilizer label.

guarantee for commercial fertilizers. Potassium in fertilizers is usually present as inorganic salts such as potassium chloride that are water soluble. When the guarantee lists potassium as soluble rather than water soluble, the interpretation is usually solubility in boiling ammonium oxalate. Guarantees for micronutrients are often expressed as a percentage by weight, and in most cases the minimum percentage guarantee lists the *total* amount present without regard to solubility criteria.

The form of nutrient expression discussed above (N, P_2O_5, K_2O) is called the *oxide* expression of fertilizer nutrients and is usually the basis on which fertilizer recommendations are made. It is also possible that fertilizer recommendations or fertilizer analysis can be given on the elemental basis (N, P, K). Conversion from one form to the other is simply a ratio of the molecular weights. For example:

$$P = \% \ P_2O_5 \text{ equivalent} \times \frac{P_2}{P_2O_5}$$

$$= \% \ P_2O_5 \text{ equivalent} \times \frac{62}{142}$$

$$= \% \ P_2O_5 \text{ equivalent} \times 0.44$$

and

$$K = \% \ K_2O \ \text{equivalent} \times \frac{K_2}{K_2O}$$

$$= \% \ K_2O \ \text{equivalent} \times \frac{78}{94}$$

$$= \% \ K_2O \ \text{equivalent} \times 0.83$$

This means that a bag of fertilizer that had an analysis of 5, 20, 20, in terms of oxide analysis would have an analysis of 5, (20 × 0.44), (20 × 0.83) or 5, 8.8, 16.6 in terms of elemental analysis. Obviously, the bag would contain exactly the same fertilizer material, but the grower or homeowner may not readily recognize the difference unless careful attention is given the basis for the fertilizer guarantee. The form of expression must be considered when comparing costs of different fertilizer sources. The cost of a mixed fertilizer such as 5, 10, 10 is a reflection of the nutrient sources used to manufacture the fertilizer (see Table 16-1). Generally speaking, the cost per unit of fertilizer constituent is greatest for nitrogen, followed by phosphorus, and least for potassium.

Total consumption of nitrogen fertilizers nearly tripled from 1960 to approximately 1980 (see Figure 16-4). Use of nitrogen fertilizers has not increased

Figure 16-4 Plant nutrient consumption in the United States from 1960–1988. (Data from *Commercial Fertilizers*, TVA Bulletin No. Y-211, Muscle Shoals, Ala., 1989.)

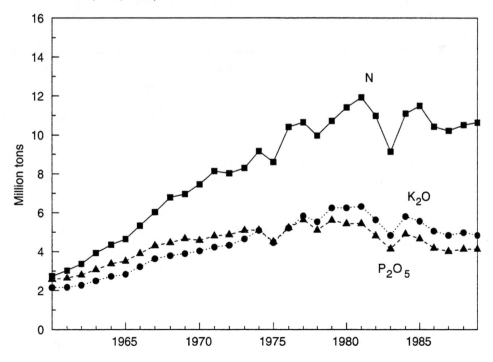

appreciable, however, since 1980, and increased concern about groundwater pollution with excessive nitrogen applications will likely reinforce the trend for a leveling off or perhaps somewhat decreased usage of nitrogen fertilizers. Annual
consumption rates for nitrogen fertilizers since the late 1970s have generally been
at about the 10 million ton level. Consumption rates for phosphorus and potassium
fertilizers have been approximately one-half that for nitrogen (Figure 16-4). Year-
to-year fertilizer consumption is altered somewhat by government programs to take
land out of production. The noticeable dip in fertilizer consumption in 1983, for
example, is due in part to a major government-sponsored program to reduce surplus
grain inventories by subsidizing farmers for not planting cash grain crops, especially corn and wheat.

16.1.5 Sources of Micronutrients

The role and function of micronutrients were discussed earlier. Micronutrients are
also essential elements, although only small amounts are needed to satisfy the needs
of plants. It should be emphasized, however, that if the soil does not contain
plant-available iron, manganese, zinc, copper, boron, molybdenum, or chlorine
plants will not survive. The desirable range of micronutrient availability in soils is
not very large. If a soil is deficient in a micronutrient, excessive fertilizer applications can quickly result in toxic levels. The addition of micronutrient fertilizers is
recommended only where the need has been documented and when the desirable
amount to be added is known. The focus of this discussion is on fertilizer sources
of these micronutrients that can be added to soils, if and when needed.

Iron.   Iron can be supplied to soils from inorganic or organic sources. Ferrous sulfate is one of the most widely used fertilizers for the treatment of iron
deficiencies (see Table 16-4). Ferrous sulfate is water soluble and can be applied
not only to the soils, but directly to the plant. Iron deficiencies occur most often on
soils with high pH values, and the addition of ferrous sulfate is often ineffective on
these soils. Other sources of inorganic iron include carbonates, ammonium sulfates,
and iron frits. Iron frits are a by-product of the metal industry containing significant
quantities of iron, but in a less soluble form. Iron can also be added to soils in
organic forms, such as animal manures or chelates. Several different forms of
organic chelates are listed in Table 16-4. Chelates are water-soluble organic compounds that bind metals, such as iron, preventing reaction of the micronutrient with
the soil. This keeps the micronutrient in a plant-available form. The structure and
function of chelates such as EDTA were discussed in Chapter 15. Chelates are an
effective means of supplying iron in calcareous soils and are also used for other
micronutrients, such as copper, manganese, and zinc.

Manganese.   Manganese can also be supplied to soils as an inorganic
sulfate, manganese sulfate. As is the case with iron, soil deficiencies of manganese
are more likely on high-pH soils, where Mn^{2+} can react with the soil and be made

TABLE 16-4 Fertilizer Sources of Iron

Source	Composition	% Iron
Inorganic materials		
Ferrous sulfate	$FeSO_4 \cdot 7H_2O$	20
Ferric sulfate	$Fe_2(SO_4)_3 \cdot 4H_2O$	20
Ferrous carbonate	$FeCO_3 \cdot H_2O$	42
Ferrous ammonium sulfate	$(NH_4)_2SO_4 \cdot FeSO_4 \cdot 6H_2O$	14
Iron frits	Variable formula	40
Organic sources		
FeDTPA	Chelate	10
FeEDTA	Chelate	9–12
FeEDDHA	Chelate	6
Manures	Variable	Variable

unavailable. Chelated forms of manganese are much better sources of manganese on these soils. Manganese, as well as other micronutrients, is sometimes more effective when applied in combination with phosphorus fertilizers, particularly when applied in a soil band. Animal manures are also a source of manganese and often prevent manganese deficiency when applied regularly. The amount of manganese in manures is variable and depends on the kind of animal manure used, as well as the diet of the animals.

Zinc. Several sources of zinc, including sulfates, oxides, chelates, frits, and manures, can be used to fertilize soils. Zinc sulfate is the primary material used and can be applied as a soil amendment or directly to the plant as a foliar spray. As is the case for iron and manganese, soil application of zinc sulfate to calcareous soils is not very effective. Under these soil conditions a chelated form of zinc is a much better choice. Plant residues and animal manures serve to recycle zinc in soils and can provide adequate amounts of this micronutrient in many situations.

Copper. Copper is also available as sulfates, oxides, and other inorganic mineral forms, but the most common fertilizer form is hydrated copper sulfate. Precipitation of copper as insoluble compounds is a potential problem with the addition of the sulfate form in alkaline or calcareous soils. Copper can also be added as a chelated form such as $Na_2CuEDTA$. As with other micronutrients, the addition in a chelated form keeps the nutrient plant-available because it limits precipitation or adsorption by the soil.

Boron. The primary source of fertilizer boron is hydrated sodium tetraborate, borax ($Na_2B_4O_7 \cdot 5H_2O$). The level of hydration may vary from no water of hydration to 10 waters of hydration in borax. Borax is soluble in water and available to plants. Less that 1 kg/ha of boron is needed to correct deficiencies in coarse-textured soils for most plants, and the addition of excess boron can be toxic. Rates of 2 to 3 kg/ha may be used for alfalfa in silt loam or silty clay loam soils. Boron can also be supplied as frits and is available in low concentrations from fly ash, collected from industrial smokestacks.

Molybdenum.   Molybdenum can be supplied in fertilizer form as sodium molybdate ($Na_2MoO_4 \cdot H_2O$) or ammonium molybdate (($NH_4)_6Mo_7O_{24} \cdot 4H_2O$). The amounts of molybdenum needed to correct soil deficiencies are often so small that it is more practical and cost-effective to add the molybdenum as a seed coating rather than directly to the soil. Molybdenum is also added as an impurity in phosphorus fertilizers. Unlike most of the fertilizer micronutrients discussed, molybdenum is most available in alkaline soils or soils with a high pH. Leguminous crops, such as alfalfa, are most responsive to soil or seed additions of molybdenum.

Chlorine.   Although chlorine is an essential micronutrient, it is a fairly ubiquitous element and almost never added to soils as a fertilizer treatment. It is added superficially to many soils as potassium chloride when potassium fertilization is required. Native chloride easily replaces that lost by leaching or used by plants in most soils. From a soils standpoint, excesses of chloride, rather than deficiencies, are much more likely to be a problem.

16.2 LIMING

Soils in the eastern half of the United States have formed in humid climates, where rainfall amounts are large enough to result in leaching of basic cations (i.e., Ca^{2+} and Mg^{2+}) out of the soil. The result of many years of leaching are soil pH values that are lower than ideal and lower availability of many plant nutrients. Soil pH values can be increased by replacing the acidic cations, H^+ and Al^{3+}, with basic cations, usually Ca^{2+} and Mg^{2+}, in a procedure known as liming.

Liming results not only in increased soil pH values, but also in increased availability of Ca^{2+} and Mg^{2+}. Liming and fertilization with other essential elements are complementary practices of good stewardship of soils. Liming of acid soils increases the availability of most soil nutrients and thus increases plant growth and uptake. Liming acid soils also results in increased microbial activity and the more rapid mineralization of soil organic matter. Liming increases both the availability and plant uptake of nutrients, resulting in increased plant growth and production. This creates a greater need for fertilizers to maintain adequate soil reserves of plant nutrients. Limestone is relatively inexpensive in relation to the fertilizer nutrients and should be one of the first soil amendments considered in any production program.

16.2.1 Liming Materials

The most commonly used liming materials are salts of strong bases and weak acids, such as calcium or magnesium carbonates. Limestone rock that is mined and ground for liming purposes usually consists of calcium and magnesium carbonates. Rock containing mostly $CaCO_3$ is called *calcitic limestone*, while rock containing mostly $CaMg(CO_3)_2$ or $CaCO_3$ plus $MgCO_3$ is called *dolomitic limestone*. The ratio of calcite to dolomite in natural limestone deposits varies and is largely a reflection of environmental conditions at the time of formation. Most deposits are a mixture of both calcium and magnesium carbonates. From a practical standpoint both are effective agricultural liming materials because they (1) increase soil pH; (2) supply Ca and Mg, which are often deficient in acid soils; (3) promote favorable soil structure; and (4) create a favorable environment for microorganisms.

Other suitable liming materials consist of oxides, such as calcium oxide (CaO) and magnesium oxide (MgO). Commercial oxides are referred to as *burned lime*, *quicklime*, or simply *oxides*. Oxides are produced by heating (burning) carbonates in large kilns to convert the carbonate forms to oxide forms. The oxide forms are caustic and more difficult to handle and store. Exposure to air or water tends to result in the formation of hydroxides, or a return to the original carbonate forms. Oxide forms have higher neutralizing powers per unit of weight, as discussed later.

Hydroxides of lime are sometimes referred to as *slaked* or *hydrated lime* and are produced by the addition of water to oxide forms. Calcium hydroxide ($Ca(OH)_2$) and magnesium hydroxide ($Mg(OH)_2$) are caustic materials that must be stored in a sealed or bagged form to prevent reaction with carbon dioxide or water vapor in the atmosphere. Although they are very suitable liming materials, hydroxides of lime are usually more expensive to purchase, more difficult to handle, and because of these problems, are less often applied to agricultural soils.

Liquid lime is a by-product of water treatment plants and can be used as a suitable liming source. The chemical nature of "liquid lime" is mostly finely divided calcium carbonate suspended in water. Fluid lime can also be made directly by suspending finely ground limestone in water. Fluid or liquid lime can be applied along with nitrogen fertilizers, eliminating extra trips across the fields. Liquid or fluid limes are both effective in raising soil pH, but are generally more variable in composition than dry liming materials from a single quarry.

16.2.2 Determination of Lime Requirement

Calcium carbonate equivalent. The effectiveness of liming materials in neutralizing soil acidity is dependent on their chemical composition. One mole of each of the liming materials, except dolomite, in Table 16-5 will neutralize 2 mol of H^+ ions. Each mole of dolomite will neutralize 4 mol of H^+ ions. For example:

$$CaCO_3 + 2H^+ \rightarrow Ca^{2+} + CO_2(gas) + H_2O \qquad [16\text{-}2]$$

$$Ca(OH)_2 + 2H^+ \rightarrow Ca^{2+} + 2H_2O \qquad [16\text{-}3]$$

$$CaO + H_2O \leftrightarrow Ca(OH)_2 + 2H^+ \rightarrow Ca^{2+} + 2H_2O \qquad [16\text{-}4]$$

$$CaMg(CO_3)_2 + 4H^+ \rightarrow Ca^{2+} + Mg^{2+} + 2CO_2(gas) + 2H_2O \qquad [16\text{-}5]$$

On a weight basis the liming materials are very different. One hundred grams of $CaCO_3$ will neutralize the same amount of acidity as 84.3 g of $MgCO_3$ or 56 g of CaO. The differences are due to the materials' molecular weights. Calcium carbonate has a molecular weight of 100, $MgCO_3$ a molecular weight of 84.3, and CaO a molecular weight of 56.

    An equivalent weight of a liming material is equal to the amount of material required to neutralize 1 mol (1 g) of H^+ ions. For $CaCO_3$, an equivalent weight is equal to its molecular weight divided by 2. For dolomite, an equivalent weight is equal to its molecular weight divided by 4. Table 16-5 lists the molecular and equivalent weights of common liming materials. It is common practice to express the neutralizing power of a liming material in terms of the amount of calcium carbonate that it is equivalent to; this is known as its *calcium carbonate equivalent* (CCE). Since one equivalent of any liming material is chemically equal to one equivalent of any other liming material, 42.1 g of magnesium carbonate will neutralize the same amount of soil acidity as 50 g of calcium carbonate and the CCE of magnesium carbonate is simply the ratio of equivalent weights, or $50/42.1 \times 100 = 119$. Thus the calcium carbonate equivalent of any material can be found by dividing 50 by the equivalent weight of the material in question. Magnesium oxide, for example, has a calcium carbonate equivalent (CCE) of 248, meaning that 100 g of this material has the same theoretical neutralizing power as 248 g of pure calcium carbonate. The relative effectiveness of magnesite, dolomite, slaked lime, and burned lime to calcite is shown in Table 16-5.

    Calcium carbonate equivalent is a common way of comparing liming materials, but it would be equally valid to express all liming materials as a calcium oxide equivalent by dividing 56 (the equivalent weight of calcium oxide) by the equivalent weight of the material in question. Similarly, the magnesium oxide equivalent of any liming material can be obtained by dividing 40.3 (the equivalent weight of magnesium oxide) by the equivalent weight of the material in question. Sometimes lime is marketed with a *conventional oxide guarantee*, which simply means that all forms of calcium in the limestone are converted to the calcium oxide equivalent, and all forms of mag-

TABLE 16-5   Standard Liming Materials and Calcium Carbonate Equivalents

Liming material	Common name	Molecular weight	Equivalent weight	Calcium carbonate equivalent
$CaCO_3$	Calcite	100	50	100
$MgCO_3$	Magnesite	84.3	42.1	119
$CaMg(CO_3)_2$	Dolomite	184.3	46.1	108
$Ca(OH)_2$	Slaked lime	74	37	135
$Mg(OH)_2$	Slaked lime	58.3	29.2	171
CaO	Burned lime	56	28	179
MgO	Burned lime	40.3	20.2	248

nesium are converted to magnesium oxide equivalent. The calcium oxide and the magnesium oxide equivalent can be summed to arrive at the *conventional oxide equivalent* of that liming material. In reality, most liming materials actually contain calcite and/or dolomite, regardless of the equivalent expression used for marketing.

Effective calcium carbonate equivalent. While CCE allows a comparison of the theoretical neutralizing power of limestone, the rate of dissolution of the limestone particles must also be considered. If limestone is too coarsely ground, its reaction will be so slow that it will be ineffective in increasing crop yields. Limestone that will react with the soil in a three-year period must be ground to pass a 60-mesh sieve. A 60-mesh sieve is one that has 60 dividing wires, and thus 60 openings per linear inch or 3600 openings per square inch. Each opening is approximately 0.21 mm across. The following is a measure of relative effectiveness established by the Iowa Agricultural Limestone Act of 1967 and representative of the principles followed by laws regulating lime sales in several states.

percent of material coarser than 4 mesh	$\times 0$	= % effectiveness
percent of material between 4 and 8 mesh	$\times 0.1$	= % effectiveness
percent of material between 8 and 60 mesh	$\times 0.4$	= % effectiveness
percent of material finer than 60 mesh	$\times 1.0$	= % effectiveness
total percent effective in the first 3 years		= % effectiveness

The effectiveness of a limestone can be estimated from the equations above. For example, a limestone analysis may reveal that all of the sample will pass through a 4-mesh sieve, 10% is retained on an 8-mesh sieve, 30% is retained on a 60-mesh sieve, and 60% passes through a 60-mesh sieve. The percent effectiveness of the limestone, based on fineness, can be calculated as follows:

coarser than 4 mesh	$0\% \times 0$	= 0%
between 4 and 8 mesh	$10\% \times 0.1$	= 1%
between 8 and 60 mesh	$30\% \times 0.4$	= 12%
finer than 60 mesh	$60\% \times 1.0$	= 60%
total effectiveness in 3 years		= 73%

A combination of CCE and fineness determines the *effective neutralizing value* (ENV) or *effective calcium carbonate equivalent (ECCE)* of a limestone. If a limestone with 73% effectiveness based on fineness had a CCE of 90%, the ECCE would be 0.9×0.73, or 65.7% ECCE. Other terms that are used by different states for expressions of limestone quality are *effective neutralizing power* (ENP), *effective neutralizing material* (ENM), and *effective neutralizing index* (ENI). In some states lime is marketed on the basis of CCE with a fineness guarantee implied.

Specific limestone requirement. Assume that a soil requires 5.8 mt $CaCO_3$/hfs to change its pH from 5.5 to 6.5 (see Chapter 9). How much limestone is required if the limestone contains 70% $CaCO_3$ and 5% $MgCO_3$ and has an effectiveness based on fineness of 80%?

$$CCE = (70\% \ CaCO_3) + (5\% \ MgCO_3 \times 50/42.1) = 75.94\% \qquad [16\text{-}6]$$
$$ECCE = 0.759 \times 0.8 = 0.607 \ or \ 60.7\%$$
$$lime \ requirement = 5.8 \ mt \ CaCO_3/hfs \ 0.607 = 9.56 \ mt \ limestone/hfs$$

Estimated lime requirement. The lime requirement of soils can also be estimated using several different techniques, rather than calculated from changes in percentage base saturation and cation-exchange capacities. One technique involves mixing a sample of the soil to be limed with a buffer solution. When the soil is mixed with a buffer solution, the active and reserve acidity in the soil will alter the pH of the buffer. A sample of the buffer solution has previously been titrated with a standard acid, so that the amount of H^+ ions required to change the buffer's pH a given number of units is known. If the soil that is added to the buffer solution lowers the pH of the buffer by one unit, it can be estimated that the milliequivalents of potential acidity in that soil are the same as the milliequivalents of standard acid required to lower the buffer's pH one unit in the titration. Thus the amount of $CaCO_3/hfs$ needed to lime the soil to the desired pH can also be estimated. This value can then be corrected for the ECCE of the actual liming material and a lime recommendation made. The buffer method is used, where applicable, because neither CEC nor the pH-base saturation curve for the soil in question have to be known.

Some states use incubation techniques where soil samples with different pH values, textures, and organic matter contents are equilibrated with a standard limestone (known ECCE). The pH value, texture, and color, as an estimator of organic matter content, are determined for the soil to be limed and compared with the results, usually charts, of the incubations. From this comparison a lime requirement in terms of the standard limestone can be determined. These results can then be converted into actual lime requirements using the ECCE of the actual liming material.

Benefits of liming. The obvious reason for liming acid soils is increased crop yields. Most crops respond well to increases in pH from pH 5.0, or less to pH 6.0 or higher, although the response will vary greatly with the individual crop and other soil chemical properties. Some plants, such as azaleas, blueberries, and pin oaks, require acid soils for optimal growth. However, the majority of plants exhibit increased production when acid soils are limed. Plants respond to liming for a number of reasons including physical, chemical, and biological changes.

Increased Ca and Mg can result in improved physical structure of the soil. While this effect may be largely indirect and difficult to measure, it is, nonetheless, helpful. With liming of soils there is an obvious increase in soil solution pH. This means that the concentration of hydrogen ions decreases while the concentration of hydroxyl ions increases. As exchangeable hydrogen is replaced with basic cations, the percent base saturation will also increase. Higher soil pH also results in lower solubility of aluminum, iron, and manganese, which can be toxic. It also increases the availability of phosphates, which are often deficient, and molybdates in soils. Liming also increases plant-available calcium and magnesium. These chemical effects are a positive result of liming.

Liming stimulates microbial activity. This can increase the rate of mineralization of nutrients such as nitrogen and sulfur. Higher soil pH is especially beneficial to bacteria involved in nitrogen fixation by legumes, as well as in nonsymbiotic nitrogen fixation. Thus the benefits of liming acid soils are reflected in physical, chemical, and biological changes in the soil environment.

Overliming. Addition of limestone, as with any other soil amendment, must be carried out judiciously. Soil conditions and the need for liming can vary considerably in the same field, so that different areas of that field may need to be limed differently. If lime is applied uniformly across a field containing both acidic and calcareous soils, the lime applied in excess to the already calcareous areas can reduce the availability of iron, phosphorus, manganese, boron, and zinc. The net result is not only a waste of capital for limestone purchase, but reduced crop yields as well. In general, acidic soils should not be limed above a pH of 7.0.

STUDY QUESTIONS

1. What is a fertilizer?
2. Identify the major nitrogen fertilizer materials. What form of N is initially released into the soil by each fertilizer?
3. Why do many nitrogen fertilizers result in increased soil acidity?
4. Why are slow-release fertilizers often more desirable than standard materials?
5. What is the difference between normal and concentrated superphosphates?
6. Under what soil conditions would untreated "raw" rock phosphate be a suitable fertilizer?
7. Why is potassium referred to as potash?
8. What is a mixed fertilizer?
9. Which would neutralize the greatest amount of soil acidity, 1 mol of CaO, 1 mol of $CaCO_3$ or 1 mol of $Ca(OH)_2$?
10. Which would neutralize the greatest amount of soil acidity, 1 ton (2000 lb) of CaO, 1 ton of $CaCO_3$ or 1 ton of $Ca(OH)_2$?
11. Calculate the CCE of a limestone that contains 62% $CaCO_3$ and 9% $MgCO_3$.
12. Express the phosphorus and potassium contents of an 8, 8, 8 fertilizer on an elemental basis.
13. Why does ammonium sulfate tend to increase soil acidity much more markedly than does ammonium nitrate?

17

Soils and Environmental Quality

Few issues in recent years have attracted such widespread attention or concern as has the quality of our environment. Maintaining and protecting the quality of our atmospheric, soil, and water resources is necessary for the survival of human civilization. Soil is a major component of the environment and an essential natural resource. When soil is lost by erosion or polluted, not only is soil degraded but other portions of the environment can be affected as well. Sediment from eroded soils and chemicals leached out of the soil or carried on sediment particles can lessen the quality of air, water and biological resources. In this chapter we briefly discuss selected potential pollutants and their reactions with soils. Erosion, salinization, desertification, and other forms of soil degradation are discussed in other chapters.

17.1 AGRICULTURAL PESTICIDES

The need for pesticides has been exemplified by historical examples of suffering and loss of human lives and mass destruction of crops by diseases and insects. These include the nineteenth-century Irish potato famine, caused by a fungus, resulting in the starvation and death of a million persons, the millions of lives lost to malaria-carrying mosquitos, and the abandonment of early attempts to construct the Panama Canal because of yellow fever. Food production has always been limited by weeds that compete with the crop for water, sunlight, and nutrients. Weeds also interfere with crop harvest, contaminate the harvested grain, and harbor insects and diseases.

While the use of crop protection chemicals has been recorded for centuries, pesticides were introduced for agricultural use mainly in the second half of the nineteenth century. Early pesticides were preparations containing salts of heavy and transition metals, such as lead, arsenic, copper, and zinc; elemental compounds, such as sulfur; and naturally produced plant compounds, such as nicotine and pyrethrum. These pesticides were used successfully for controlling or reducing disease and insect damage in crops. The first synthetic pesticides were introduced before 1900, but 2,4-D, a herbicide, and DDT, an insecticide, which were introduced in the 1930s and 1940s marked the beginning of a technological revolution and heavy dependence on pesticides. Pesticides have rapidly evolved as valuable aids to world agricultural production. It is estimated that insects, weeds, plant diseases, and nematodes account for losses exceeding $20 billion in the United States and exceed $100 billion worldwide and would be much higher without pesticides. Pesticides are important in providing high-quality food and fiber at reasonable cost to an ever-increasing world population (see Chapter 18).

Pesticides can be defined as substances that eliminate or inhibit the growth and/or reproduction of species considered to be pests. Pesticides can be categorized according to the type of pest they target, with the main groups being herbicides, used to control weed species; insecticides, used to control insects; fungicides, used to control fungal diseases; and others, which include nematicides and rodenticides. The use of pesticides is extensive (Table 17-1). In the United States over 1 billion pounds of active ingredients, or 4 pounds of pesticides for every man, woman, and child, were used in 1985. This huge volume of pesticides represents over 960 different active chemicals. These active ingredients were used to make over 25,000 formulations or products. This huge number of compounds and their diversity is illustrated by Table 17-2, which lists major chemical classes, subgroups, and a few examples of each.

Agriculture is by far the largest user of pesticides (Table 17-1). Herbicides account for approximately one-half of the volume of pesticide usage in the United States. Approximately 90% of the herbicides and insecticides used for agriculture are accounted for by four major crops: corn, cotton, soybeans, and sorghum. Thus the usage of herbicides is heavily weighted to the major U.S. crop-producing regions. Industrial, commercial, and government use accounts for a rather sizable

TABLE 17-1 Estimated Annual Use of Pesticides in the United States, 1985 (Millions of Pounds of Active Ingredients)

	Herbicides	Insecticides	Fungicides	Other	Total
Agriculture	525	225	51	60	861.0
Nonagricultural[a]	115	40	21	0.1	176.1
Home and garden	30	35	12	0.1	75.1
Total	670	300	84	60.2	1112.2

Source: EPA, Office of Pesticides and Toxic Substances, *Agricultural Chemicals in Ground Water: Proposed Strategy*, 1987.
[a]Includes industrial, commercial, and governmental uses but not home and garden.

TABLE 17-2 General Pesticide Classes and Subgroups with Representative Examples and Some General Characteristics

Class—Subgroup	Examples	Comments
I. Insecticides		
Organochlorines	Methoxychlor, aldrin	Highly chlorinated, persistent
Organophosphates	Malathion, methyl parathion	Acutely toxic, nonpersistent
Carbamates	Carbaryl, methomyl	Inhibit cholinesterase
Organosulfurs	Tetradifon	Sulfur-containing miticides
Formamidines	Chlordimeform, amitraz	For carbamate-resistent pests
Amidinohydrazones	Pyramdron	Ant and cockroach control
Dinitrophenols	Dinoseb, dinitrocresol	Wide range of toxicity
Botanicals	Nicotine, rotenone, pyrethrum	"Natural" (derived from plants)
Synthetic pyrethroids	Permethrin, fenvalerate	Effective at very low rates
Fumigants	Methyl bromide, chlorothene	Usually halogenated
Repellents	Diethyl toluamide	Human insect repellants
II. Herbicides		
Phenoxyaliphatic acids	2,4-D, MCPA	Growth-hormone-like action
Substituted amids	Propanil, alachlor, metolachlor	Widely used
Diphenyl ethers	Diclofop methyl, fluzifop-butyl	Contact herbicides
Nitroanilines	Trifluralin, oryzalin, ethalfluralin	Soil incorporated
Substituted ureas	Linuron, diuron, fluometuron	Photosynthesis inhibition
Carbamates	Propham, chlorpropham	Esters of carbamic acid
Thiocarbamates	EPTC, cycloate, butylate	Sulfur containing
Heterocyclic nitrogens	Atrazine, simazine, metribuzin	Triazine herbicides
Aliphatic acids	Dalapon, dicamba, DCPA	Carbon chain acid
Phenol derivatives	Dinoseb, DNOC, PCP	Highly toxic, nonselective
Substituted nitriles	Dichlobenil, bromoxynil	C≡N or cyanide grouping
Bipyridylium	Diquat, paraquat	Contact herbicides
Cineoles	Cinmethylin	Disrupt growing tissue
Miscellaneous	Glyphosate, sithoxydim, acrolein	Varied structure and functions
III. Fungicides and bactericides		
Dithiocarbamates	Maneb, zineb, nabam	Amino acid inhibitor
Thiazoles	Ethazol	Degraded easily
Triazines	Anilazine	Structure similar to herbicides
Substituted aromatics	PCP, chloroneb	Benzene derivatives
Dicarboximides	Captan, folpet, captifol	Sulfur containing
Oxathiins	Carboxin, oxycarboxin	Systemic fungicides
Benzimidazoles	Benomyl, thiabendazole	Broad spectrum of diseases
Pyrimidines	Dimethirimol, ethirimol	Powdery mildew control

TABLE 17-2 (cont.)

Class—Subgroup	Examples	Comments
Organophosphates	IBP, fosetyl	Systemic action
Acylalanines	Metalaxyl, furalaxyl	Foliar, soil, or seed treatment
Triazoles	Triadifefon, bitertanol	Sterol inhibition
Piperazines	Triforine	Used on fruit and vegetables
Imides	Iprodione, vinclozolin	Selective fungicides
Pyrimidines	Fenarimol	Broad-spectrum fungicides
Carbamates	Propamocarb	Soil and foliar
Dinitrophenols	Dinocap	Uncouple oxidative phos-phorylation
Quinones	Chloranil, dichlone	Affect respiration in fungi
Organotins	Fentin hydroxide	Phytotoxic and fungicidal
Aliphatic nitrogens	Dodine	Control apple and pear scab
IV. Nematicides		
Halogenated hydrocarbons	DBCP, methyl bromide	Affect nervous system
Organophosphates	Phorate, fenamiphos	Similar to insecticides
Isothiocyanates	Vorlex, dazomet	Affect nematodes, fungi, weeds
Carbamates	Aldicarb, carbofuran	Similar to insecticides
V. Rodenticides		
Coumarins	Warfarin, brodifacoum	Anticoagulants
Indandiones	Pindone, chlorophacinone	Anticoagulants

Source: G. W. Ware, *Fundamentals of Pesticides: A Self-Instruction Guide*, Thompson Publications, Fresno, Cal., 1986.

amount of herbicides and insecticides. In addition, pesticides for lawn weed control; insect powders for pets; sprays and powders for controlling lawn and garden insects; aerosols for control of flies, mosquitoes, ants, roaches, mites, and termites; rodent baits; and pesticides to protect clothing in storage are used by over 90% of U.S. households.

There is increasing public awareness of the potential problems associated with the use and misuse of pesticides. Continued use of specific chemicals have led to the development of resistance, particularly among insects, and the need to develop new types of pesticides. Pesticides are valuable because they are lethal to target organisms, but less than 1% of the pesticides applied to the soil reaches the target organism. The remaining 99% represents a potential for the contamination of soils, water supplies, and the food we eat. In addition, some pesticides are not readily biodegradable and may persist for long periods of time in the environment. The lack of biodegradability increases the possibility of transport of that chemical into water, or through the natural food chain. Therefore, we must consider the costs of pesticide use, to human health and safety, and to the environment, as well as their benefits to agriculture. The fate of a pesticide in the environment is determined primarily by the chemical structure of the particular pesticide.

17.2 PESTICIDE REACTIONS WITH SOILS

Herbicides and insecticides account for over 85% of the organic compounds used as pesticides. Many of these chemicals are applied directly to the soil or eventually contact the soil; hence it is necessary to consider the interactions of these compounds with soils. The reactions with and movement of synthetic organic compounds through soils cannot be generalized because of the diverse chemical characteristics of individual compounds and because of the wide range in physical, chemical, and biological properties of soils. Six major types of potential soil reactions with pesticides are shown in Figure 17-1. These include plant uptake, volatility, adsorption, leaching, nonbiological degradation, and microbial degradation.

Plant uptake.   Crop removal is one avenue of pesticide removal from soils. This includes removal by the target weed species or by some nontarget crop species. Upon plant uptake the chemical structure of some pesticides is altered by the plant's metabolism, while in other cases the plant may only accumulate the pesticide in its unaltered form.

Volatilization.   The magnitude of pesticide loss by volatilization ranges from insignificant to more than 50% of the pesticide applied. The amount of volatilization depends on the chemical nature of the compound. Compound properties such as vapor pressure, water solubility, number, kind, and position of functional groups (groups such as –COOR, –NHR, –OH, and –NH$_2$) are all important in determining the volatility of a pesticide. For pesticides with very high vapor pressures, such as methyl bromide, volatilization becomes the predominant pathway by which the pesticides are lost from soil. The trifluralin herbicides are easily volatilized and must be incorporated into the soil to prevent loss and to remain effective. Volatilization during an entire cropping season can be important even for pesticides that are considered only slightly volatile. For example, annual losses of 3 to 18% for soil-applied chlorinated insecticides, which are only slightly soluble, have been demonstrated. Since vaporization does not necessarily imply degradation, this may be and often is a mechanism that results in the long-range transport of pesticides via the atmosphere. Volatilization creates the possibility of contaminating nontarget areas or organisms, sometimes at great distances from the point of the original application.

Soil adsorption.   Adsorption to soil particles has a great influence on the behavior of pesticides in soils. Of the many potential soil reactions shown in Figure 17-1, adsorption and degradation are the two most important processes determining

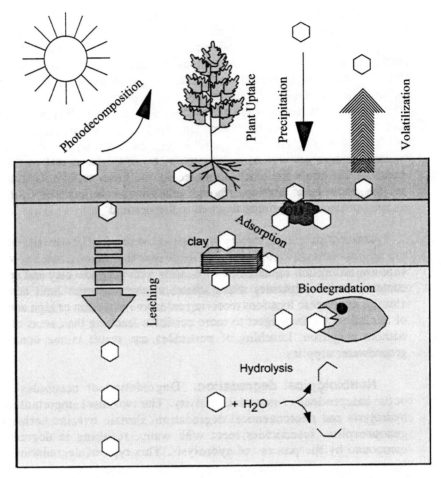

Figure 17-1 Major types of reactions of pesticides in soils.

the fate of a pesticide. Adsorption removes the pesticide from the soil solution, resulting in a reduction in leaching, volatilization, and even bioavailability of the compound. The two most active adsorbers in soils are clays and organic matter (humus). Since clays are negatively charged, they are important in the adsorption of positively charged molecules, that is, cationic pesticides. Clays are also important in the adsorption of weakly basic compounds which can protonate, that is, accept a hydrogen ion forming a cation. Pesticides that are acidic, that is, have negatively charged functional groups, could react with anion-exchange (positively charged) sites on clays or organic matter. However, the anion-exchange capacities of most soils are low and anionic pesticides are very weakly adsorbed and easily displaced. By far the largest group of pesticides are the nonionic pesticides. These are neutral molecules that range in polarity and water solubility from highly polar and therefore very soluble in water, to very nonpolar and therefore very insoluble in water. Studies have shown that the more polar neutral pesticides can be sorbed by swelling clay minerals, but the extent and the energy of adsorption are much less than with pesticides that occur as organic cations. Nonpolar neutral pesticides have been shown to be strongly sorbed by hydrophobic sites associated with soil organic matter. The amount of adsorption of neutral pesticides has been shown to be primarily a function of the organic matter content of the soil; in fact, recommended application rates for herbicides are often based on the organic matter content of the soil (see Chapter 8). Generally speaking, the lower the water solubility of a compound, the stronger will be its adsorption by soil organic matter. Also generally speaking, the larger the pesticide molecule, the lower will be its water solubility, and the greater its adsorption, although solubility can be modified by the nature and number of functional groups attached to the pesticide.

Leaching.   Leaching of pesticides is the result of downward movement and loss of gravitational water in soils. Conditions that favor leaching would be soils with low adsorption capacities, that is, soils with very low clay and organic matter contents, and/or pesticides whose chemical characteristics limit adsorption. Obviously, geographic locations receiving extensive irrigation or high annual amounts of rainfall would be subject to more pesticide leaching than areas of low rainfall without irrigation. Leaching of pesticides can result in the contamination of groundwater supplies.

Nonbiological degradation.   Degradation of pesticides in soils can occur independent of microbial activity. The two most important reactions are hydrolysis and photochemical degradation. Certain triazine herbicides and organophosphate insecticides react with water, resulting in degradation of the compound by the process of hydrolysis. This type of degradation reduces the long-term risk of leaching and groundwater contamination by the parent compound but may result in contamination of the groundwater by the degradation products. For some pesticides degradation products are nontoxic, but for others the degradation products may even be more toxic than the parent compound. Other pesticides absorb light energy, resulting in isomerization, substitution, oxidation, or other chemical changes. Photochemical reactions have been reported for a wide range of pesticides, including chlorinated insecticides, benzoic and phenylacetic acids, triazines, ureas, and dinitroanaline herbicides. The major limiting factor in photodegradation is the ability of the sunlight to reach the pesticide once it is incorporated in the soil or moved by leaching below the soil surface.

Biological degradation.   Soil microorganisms are essential in removing pesticides from the environment. Without the natural and active process of biodegradation by a wide variety of heterotrophic organisms, residual buildup of pesticides would be an overwhelming problem. Organisms metabolize pesticides

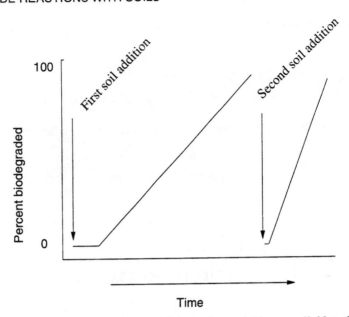

Figure 17-2 Effect of multiple additions of a pesticide to a soil. Note the shorter lag time and faster rate of decomposition for the second addition.

either directly as a source of carbon, energy, or nutrients, or through a process known as *cometabolism*. In the process of cometabolism, the chemical is degraded even though it does not serve as a major source of energy for the organism. Biological degradation of pesticides is favored by environmental factors that support the microbial population as a whole. Degradation is enhanced by warm temperatures, by increased soil water levels, and by higher soil organic matter contents.

Since organisms can use many pesticides as an energy or nutrient source, the population degrading a specific chemical may build up somewhat in the soil after the first addition of that chemical to a soil. A subsequent addition may result in more rapid degradation than the first. To a limited extent soils can be "trained" to degrade specific compounds (Figure 17-2). Note that there is both a shorter lag time prior to degradation for the second soil addition and a more rapid rate of breakdown once the process begins.

It is also important to distinguish between degradation and detoxification. Microbial changes in the structure of a chemical do not necessarily render it nontoxic. Occasionally, a compound will undergo a microbial transformation that would produce a degradation product that is just as toxic or even more so than the original compound and in some cases the degradation product may be resistant to further degradation. In general, extensive degradation results in detoxification of pesticides. Examples of some of the initial reactions involved in the microbial degradation of pesticides include the following (note that R represents the major portion of the organic molecule; only the functional groups are emphasized in these examples):

1. Addition of a hydroxyl group:

$$RCH_3 \rightarrow RCH_2OH$$

2. Oxidation of an amino group or reduction of a nitro group:

$$\frac{RCH_3}{\text{reduced form}} \leftrightarrow \frac{RNO_2}{\text{oxidized form}}$$

3. Removal or addition of a methyl group:

$$RN(CH_3)_2 \leftrightarrow RN(CH_3H) + CH_2 \leftrightarrow RNH_2 + CH_2$$

4. Removal of a chlorine atom:

$$RCH_2Cl + H_2O \rightarrow RCH_2OH + HCl$$

5. Hydrolysis:

$$RCOR' + H_2O \rightarrow RCOH + R'OH$$

These initial reactions may be followed by further degradation. For pesticides containing a benzene ring further degradation would include transformations that would open the ring and allow complete degradation.

Rate of disappearance in soils.   Almost all synthetic pesticides that are added to soils disappear with time. The rate of disappearance is a function of all the chemical and biochemical reactions that affect the pesticide and is a measure of the persistence of the pesticide in the environment. Some pesticides, such as organophosphate insecticides or the contact herbicide glyphosate, are biologically active in soils for only a few days under normal environmental conditions. Others, such as the triazine herbicides, usually persist at biologically active levels for 3 to 12 months. However, concentrations of triazine herbicides less than those required for biological activity may be detected in soils for 1 or 2 years after application, or even longer if environmental conditions do not favor breakdown. Some pesticides, such as DDT or dieldrin, persist in soils for extended periods of 2 to 5 years or longer. Usage of these very persistent pesticides is decreasing and many have been banned from use. Most pesticides degrade in a period of one year or less under normal soil and environmental conditions, so that repeated usage at recommended rates should not result in appreciable soil buildup or residues (Table 17-3). Persistence of pesticides beyond the target period of their intended use results in potential carryover problems. For example, carryover of a triazine herbicide used to control weeds in corn may devastate an oat crop planted on the same field the next cropping season. From an environmental point of view, persistent pesticides pose the greatest potential for movement from the point of application by the processes of leaching,

TABLE 17-3   Persistence of Biologically Active Levels of Selected Pesticides under Average Field Conditions

Transient, 1 month or less	Short term, 1–3 months	Seasonal, 3–12 months	Extended, over 1 year
Acrolein	Butylate	Alachor	Arsenic
Carbaryl	Chlorpropham	Atrazine	Borate
Dalapon	EPTC	DCPA	Dieldrin
2,4-D	PCP	Diuron	DDT
Dinoseb	2,4,5-T	Dicamba	Terbacil
Glyphosate	Cycloate	Linuron	Chlorodane
Malathion		Metribuzin	
Methyl bromide		Simazine	
Propanil		Trifluralin	
Diazinon		Bromoxynil	

Source: F. Matsumura, and C. R. Krishna Murti, *Biodegradation of Pesticides.* Plenum Press, New York, 1982.

soil erosion, or volatilization. Clearly, from a long-range perspective, desirable pesticides are those pesticides with short residence times.

17.3   GROUNDWATER AND CHEMICAL CONTAMINATION

A critical renewable, although sometimes very slowly renewable resource that has attracted national attention in the late 1980s and early 1990s is the nation's groundwater supply. The total amount of groundwater withdrawn daily has been in excess of 80 billion gallons in recent years (Figure 17-3). While considerably more groundwater is used for irrigation than for industrial, urban, and rural water needs, it is the quality of water for domestic use that is of primary concern. In several states over 90% of the population depends on groundwater as a partial source of water for domestic use (Table 17-4). In many rural areas, groundwater is the only available source other than transporting water considerable distances.

Groundwater is defined as water found below the earth's surface in geological formations called *aquifers*. The soil surface is for the most part in the unsaturated zone (*vadose zone*) and therefore does not represent a source of groundwater for wells. However, it is through this unsaturated zone that recharge water for aquifers must pass on its way to the saturated groundwater zone (Figure 17-4). Many aquifers are confined areas sandwiched between impermeable layers. These confined aquifers may reach the soil surface and be recharged at some point far removed from the site where a well may be drilled. Many states have strict regulations governing land use in recharge areas of confined aquifers. The rate of water movement into an aquifer and through an aquifer depends greatly on the per-

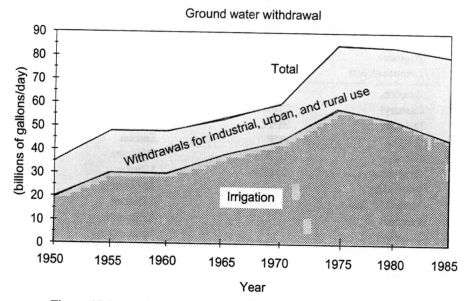

Figure 17-3 Total groundwater withdrawal for the years 1950 to 1985. (EPA, *Agricultural Chemicals in Ground Water: Proposed Pesticide Strategy*, Office of Pesticides and Toxic Substances. Washington, D.C., 1987.)

meability of the parent material and may vary from less than 1 m/yr to more than 1000 m/yr. The problem then is the possibility that pesticides or other chemicals may move with recharge water and contaminate an aquifer that is used as a source of water for rural farm families or urban populations. Pesticides can enter the groundwater as the result of a point source of pollution, such as an accidental spill of chemical in the aquifers recharge area or into a stream that feeds into a groundwater aquifer. Groundwater contamination could also be the result of non-point pollution, such as agricultural application of a pesticide that was persistent enough and mobile enough to pass through an unsaturated zone into a saturated zone used as a groundwater source.

For example,[1] in 1979, the pesticide DBCP (dibromochloropropane) was found in California wells and the pesticide aldicarb was found to have contaminated a groundwater supply in New York. By 1986, a total of 19 different pesticides had been detected in groundwater. In approximately half of the states, nonpoint sources were suspected to be the origin of the groundwater contamination, and more are discovered each year. While the measured levels are often very low (less than 5 ppb in many cases), the presence of any detectable level of the pesticide in groundwater is raising concern about the use of these pesticides and their long-range implications.

[1]EPA, *Agricultural Chemicals in Ground Water: Proposed Pesticide Strategy*, Office of Pesticides and Toxic Substances, Washington, D.C., 1987.

TABLE 17-4   Percentages of Population Relying on Groundwater For Domestic Use

Percent of state population	States	Percent of state population	States
90+	Arizona	40-49	Georgia
	Florida		Minnesota
	Hawaii		New Jersey
	Idaho		New York
	Mississippi		Ohio
	Nebraska		Pennsylvania
	Nevada		Virginia
	New Mexico		
		30-39	Alabama
80-89	South Dakota		Connecticut
			Massachusetts
70-79	Delaware		Missouri
	Iowa		North Carolina
	Maine		Oklahoma
			Oregon
60-69	Alaska		
	Indiana	20-29	Colorado
	Kansas		Kentucky
	South Carolina		Rhode Island
	Washington		
	Wisconsin	<20	Maryland
	Utah		
50-59	Arkansas		
	California		
	Illinois		
	Louisiana		
	Michigan		
	Montana		
	New Hampshire		
	North Dakota		
	Tennessee		
	Texas		
	Vermont		
	West Virginia		
	Wyoming		

Source: EPA, *Agricultural Chemicals in Ground Water: Proposed Pesticide Strategy,* Office of Pesticides and Toxic Substances, Washington, D.C., 1987.

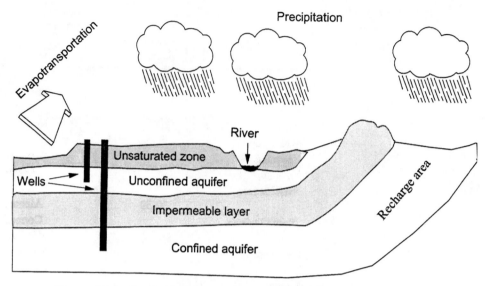

Figure 17-4 Confined and unconfined aquifers and their relationship to the unsaturated (vadose) zone, rivers, and recharge areas. (After EPA, *Agricultural Chemicals in Ground Water: Proposed Pesticide Strategy*, Office of Pesticides and Toxic Substances, Washington, D.C., 1987.)

As already discussed, the source of contamination can be accidental spills or leaks of pesticides where they are manufactured, stored, or distributed. These types of point sources usually involve very high concentrations at the source of the spill such that soil adsorption and microbial degradation are not sufficient to contain the chemical or prevent it from moving through the soil and underlying parent material. Very high concentrations of toxic chemicals may cause a soil to be nearly sterile and microbial breakdown may be too slow to result in significant degradation. Such sources of pollution generally also result in a localized plume of contamination that can be sampled and traced. When pesticides are applied to agricultural soils, the rates of application are usually low (from a few ounces to a kilogram per hectare) and quite uniform. Even though the rates per hectare are low, the total amount of pesticides applied annually exceed 1 billion pounds in the United States (Table 17-1). Thus there is the potential for groundwater contamination.

Not all pesticides have the same potential to leach through soils and contaminate groundwater. First, many are very water insoluble and bind very tightly to soil organic matter, preventing movement from nonpoint sources. Also, many pesticides degrade in short time periods (a few weeks or less), preventing them from reaching groundwater sources. The conditions favoring groundwater contamination include one or more of the following; (1) point sources involving concentrated amounts, such as accidental spills; (2) very persistent pesticides; (3) applications on soils with very low organic matter or clay contents, that is, soils with low adsorption capacities; (4) areas of high rainfall amounts or extensive irrigation; and

(5) areas where the saturated zone is very close to the soil surface. State and federal governments are making a concentrated effort to reduce the risk of contaminated groundwater supplies. These efforts include greater regulation of pesticide use, lower rates, encouragement of less persistent pesticides, regulation and education of pesticide handlers and applicators, and research programs to reduce the dependency of growers on pesticides.

In addition to organic pesticides, another chemical contaminant found more and more frequently in groundwater supplies is nitrate (NO_3^-). Since nitrate is an inorganic anion, it would be expected to move fairly rapidly through unsaturated zones into aquifers. In a random survey by the Environmental Protection Agency (EPA)[2] about 3% of the wells sampled that were supplying rural homes had nitrate concentrations exceeding the EPA drinking water standard of 10 mg/liter of NO_3^-–N. The levels and percentage of contaminated wells varies considerably with geographic area and with soil and geologic conditions. Several states have considered legislation that would regulate the rates of nitrogen fertilizer used for agricultural production as a means of groundwater protection. As with pesticides, nitrates can also enter groundwater supplies from point sources such as industrial spills or runoff from feedlots or confinement livestock operations (see Figure 17-9). Wells are sometimes drilled in very close proximity to livestock feeding operations and are not adequately protected from contamination by the feedlot runoff. Sandy-textured soils with low water-holding capacities pose the greatest potential for nitrate leaching.

17.4 SOILS AND WASTE DISPOSAL

Land disposal of sewage sludges is an option that has received increasing attention by many municipalities in the United States and throughout the world. This is in part because alternative methods such as incineration, ocean dumping, or landfills may not be suitable, adequate, or cost-effective. Total U.S. sludge production was estimated to be approximately 7 million tons in 1982[3] and is increasing. Approximately 25% of municipal sludge generated is disposed of by some form of land application. Land application of sludge can be classified in four major categories.

1. Agricultural utilization as a source of fertilizer nutrients, as well as improving the physical characteristics of the soil
2. Forest utilization as a means to enhance forest productivity
3. Land disposal where the primary goal is to dispose of the sludge rather than improving the physical soil conditions or as a source of nutrients

[2]*Groundwater Protection*, The Conservation Foundation, Washington, D.C., 1987.

[3]EPA, *Process Design Manual: Land Application of Municipal Sludge*, EPA-625/1-83-016, 1983.

4. Land reclamation where sludge is used to reclaim strip-mined areas and other disturbed sites

Municipal sewage sludge contains essential elements often applied to soils as fertilizers. Table 17-5 shows the average content of nitrogen, phosphorus, potassium, and sulfur in aerobically and anaerobically digested sludge. A common sludge treatment practice used by municipalities is anaerobic digestion. The resulting material is a dark-colored and reasonably stable humuslike material that can be transported and land applied. As shown in Table 17-5, the nutrient content of either aerobically or anaerobically digested sludges can vary widely. This rather wide range and unpredictable content of nutrients is one of the limitations of sludge application as a fertilizer. The median available N content for anaerobically digested sludge (ammonium plus nitrate) is about 1600 mg/kg, while the median total phosphorus content is about 3.0%, or 30,000 mg/kg, of which about 80% is in inorganic forms and readily available for plant uptake. This means that if enough sludge were applied to meet the nitrogen needs of a crop, excess phosphorus would be added. Thus, when application rates for land application of sludge are determined, these should be based on rates where no plant nutrient is applied in excess of the level needed by the crop nor in excess of levels that are environmentally safe.

In addition to the major plant nutrients, such as nitrogen, phosphorus, and potassium, sewage sludges often contain trace metals such as cadmium, mercury, lead, zinc, copper, and nickel. Some of the trace metals are plant nutrients that are needed in small amounts, but all trace metals can be toxic if present in large amounts. All "native" soils contain trace metals even when they have not been treated with sludge or contamination by other sources. Table 17-6 gives the minimum, maximum, and median values for Cd, Pb, Zn, Cu, and Ni, for soils with no known source of contamination. The concentrations in these "uncontaminated" soils varies considerably and all soils sampled had measurable concentrations of these metals. Similarly, the presence of low levels of metals in "uncontaminated" plants

TABLE 17-5 Nutrient concentrations (dry weight basis) in aerobically and anaerobically digested sludges

Component[a]	Anaerobically digested			Aerobically digested		
	Range	Median	Mean	Range	Median	Mean
Total N (%)	0.5–17.6	4.2	5.0	0.5–7.6	4.8	4.9
NH_4^+ - N (mg/kg)	120-67,600	1600	9400	30-11,300	400	950
NO_3^- - N (mg/kg)	2–4,900	79	520	7–830	180	300
Total P (%)	0.5–14.3	3.0	3.3	1.1–5.5	2.7	2.9
Total S (%)	0.8–1.0	1.1	1.2	0.6–1.1	0.8	0.8
Total K (%)	0.02–2.64	0.30	0.52	0.08–1.10	0.39	0.46

Source: Chongrak Polprasert, *Organic Waste Recycling*, John Wiley & Sons, New York, 1989.
[a]Number of samples analyzed for individual components varies from 8 to 86 different sewage samples.

TABLE 17-6 Concentration of Selected Trace Metals in Soils with No Known Point Source of Contamination (µg/g)

Metal	Minimum	Maximum	Median
Cd	0.01	2.3	0.20
Pb	0.2	4,109	11
Zn	1.5	402	57
Cu	0.3	735	19
Ni	0.3	269	24

Source: Sommers et al., in *Land Application of Sludge*, A. L. Page et al. (eds.), Lewis Publishers, Chelsea, Mich., 1987.

would be expected based on the fact that almost all soils have measurable levels of trace metals. The concentration of Cd, Zn, and Pb in several crops, including vegetables, grain, and root crops, is shown in Table 17-7. As was the case for soils, the concentration varies widely and accumulation of a particular metal, cadmium for instance, is much different for rice grain (about 5 µg/kg) than for leafy vegetables such as lettuce (435 µg/kg) or spinach (800 µg/kg).

The amounts of trace metals potentially added to soils as sewage sludges are shown in Table 17-8. This table shows the average concentrations in sludges from Illinois communities of population less than 200,000 and from Chicago, which has the largest metropolitan municipal sewage treatment facility in the United States. Sewage sludge from Chicago is influenced more by industrial wastes that have a

TABLE 17-7 Concentration of Selected Metals in the Edible Portion of Plants Grown on Soils with No Known Point Source of Pollution

Crop	Cd (µg/kg)[a]			Zn (mg/kg)			Pb (µg/kg)		
	Min.	Max.	Med.	Min.	Max.	Med.	Min.	Max.	Med.
Lettuce	34	3800	435	13	110	46	36	1700	190
Spinach	160	1900	800	17	200	43	240	2300	530
Potatoes	9	1000	140	5.1	35	15	1	2200	25
Wheat	5	220	36	11	76	29	1	770	21
Rice	<1	250	5	7.7	23	15	<1	80	5
Sweet corn	0.5	230	86	28	55	25	7.6	260	9
Field corn	<1	350	4	12	39	22	<1	3600	6
Carrots	15	1200	160	3.8	61	20	10	720	55
Tomatoes	45	790	220	12	35	22	<1	460	27
Soybeans	1	1200	45	32	70	45	3	350	36

Source: Sommers et al., in *Land Application of Sludge*, A. L. Page et al. (eds.), Lewis Publishers, Chelsea, Mich., 1987.
[a]Min., minimum; max., maximum; and med., median values.

TABLE 17-8 Metal Content (μg/g, Dry Weight Basis) of Sewage Sludge from Chicago and Illinois Communities with Populations of Less than 200,000

Metal	Chicago			Communities of <200,000		
	Minimum	Maximum	Median	Minimum	Maximum	Median
Zn	1,670	4,850	2,630	338	14,900	1,430
Cd	120	312	197	3	3,410	12
Cu	679	2,270	1,380	117	4,060	760
Ni	186	840	343	0	1,650	70
Pb	304	1,160	680	58	1,300	267
Hg	0.8	7.5	3.2	0.5	15.6	2.5

Source: T. M. McCalla, J. R. Peterson, and C. Lui- Hing, in *Soils for Management of Organic Wastes and Waste Waters*, L. F. Elliot and F. J. Stevenson (eds.), SSSA, ASA, CSSA, Madison, Wis., 1977.

somewhat higher trace metal content. This is reflected in higher median values for the trace metals shown in Table 17-8. The actual amount of any trace metal applied to a specific land area can only be determined if analysis is performed on individual batches of sludge. The effect of soil-applied sludges on concentrations of metals in soils, barley leaves, and barley grain is shown in Table 17-9. These data are based on results from 13 states where crops were harvested for five consecutive years. Treatments were no applied sludge, a one-time application of 100 metric tons/ha the first year only, and an annual application of 20 metric tons/ha for each of the 5 years of the experiment. The levels of metals extracted from the soil were increased

TABLE 17-9 Effect of a One-time Initial Application of 100 mt/ha or 20 mt/ha Each Year for Five Consecutive Years on the Average Soil and Plant Content of Cd, Zn, Cu, and Ni for Five Growing Seasons in 13 States and 15 Locations

Extract	Sludge Rate	Cd	Zn	Cu	Ni
Soil extract	0	0.19	3.7	1.9	1.5
	20 mt/ha/yr	2.17	32.2	10.2	3.8
	100 mt	3.45	53.0	17.7	6.6
Barley leaf	0	0.31	28.2	9.5	1.5
	20 mt/ha/yr	0.51	38.4	10.6	1.8
	100 mt	0.76	47.2	10.8	1.9
Barley grain	0	0.11	39.6	6.7	1.1
	20 mt/ha/yr	0.24	53.4	7.7	1.3
	100 mt	0.36	60.5	7.3	1.6

Source: Sommers et al., in *Land Application of Sludge*, A. L. Page et al. (eds.), Lewis Publishers, Chelsea, Mich., 1987.

dramatically where a one-time application of 100 mt/ha was added when compared to nonamended soils. The levels of extracted metals depends on soil properties, such as pH, organic matter content, and CEC. The pH values of the soils used in this study ranged from 5.2 to 8.2. Metals are generally bound less tightly and hence are more plant available in acid soils than in neutral or basic soils. Molybdenum is an exception to this rule, in that molybdenum is more available in higher pH soils. The level of metals in barley leaves and grain were also higher where sludges were applied at the rate of 20 metric tons/ha than in the control (Table 17-9). Additional research has shown that when sludge is applied at rates to satisfy the N requirement of the crop, the Cd and Zn contents of plant tissue remain at nearly constant levels with successive sludge applications. EPA guidelines require that the annual rate of Cd addition to soils should not exceed 0.5 kg/ha and cumulative loadings for Cd should not exceed 5 kg/ha for soils with a CEC of less than 5 $cmol_c$ kg^{-1}, 10 kg/ha for soils with CEC in the range of 5 to 15 $cmol_c$ kg^{-1} and 20 kg/ha of Cd for soils with CEC exceeding 15 $cmol_c$ kg^{-1}.

Another challenge to the safe use of sludge on agricultural land is the presence of trace but variable amounts of natural or synthetic organic compounds. Literally thousands of organic compounds can be potentially found in sludge, and analysis for all would be impossible. To check for potential contamination by organic compounds, sludges can be analyzed according to a predetermined list of specific organic chemicals, such as the priority pollutants list[4] developed by the EPA. Sludges can also be analyzed for specific suspected chemicals based on some knowledge of the discharge by users in a specific location. Finally, short-term bioassays can be used to test sludges for mutagenicity. To keep organic chemical loadings in perspective, application rates of approximately 100 mt/ha of sludge containing 10 mg/kg of a pesticide would be necessary to add amounts of organic compounds comparable to quantities of pesticides commonly added to soils in agricultural operations. An important difference between the presence of trace amounts of organics and trace amounts of metals is their persistence in soils. The half-lifes of even the most persistent organic pesticides seldom exceed 10 or more years, while some organic pesticides have a half-life of only a few days. By contrast, the residence time for most metals is estimated to be on the order of a few thousand years. It should be noted that no adverse effects on the growth of crops have been observed when sludges containing less that 10 mg/kg of organics are applied to soil at rates needed to meet the nitrogen requirement of the crop.

17.5   LAND DISPOSAL OF DOMESTIC AND INDUSTRIAL TRASH AND ANIMAL WASTES

Soils have long been and will probably continue to be the receptor of society's wastes. Three somewhat different problems are found for the land disposal of

[4]EPA, *Summary of Environmental Profiles and Hazard Indices for Constituents of Municipal Sludge*, Office of Water Regulations and Standards, Washington, D.C., 1985.

domestic wastes (trash), the disposal of hazardous industrial wastes, and land disposal of animal manures.

Domestic wastes. Trash collection in the United States is a huge and growing concern. Not only has the U.S. population grown by about 40 million people from 1970 to 1990, but the amount of solid waste generated per person has increased by approximately 50%. It has been estimated that every man woman and child generates approximately 2 kg of trash per day. This totals over 200 million tons of municipal solid waste or domestic solid waste each year. On the average about one-half of this solid waste is in the form of paper and glass (Table 17-10). Enough paper is buried in landfills each year to produce all the newspapers in this country. Food wastes, iron-containing metals, plastics, and wood also contribute significantly to the total. The huge volume of this waste places increased pressure on scarce land resources and increases environmental risks to soil and water resources. Open burning of wastes in dumps was used to reduce the volume of trash until the 1960s, when air pollution laws gradually ended this practice. Most municipal solid waste is now disposed of in landfills (Figure 17-5). The basic procedure is to find a natural ravine or valley or to dig a hole or trench in which to bury solid waste. The waste is covered with soil and when the landfill is completed it is often capped, seeded with grasses and used as a park or for other public recreational use. Suitability for construction on these sites is limited because of land settling. A major problem with landfills is the potential for leaching and groundwater contamination. Most municipal landfills are not designed to dispose of pesticides, fossil fuels, and toxic industrial chemicals, but it is almost impossible to keep homeowners from discarding these potential pollutants. With such potential pollutants buried deep in the regolith and with little or no thought given to protecting the landfill from

TABLE 17-10   Composition of Domestic Waste

Component	percent
Paper	41
Food waste	21
Glass	12
Ferrous metals	10
Plastics	5
Wood	5
Rubber and leather	3
Textiles	2
Aluminum	1
Other	0.3

Source: B. J. Nebel, *Environmental Science: The Way the World Works*, 2nd ed., Prentice Hall, Englewood Cliffs, N.J., 1987.

Figure 17-5   Sanitary landfill in DuPage County, Illinois. Solid waste is trucked to the site, dumped, and covered with clay material. Bottom and sides of pit are packed with clay to form an impervious seal to prevent groundwater pollution. Ultimate use of this landfill will be a ski hill. (Courtesy of USDA Soil Conservation Service.)

leaching by rainwater, groundwater contamination is a serious concern. Because potential dump sites in which landfills are established are not protected against potential leaching, proposed sites for landfill development are often rejected by local political pressure. This is a very serious problem, as many cities face the reality that current landfill needs will exceed the present landfill capacity in the next few years to a decade with no new sites available. The lack of available landfills and acceptable technology is a major driving force for programs to recycle many components of domestic wastes.

Industrial wastes.   An even more serious problem is the safe land disposal of toxic chemical wastes. Federal and state laws protecting air and water resources from toxic chemicals means the vast majority of industrial wastes will be disposed of by land disposal methods. Three methods of land disposal used are (1) deep-well injection, (2) surface impoundments, and (3) landfills.

Deep-well injection is based on the principle that dry porous layers of strata exist far beneath the soil surface that are isolated by impermeable layers of clays or consolidated bedrock. A carefully sealed well is drilled into this dry layer and hazardous waste liquids are pumped into this porous strata as a permanent containment facility (see Figure 17-6). Some of the potential problems arise with spills or

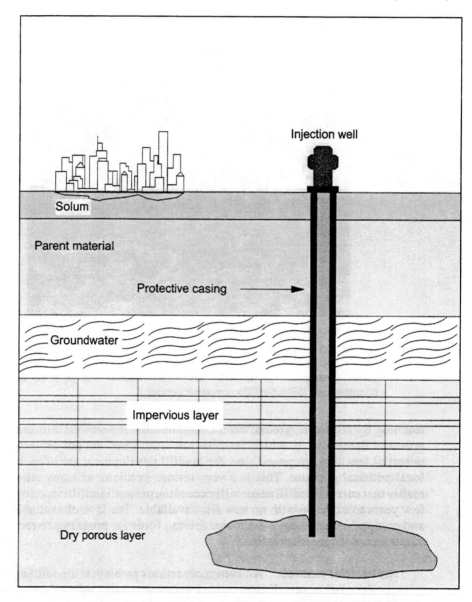

Figure 17-6 Deep-well injection disposal of hazardous wastes. Well is drilled into a dry porous layer that is separated from groundwater by a thick impervious layer. (After B. J. Nebel, *Environmental Science: The Way the World Works*, 2nd ed., Prentice Hall, Englewood Cliffs, N.J., 1987.)

leaks at the surface, corrosion of the well casing, or failure of the "impermeable" layers of clay or consolidated bedrock.

Surface impoundments are another method used to dispose of wastewaters containing relatively low levels of hazardous chemicals. A contained pond is built by lining with impervious clays topped by plastic liners (Figure 17-7). If properly

Figure 17-7 Surface impoundment used to dispose of wastewaters. Evaporation of water concentrates the hazardous wastes, reducing the volume. A plastic liner and an impermeable clay liner are used to prevent groundwater contamination. (After B. J. Nebel, *Environmental Science: The Way the World Works*, 2nd ed., Prentice Hall, Englewood Cliffs, N.J., 1987.)

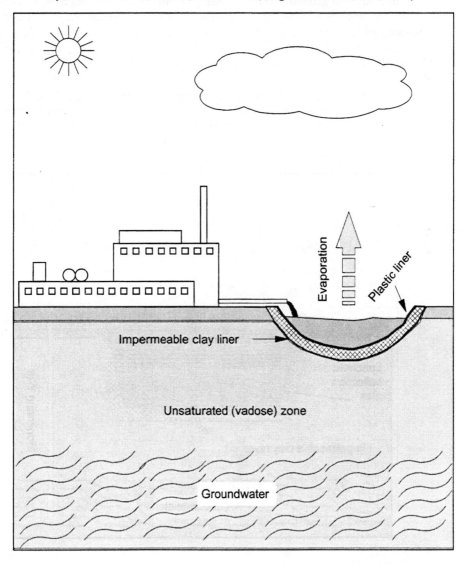

designed and maintained, this creates a sealed pond that evaporates large amounts of water while concentrating and containing the hazardous waste. Potential problems exist in locating clay sources that are truly impervious and the possibility of deterioration of failure of plastic liners. In addition, excessive discharge could lead to overflow of the containment pond.

Landfills for hazardous wastes are similar to solid waste disposal sites but are designed to be protected from leaching by clay and plastic liners and by a leachate collection field and pumping system to remove any toxic leachate (Figure 17-8). The success of this disposal method once again depends on the physical properties of the clay and plastic liners, as well as the maintenance and operation of leachate removal system. Soil and groundwater contamination are the major concerns, and continuous site monitoring is a long-term obligation. Disposal of hazardous wastes continues to be a major environmental problem. More effort is needed to find efficient, secure, and long-term solutions to land disposal or alternative methods, such as biodegradation, incineration, and reduced levels of hazardous waste generation.

Figure 17-8   Landfill for hazardous waste disposal. Clay and plastic liners and caps are used to prevent leaching by precipitation. A leachate collection system is used to collect and remove any toxic leachate. (After B. J. Nebel, *Environmental Science: The Way the World Works*, 2nd ed., Prentice Hall, Englewood Cliffs, N.J., 1987.)

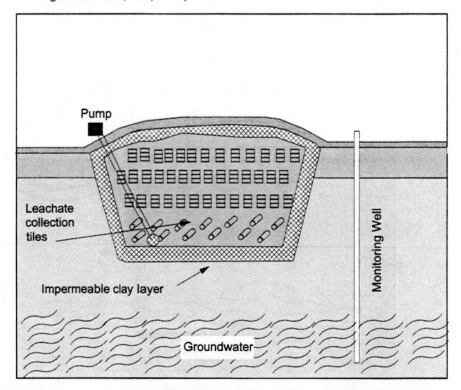

Animal manures.   The livestock and poultry industry generates large amounts (in excess of 1.7 billion tons/year) of animal manures. These manures are routinely and successfully applied to agricultural land. Application of these organic materials improves the tilth of the soil by increasing the organic matter and hydraulic conductivity and by decreasing the bulk density. Exchangeable cations and soil nutrient levels are also increased. The major inorganic nutrients are nitrogen, phosphorous, and potassium. The mineral composition of manures from a nutrient standpoint vary with the type of livestock and the age of the livestock. The average composition of several livestock manures is given in Table 17-11. Manure can be applied to land as a slurry or semisolid by surface application, or injection. Not all of the nitrogen in manures is available for crop use. Thirty to as much as 90% of the nitrogen in manures can be lost as NH_3(gas) or N_2(gas) in storage and handling or as NH_3(gas) in surface spreading. Nitrates released by mineralization of manures may be lost by leaching or by denitrification as N_2(gas). When animal manures are applied at reasonable rates to cropped soils, their use is both environmentally acceptable and an economic advantage to growers. There has been increased emphasis in the late 1980s and early 1990s on the concept of sustainable agricultural systems that encourage the increased use of manures as a nutrient source reducing the need for chemical fertilizers.

Mechanized livestock production, however, often results in very concentrated livestock operations such as feedlot operations of 100,000 cattle or dairy operations of 2000 animals or more (Figure 17-9). In such concentrated operations there may not be enough land within a reasonable proximity (hauling distance) to dispose of the manures without the risk of excess nitrates in the soil, excess salts from the manures, or imbalance of applied nutrients. There is also the risk in any open livestock operation of feedlot runoff, resulting in the potential for nitrate pollution of groundwater (Figure 17-10). In addition to the nitrate pollution potential, feedlot runoff into streams and lakes can result in microbial breakdown of these organics,

TABLE 17-11   Manure Production and Characteristics for Farm Animals on a Yearly Basis

	Assumed avg. wt. (lb)	Daily manure production (percent body wt.)	Annual prod. (tons/ animal)	Annual prod. (ft^3/animal)	Mineral content (lb)		
					N	P_2O_5	K
Dairy cattle	1,500	8.8	24.1	756	197.1	54.1	83.7
Beef cattle	800	6.0	8.76	276	136.0	21.0	31.5
Hens	5	5.9	0.054	1.72	1.81	1.46	0.67
Pigs	100	5.0	0.915	28.4	14.7	6.6	3.7
Sheep	100	3.7	0.675	21.2	12.3	4.3	8.9

Source: A. C. Dale, Properties of animal wastes, in *Proceedings of Livestock Waste Management Conference*, Department of Agricultural Engineering. University of Illinois, Urbana, Ill., 1972.

Figure 17-9   Overview of a large cattle feeding operation in Montana. The pens contain 25,000 cattle at the time of the photograph. (Courtesy of USDA Soil Conservation Service.)

Figure 17-10   Manure and nutrient-laden runoff from a Minnesota farm have washed into an otherwise clear flowing stream following two days of heavy rain. (Courtesy of USDA Soil Conservation Service.)

Figure 17-11   Fish kill near Colorado Springs, Colorado, due to oxygen deficiency. (Courtesy of USDA Soil Conservation Service.)

resulting in an increased *biological oxygen demand* (BOD). If the BOD is high enough, there will not be enough dissolved oxygen left in the water to support the needs of fish, and a fish kill can result (Figure 17-11). To reduce these environmental risks, animal waste disposal is increasingly regulated by local, state, and federal governments.

17.6   ACID RAIN

Acid rain has been and continues to be a topic of considerable environmental interest in the United States and around the world. The focus of this discussion will be the effects of acid rain on soils. To discuss the effect of acid rain on soils, let's briefly define acid rain and its causes. Acidity, or pH, as discussed in Chapter 9, is the negative log of the hydrogen ion concentration. At pH 7 the concentration or activity of hydrogen ions would be 10^{-7} mol/liter. Because pH is logarithmic, rainfall of pH 6 would have 10 times the hydrogen ion concentration of rainfall of pH 7, while rainfall of pH 5 would have 100 times the concentration of hydrogen ions found in pH 7 rain. Unpolluted rainfall that is in equilibrium with clean air would not have a pH of 7, but rather a pH of approximately 5.6. A pH of 5.6 occurs because air naturally contains carbon dioxide, which dissolves with water to form carbonic acid, a weak acid (see equations [9-19] and [9- 20]). Acid rain is therefore defined as rainfall that is more acidic than "clean rain," that is, rainfall with a pH of less than 5.6.

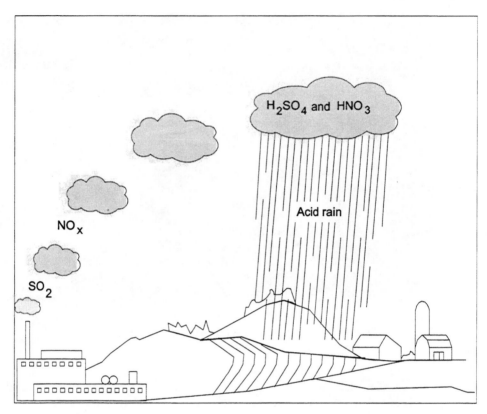

Figure 17-12   Acid precipitation results when fossil fuel use produces SO_2
and NO_x that are oxidized to sulfuric and nitric acids in the atmosphere.

Rainfall, particularly in the eastern United States, often has pH values less
than 5.6. The lower values are associated with increased sulfate and nitrate concentrations in the rainfall. Lower acidity and increased sulfate and nitrate concentrations in rain indicate that the major constituents resulting in acid rain are sulfuric
and nitric acids. These acids originate from sulfur and nitrogen oxides in the air that
are the result of human activities. These anthropogenic contributions of sulfur are
primarily in the form of sulfur dioxide, and the major source of the sulfur dioxide
is the burning of fossil fuels, especially coal. Nitrogen oxides are also added to the
air by the burning of fossil fuels, including automobile exhausts and other industrial
sources. The conversion of these gases to sulfuric and nitric acids in the atmosphere
results in the formation of acid rain (Figure 17-12).

There is an extensive monitoring network called NADP (NADP/NTN), National Atmospheric Deposition Program/National Trends Network,[5] that collects

[5]National Atmospheric Deposition Program, *NADP/NTN Annual Data Summary:
Precipitation Chemistry in the United States, 1988*, Natural Resource Ecology Laboratory,
Colorado State University, Fort Collins, Colo., 1989.

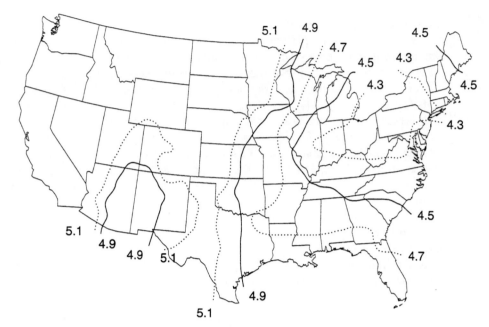

Figure 17-13 The weighted mean annual (1988) pH for precipitation.
(Data from the *NADP/NTN Annual Data Summary, Precipitation Chemistry
in the United States*, Natural Resource Ecology Laboratory, Colorado State
University, Fort Collins, Colo.)

precipitation samples every week from approximately 200 U.S. monitoring stations.
These samples are analyzed weekly for pH and major cations and anions. Figure
17-13 shows the distribution of rainfall pH in the United States. It is clear that
rainfall pH is much lower east of the Mississippi than in the west and also lower in
the northeastern states than in the southeastern states. Rainfall in the west, in
general, has higher pH values because of lower emissions of sulfur and nitrogen
oxides and also because alkaline soils in the western United States provide dust
particles that tend to raise the pH of rainfall in the air.

Rainfall results in leaching during soil formation and thus has a direct effect
on soil acidification. Soil acidification mainly results from the loss of basic constituents, such as calcium and magnesium, and their replacement with hydrogen
ions from rainfall. Processes involving soil organisms, such as nutrient uptake and
organic matter decomposition, can also lower soil pH. Soil acidification is a natural
process; however, the real question is how increased rainfall acidity, that is, increased hydrogen ion input from anthropogenic activities, can accelerate the natural
process of soil acidification.

The extent to which soils are affected by or sensitive to acid precipitation
depends on the properties of the soil. Soils that are considered to have the greatest
sensitivity to acid rain would be those soils with low buffer capacities (cation-

exchange capacity of less than 6.2 $cmol_c$ kg^{-1} in the top 25 cm) and which are only slightly acidic, that is, pH values in the range of 5.5 to 6.5. Soils that are highly weathered and extremely acid, that is, that have pH values of less than 5, are not likely to become perceptibly more acid due to exposure to acid rain, although increased acid input from acid rain will result in the release of additional aluminum ions (see equation [9-24]). Soils that are considered to be nonsensitive to acid rain include calcareous soils, which contain calcium or magnesium carbonates and have pH values in the range 7.2 to 8.5, soils that have high buffering capacities (cation-exchange capacities greater than 15.4 $cmol_c$ kg^{-1} in the top 25 cm), and those soils that are frequently renewed by soil deposition, that is, soils in landscape positions that are accumulating eroded soil material from other landscape positions, for example, Cumlic Haplaquolls.

Soil pH dictates many of the chemical properties of a soil, including nutrient availability, solubility of toxic elements, buffer capacity, dominant buffer mechanism, and level and type of microorganism activity. A discussion of the effect of acid rain on the pH of soils clearly merits a distinction between highly managed cultivated soils, typical of agricultural production, and unmanaged, uncultivated soils, typical of most forest, rangeland, and natural areas. Most cropped soils in humid regions are periodically limed to correct for acidification due to the addition of nitrogen fertilizers, for acidity resulting from symbiotic N fixation, as well as for acid rain inputs. The H^+ ion input[6] from 100 cm of pH 4.0 acid rain is approximately 1 kEq/ha/yr. This compares to approximately 12 kEq/ha/yr from nitrogen fixation by a legume crop, such as alfalfa, with an annual yield of 10 tons/ha. The largest source of acidity for cultivated soils is from the application of nitrogen fertilizers. Theoretical calculations show that the application of 195 kg/ha of anhydrous ammonia, such as used for corn production, results in a theoretical acidity equal to approximately 25 years of acid rain of pH 4.3. As mentioned previously, cultivated soils in humid regions are periodically limed to correct the buildup of acidity, regardless of the source. Thus, although acid rain does contribute slightly to increased soil acidity of managed soils, soil pH is not changed dramatically or permanently by rainfall.

Some research has indicated small pH changes in soils of approximately two-tenths of a pH unit by acid rain over a period in excess of 130 years for unmanaged soils. Other studies have not been able to measure soil pH changes due to acid rain in well-buffered soils or in soils amended by normal agricultural practices. Hence these soils are not likely to be harmed by acidic deposition. The greatest potential for acid rain to affect soil pH would be for poorly buffered range or forested soils that are not amended by the addition of limestone. The long-term effect of exposure of these soils to acid rain would include the removal of basic cations and accelerated acidification, but studies to document this effect over decades or centuries are not available. These effects are difficult to document,

---

[6]A kEq is defined as the amount of acidity equivalent to 1000 mol (1000 g) of hydrogen ions.

partially because of the high buffer capacities of even these soils (see Chapter 9). This means that even a very "sensitive soil" with a cation exchange capacity of 5 $cmol_c$ kg^- has a buffer capacity in the surface 25 cm of soil of approximately 150 times the annual hydrogen ion input from acid rain. In addition, soils of very low buffer capacity are often already quite acidic, making them less sensitive to the effects of acid rain. While acid rain affects the pH of soils, it is considered a minor affect, even for unmanaged soils.

As discussed in Chapter 9, acidic soils have a greater amount of exchangeable aluminum and hydrogen than do soils with higher pH values. If acid rain significantly lowers soil pH values, this would be the result of increased loss of basic cations and their replacement with H^+ and Al^{3+} ions. The increased Al^{3+} levels on the exchange complex would be the result of the soil buffering acidity added from acid rain (see equation [9-24]). Increased levels of aluminum are probably the most significant potential effects of acidic deposition because aluminum influences terrestrial plant growth and life in aquatic systems. In cultivated soils, however, the additional effect of acid rain on aluminum release is negated by limestone additions to correct the pH of the surface soils. Liming can also be used to neutralize the low pH values of affected lakes.

Other potential effects of acid rain are on soil microbial activity and associated biochemical reactions. Most research studies examining the effects of acid rain on microbial activities have been conducted in the laboratory using accelerated rates of simulated acid rain. This is because the rate of soil acidification by natural precipitation is too slow to allow meaningful comparisons in the field. Some studies have shown that the numbers of protozoa and mycorrhizae fungi are greater in plots treated with normal rainfall than in plots receiving rainfall with higher or lower pH values. Even though organism activity has been altered slightly by rainfall pH, these changes are not reflected in decreases in crop production. In addition, soil is very resilient and there is no evidence that any changes in microbial populations or numbers resulting from acid rain events, particularly in the surface few centimeters of soil, are permanent.

While soil pH can have a marked effect on symbiotic nitrogen fixation, evidence for the effects of acid rain on this process are also inconclusive. Some studies have shown that inhibition of rhizobia on soybeans with addition of simulated rains of pH 3.2, while other studies have shown that simulated rain treatments as low as pH 2.4 had no effect on *Rhizobium* nodulation or nitrogen fixation. Studies have also examined the effect of acid rain on the decomposition of organic residues. These studies have shown that the initial decomposition of plant residues is only slightly influenced by simulated acid rains; again this is at least partially due to the high buffering capacities of both soils and organic materials.

Acid rain does contribute to the nitrogen and sulfur balance in soils. The total annual inorganic nitrogen deposited by precipitation varies from less than 2.0 to more than 8 kg/ha/yr (Figure 17-14). Approximately 4.5 kg/ha of nitrate nitrogen would be deposited by 75 cm of precipitation with a pH of 4.2 and the average chemistry of rainfall received in the central United States. Generally, this amount

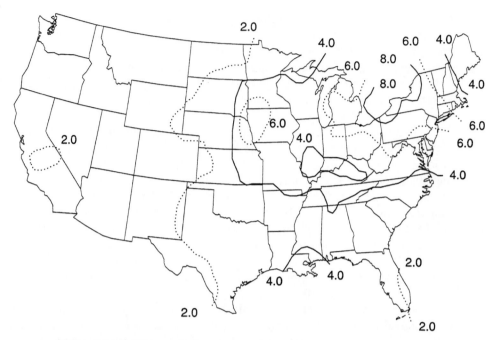

Figure 17-14   Measured nitrogen deposition (kg/ha) for 1986. (Data from the
*NADP/NTN Annual Data Summary, Precipitation Chemistry in the United
States*, Natural Resource Ecology Laboratory, Colorado State University, Fort
Collins, Colo.)

of nitrogen is small compared with that mineralized from soil organic matter and
added by fertilizers. However, this amount of nitrogen is significant for forests,
unimproved pastures, rangeland, or crops not receiving nitrogen fertilizer. The
amount of nitrogen deposited is a direct function of the pH of the rainfall. If the
mean annual pH of rainfall is increased from 4.2 to 4.6 through reductions of acid
rain, the amount of nitrate-nitrogen deposited would be reduced by approximately
one-half. On the other hand, if the mean annual pH of rainfall were reduced from
4.2 to 3.8, the amount of nitrate-nitrogen added would increase approximately
twofold.

The sulfate content of rainfall is approximately two-thirds of the hydrogen ion
content (equivalent basis) for major agricultural areas. Seventy-five centimeters of
precipitation with an annual average pH of 4.2 would contain approximately 8 kg/ha
of sulfate-sulfur. The amount of sulfur added annually by precipitation is important
to both forest plants and cultivated agricultural crops. Many soils in major crop-
producing regions do not contain sufficient plant-available sulfur to meet crop
requirements, yet sulfur deficiency symptoms are seldom observed. This is in part
because plants receive sulfate in the form of precipitation during the growing season
(see Figure 13-6), and both the soils and plants absorb sulfur dioxide gas directly

from the atmosphere. The amount of sulfur in precipitation with an average pH of 4.6 would be approximately one-half of that in rainfall of pH 4.2, while the amount of sulfur in precipitation of pH 3.8 would be approximately twice that of pH 4.2. Reduced emissions of sulfur would reduce acid rain and sulfur input to soils, both in the form of sulfate in rainfall and by direct absorption of sulfur dioxide by soils and crops.

In summary:

1. Acid rain does occur in major U.S. crop-producing areas.
2. The effects of acid rain on soil pH are relatively small, especially when compared to the effects of N fertilizers in cultivated cropping systems.
3. Effects of acid rain on soil biota in the field are not conclusive but probably not large or permanent.
4. Acid rain does supply some plant-available sulfur and nitrogen to the soil, although this is not the recommended procedure to maintain soil fertility.

STUDY QUESTIONS

1. What are the major classes of pesticides?
2. Explain the difference between biological and nonbiological degradation of organic compounds in the soil.
3. What are the advantages and disadvantages of persistent pesticides?
4. What factors control the movement of a pesticide through the soil?
5. Distinguish between point and nonpoint sources of pollution.
6. What are the benefits and drawbacks to the application of sludge on soils?
7. What is acid rain? What are the effects of acid rain on soils?

18

World Soil Resources

18.1 POPULATION TRENDS

Discussion of world soil resources must be coupled with a discussion of world population and its potential growth. World population trends, as well as predictions for the future, are given in Table 18-1. World population has increased from approximately 250 million people in 1000 A.D. to 5.1 billion people in 1988. Notice that it took 650 years for the world's population to double from 250 million to 500

TABLE 18-1 World Population Trends

Year	Population	Years to Double
8000 B.C.	<5 Million	
0 A.D.	200 to 250 million	
1000 A.D.	257 million	
1650 A.D.	500 million	650
1850 A.D.	1 billion	200
1930 A.D.	2 billion	80
1975 A.D.	4.079 billion	45
1988 A.D.	5.114 billion	≈ 35
Predicted 2001 A.D.	8 billion	
Predicted 2010 A.D.	10 billion	Does population level off or double?
Predicted 2050 A.D.	10.5 billion	?

Source: B. J. L. Berry, E. C. Conkling, and D. M. Ray, *Economic Geography*, Prentice Hall, Englewood Cliffs, N.J., 1987.

million, 200 years to double from 500 million to 1 billion, and then only 80 and 45 years for the next two doublings. Demographers generally feel that the world's population will double from 5.1 billion people in 1988 to approximately 10 billion people by 2010 and then stabilize at 10.5 billion people by the year 2050. Most of the population increase will occur in developing countries, although increasing life spans will result in substantial population growth in developed countries even if birthrates just replace the parents.

18.2 SOIL RESOURCES

Can the earth's soil resources provide adequate food, fiber, fuel, and wood products for the world's present and potential population and still provide areas for leisure-time activities, as well as minimize environmental pollution and disruption of biogeochemical cycles? Table 18-2 gives the total land area, including land under inland waters, of the seven continents and the major islands. Also included in Table 18-2 are the total areas of the major oceans and seas. Table 18-3 gives the total land area of the world, corrected for land under inland waters, and the 1988 distribution of land in different uses.

The Food and Agriculture Organization (FAO)[1] classifies land in to several different categories of use. *Arable land* refers to land under temporary and permanent crops. Temporary crops includes commercial row (cash) crops that are planted after each harvest (e.g., corn, soybeans, and sugarcane), temporary meadows for mowing or pasture, land under market and kitchen gardens (including cultivation under glass), and land temporarily fallow or lying idle. Permanent crops refers to land with crops that occupy the land for long periods and need not to be replanted after each harvest; this includes crops such as fruit orchards, nut trees, cocoa, coffee, figs, and grapes. It does not include trees grown for wood or timber. Permanent pasture refers to land used permanently (5 years or more) for herbaceous forage crops, either cultivated, wild prairie, or other rangeland. Forest and woodland refers to land under natural or planted stands of trees, whether productive or not, and includes land that has been cleared of trees but that will be reforested in the foreseeable future. Other land includes unused but potentially productive land, built-on areas, wasteland, parks, ornamental gardens, roads, lanes, barren land, and any other land not included in the previous categories of land use.

Examination of Table 18-3 shows that the world contains 13.08 billion hectares of land, excluding land under inland waters and in the Antarctic. The 1988 land use of the 13.08 billion hectares is: 1.47 billion hectares are classified as arable (11.27%), 3.21 billion hectares as permanent pasture (24.58%), 4.07 billion hectares as forest and woodland (31.11%), and 4.32 billion hectares as other land (33.04%). Figure 18-1 shows that of the 1.48 billion hectares of arable land, Africa contains 185.4 million hectares or 12.58% of the world's arable land. The USSR contains

---

[1]Food and Agriculture Organization, *FAO Production Yearbook*, United Nations, Rome, 1988.

TABLE 18-2   Major Land Areas and Water Areas

Earth

Mass	5.974×10^{21} metric tons	
Area	51,006,600,000 hectares	
Land	14,842,900,000 hectares	*Note:* Includes land under inland waters.
Water	36,163,700,000 hectares	

Continents and major islands	Hectares	%	Oceans and seas	Hectares	%
Asia	4,402,600,000	29.66	Pacific	16,624,100,000	45.97
Africa	3,027,100,000	20.39	Atlantic	8,655,700,000	23.93
N. America	2,425,800,000	16.34	Indian	7,342,700,000	20.30
S. America	1,782,300,000	12.01	Artic	948,500,000	2.62
Antarctic	1,320,900,000	8.90			
Europe	1,040,400,000	7.01	South China	297,460,000	0.82
Australia	768,200,000	5.18	Caribbean	251,590,000	0.70
			Mediterranean	251,000,000	0.69
Greenland	217,560,000	1.47	Bering	226,110,000	0.63
New Guinea	79,250,000	0.53	Gulf of Mexico	150,760,000	0.42
Borneo	72,550,000	0.49	Sea of Okhotsh	139,210,000	0.38
Madagascar	58,700,000	0.40	Sea of Japan	101,290,000	0.28
Baffin	50,700,000	0.34	Hudson Bay	73,010,000	0.20
Sumatra	42,730,000	0.29	East China Sea	66,460,000	0.18
Honshu	22,700,000	0.15	Andaman	56,490,000	0.16
Great Britain	21,800,000	0.15	Black Sea	50,790,000	0.14
Victoria	21,700,000	0.15	Red Sea	45,300,000	0.13
Ellesmere	19,600,000	0.13			
Celebes	17,870,000	0.12			
S. New Zealand	15,100,000	0.10			
Java	12,670,000	0.09			
N. New Zealand	11,400,000	0.08			
New Foundland	10,800,000	0.07			

Source: National Geographic Atlas of the World, 5th ed., National Geographic Society, Washington, D.C., 1981.

232.6 million hectares (15.78%); Asia, excluding USSR, contains 450.9 million hectares (30.60%); Europe contains 140.1 million hectares (9.51%); North and Central America contain 273.9 million hectares (18.58%); South America contains 142 million hectares (9.63%); and Oceania contains 48.9 million hectares or about 3.5% of the world's arable land. It should be noted that land presently used for permanent pasture, forest, or woodland or classified as other land can be used to grow crops (arable land), but this usually requires greater expenditure of resources (fertilizers, etc.) and has greater environmental costs, such as increased erosion,

TABLE 18-3 Land Use (Hectares) and Population for Various Geographical Regions of the World

Geographical area	Population	Total land	Arable land			Permanent pasture	Forest and woodland	Other land	Arable land per person	
			Temporary crops	Permanent crops	Total				Hectares	Acres
World	5,114,788,000	13,076,536,000	1,373,200,000	100,499,000	1,473,699,000	3,214,352,000	4,068,536,000	4,319,949,000	0.29	0.71
Africa	609,922,000	2,963,627,000	166,813,000	18,611,000	185,424,000	787,473,000	686,377,000	1,304,393,000	0.30	0.75
Asia (−USSR)	2,994,005,000	2,678,653,000	420,910,000	30,010,000	450,920,000	678,696,000	538,806,000	1,010,231,000	0.15	0.37
China	1,100,988,000	932,641,000	93,666,000	3,310,000	96,976,000	319,080,000	116,565,000	400,020,000	0.09	0.22
India	819,482,000	297,319,000	165,570,000	3,420,000	168,990,000	12,000,000	67,100,000	49,229,000	0.21	0.51
USSR	285,993,000	2,240,220,000	228,200,000	4,370,000	232,570,000	371,600,000	944,000,000	679,030,000	0.81	2.01
Europe (−USSR)	496,812,000	472,960,000	126,094,000	14,006,000	140,100,000	83,728,000	157,424,000	91,708,000	0.28	0.70
North and Central America	417,276,000	2,137,796,000	267,105,000	6,748,000	273,853,000	367,656,000	685,703,000	810,584,000	0.66	1.62
Canada	25,932,000	922,097,000	45,910,000	80,000	45,990,000	32,000,000	354,000,000	490,107,000	1.77	4.38
Mexico	84,884,000	190,869,000	23,150,000	1,555,000	24,705,000	74,499,000	44,080,000	47,585,000	0.29	0.72
US	264,079,000	916,660,000	187,881,000	2,034,000	189,915,000	241,467,000	265,188,000	220,090,000	0.77	1.91
South America	285,024,000	1,753,473,000	116,256,000	25,716,000	141,972,000	474,843,000	899,961,000	236,697,000	0.50	1.23
Argentina	31,536,000	273,669,000	26,000,000	9,750,000	35,750,000	142,500,000	59,500,000	35,919,000	1.13	2.80
Brazil	144,428,000	845,651,000	65,500,000	12,000,000	77,500,000	168,000,000	557,990,000	42,161,000	0.54	1.33
Oceania	25,757,000	842,827,000	47,822,000	1,038,000	48,860,000	450,356,000	156,305,000	187,306,000	1.90	4.69
Australia	16,353,000	761,793,000	46,941,000	164,000	47,105,000	436,000,000	106,000,000	172,688,000	2.88	7.11

Source: Food and Agriculture Organization of the United Nations, *FAO Production Yearbook,* 1988.

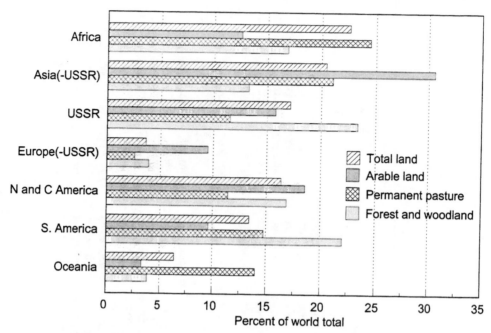

Figure 18-1 Percentage of land in different uses in various geographical regions of the world. (Food and Agriculture Organization of the United Nations, *FAO Production Yearbook*, 1988.)

destruction of watersheds, woodlands, and other important components of biogeochemical cycles.

In 1988, with a population of 5.1 billion people there was an average of about 0.29 ha (0.71 acre) of arable land per person in the world. Africa is close to the world average (Figure 18-2) with 0.30 ha per person. Africa has 11.92% of the world's population. Asia, excluding the USSR, has 58.54% of the world population with approximately 0.15 ha/person. Europe, excluding the USSR, has 9.71% of the world's population with 0.28 ha/person. The USSR has 5.59% of the world's population and has approximately 0.81 ha/person. North and Central America have 8.66% of the world's population with about 0.66 ha/person of arable land. South America is very similar to North and Central America in that it contains 5.57% of the world's population with about 0.5 ha of arable land per person. Oceania only contains 0.5% of the world's population and has 1.9 ha/person of arable land. It should be emphasized that Oceania does not represent a wealth of arable land, as it contains only 3.5% of the world's arable land.

If the world's population doubles by the year 2010, as predicted by many demographers, the amount of arable land in the world would be reduced to 0.15 ha/person. If this doubling occurs, approximately 1 billion of the 5 billion growth is expected to occur in the developed countries of the northern hemisphere. Much

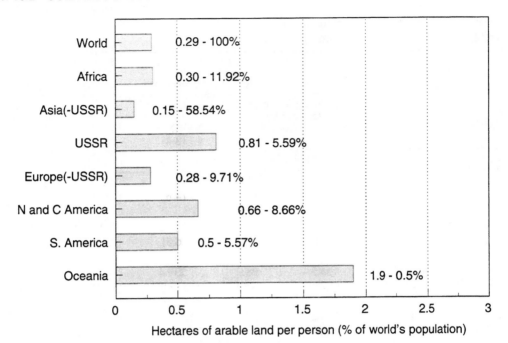

Figure 18-2   Hectares of arable land per person and percentage of world's population in various geographical regions of the world. (Food and Agriculture Organization of the United Nations, *FAO Production Yearbook*, 1988.)

of the population growth in the developed countries is expected to be due to longer life spans, as some developed areas such as Europe currently have negative birthrates, and the birthrates of many developed countries are declining. The remaining 4 billion of growth is expected to occur in the developing regions of the world. These include Africa, Asia, and South and Central America.

18.2.1 Food Production Capacity of Developing Regions

Many areas within the developing regions are having difficulty providing adequate food for present populations. These shortages may become more widespread and even catastrophic as the world population nears the 10 billion mark. The capacity of a region to feed itself is a function of many variables. These include the type of soils, the nature of the climate, the skill of the farmers, and the presence of an infrastructure to transport, store, and market the produce. The maximum capacity of a region to produce food is determined by the ability of the land to produce food, and this is limited. Limits of production are set by soil and climatic conditions and by the management applied to the land.

Table 18-4 gives the capability of climate for crop production in the developing regions of the world. For example, in Africa 9.1 million hectares (0.3% of total

TABLE 18-4 Capability of climate for crop production in developing regions [millions of hectares(percent of total land area)]

	Africa	Southwest Asia	South America	Central America	Southeast Asia
Severe Temperature Constraints					
Too hot or cold	9.1 (0.3)	113.7 (16.8)	60.8 (3.4)	0.7 (0.3)	47.7 (5.3)
No Severe Temperature Constraints					
Dry 0 growing days	846.7 (29.4)	369.7 (54.6)	81.2 (4.6)	35.6 (13.1)	39.2 (4.4)
Inadequate growing periods 1–74 days	487.9 (17.0)	72.6 (10.7)	114.6 (6.5)	62.2 (22.9)	54.6 (6.1)
Short growing periods 75–179 days	545.4 (19.0)	98.9 (14.6)	230.4 (13.0)	63.2 (23.3)	201.9 (22.5)
Long growing periods 180–365 days	969.2 (33.6)	22.5 (3.3)	1,163.5 (65.7)	109.9 (40.4)	467.8 (52.1)
Year round humid 365 days	19.8 (0.7)	—	119.7 (6.8)	—	86.4 (9.6)

Source: FAO/UNFPA, *Land Resources for Population of the Future*, Rome, 1980.

land area) are too cold or too hot to support crop production, 846.7 million hectares have no severe temperature restrictions but are too dry to support rainfed crop production, 487.9 million hectares have an inadequate growing season (1 to 74 days), 545.5 million hectares have a short growing season (75 to 179 days), 969.2 million hectares have a long growing season (180 to 365 days), while 19.8 million hectares have a year-round humid climate.

Table 18-5 gives the limitations of soils for crop production in developing regions. Soils in Table 18-5 may presently be used for crops, permanent pasture, or any other land use (Table 18-3), hence the values in Table 18-5 represent the maximum potential arable land in the various regions. In Africa, for example, 535.2 million hectares (18.6% of the total land area) has no inherent fertility limitations for crop production, 419.1 million hectares have severe fertility limitations, 98.8 million hectares have high contents of swelling clays, 64.3 million hectares of soil have salt or sodium problems, 152.9 million hectares are poorly drained soils, 376.3 million hectares are shallow to bedrock or coarse gravel, 567.5 million hectares are coarse-textured soils and are very droughty, and 459.2 million hectares are semi-desert or desert soils.

If soil and climate information are combined with information about manage-

TABLE 18-5   Limitation of Soils for Crop Production in Developing Regions [millions of hectares (percent of total land area)]

	Africa	Southwest Asia	South America	Central America	Southeast Asia
Soils with no inherent chemical or physical limitations	535.2 (18.6)	51.5 (7.6)	359.8 (20.3)	118.9 (43.8)	324.2 (36.2)
Soils with severe chemical or physical limitations	419.0 (14.6)	2.1 (0.3)	722.3 (40.8)	16.2 (6.0)	219.9 (24.5)
Heavy cracking clay soils	98.8 (3.4)	5.7 (0.8)	24.9 (1.4)	13.2 (6.5)	57.9 (3.1)
Salt-affected soils	64,3 (2.2)	53.2 (7.9)	56.5 (3.2	2.3 (0.8)	20.0 (2.2)
Poorly drained soils	152.9 (5.3)	2.6 (0.4)	179.0 (10.2)	12.7 (4.7)	75.8 (8.4)
Shallow soils	376.3 (13.1)	180.4 (26.6)	193.6 (10.9)	60.0 (22.3)	98.7 (11.0)
Coarse-textured soils	567.5 (19.7)	126.9 (18.7)	132.4 (7.5)	15.9 (5.8)	52.3 (5.8)
Semidesert and desert soils	459.2 (16.0)	230.9 (34.1)	93.9 (5.3)	31.8 (11.7)	42.7 (4.8)

Source: FAO/UNFPA, *Land Resources for Populations of the Future,* Rome, 1980.

ment, the capability of various regions and countries to produce food can be predicted. For example, FAO[2] predicts that for a low level of inputs, Zaire can support 1.3 people/ha, for a medium level of inputs, Zaire can support 5.4 people/ha and for a high level of inputs, Zaire can support 12.44 people/ha. A low level of inputs assumes only hand labor, no fertilizer or pesticide applications, no conservation measures, and cultivation of the presently grown mixtures of crops on *potentially* cultivable rainfed land. An intermediate level of inputs assumes the use of improved hand tools or draft implements, some fertilizer and pesticides, some simple soil conservation measures, and cultivation with both the present mixture of crops, as well as some use of the most calorie-protein productive crops on *potentially* cultivable rainfed lands. A high level of inputs assumes complete mechanization, full use of optimum plant genetic material, necessary farm chemicals and soil conservation measures, and cultivation of only the most calorie-protein productive crops on *potentially* cultivable rainfed lands.

18.2.2 Limitations to Food Production

At all levels of inputs there are critical levels of production that can be supported in perpetuity from any given land area. Attempts to produce food for populations

[2]FAO/UNFPA, *Land Resources for Population of the Future*, Rome, 1980.

in excess of the restrictions set by soil and climatic conditions will, in the long term, result in degradation of the land and failure of the agricultural system. *Land degradation* refers to the partial or total loss of productivity resulting from such processes as:

>   Erosion (soil loss due to wind or water)
>   Salinization (accumulation of soluble salts)
>   Alkalization (accumulation of exchangeable sodium)
>   Waterlogging
>   Depletion of plant nutrients
>   Depletion of soil organic matter
>   Deterioration of soil structure
>   Pollution
>   Desertification

Erosion is one of the most serious threats to long-term agriculture. This is especially the case when secondary soils are cultivated. These soils usually are less fertile, have lower organic matter contents, and greater slopes, all of which contribute to greater amounts of erosion than occurs with prime farmland. When these soils are cultivated with low level of inputs, erosion will be at least as much as predicted by the universal soil loss equation; when cultivated with a medium level of inputs, soil loss will be 50% of that predicted by USLE; and when cultivated with a high level of inputs, soil loss will be at acceptable soil loss levels (T values). Soil erosion is discussed in detail in Chapter 7.

Salinization, alkalization, and waterlogging occur when soils are improperly irrigated (see Chapters 5 and 10). Often the soils affected by salt or sodium accumulation or by waterlogging are not the soils being irrigated, but soils lower in the landscape. Often these were initially the most productive soils in the landscape, but have been degraded when secondary soils higher in the landscape were brought into cultivation and irrigated.

Depletion of plant nutrients and soil organic matter and the deterioration of soil structure often occurs when soils are cultivated with a low level of inputs. The soils are "mined" of nutrients by the crops and as the fertility level falls, soil organic matter is lost and soil structure deteriorates. Long-term agriculture must build and maintain a high level of soil fertility. This will encourage the accumulation of organic matter and stable structure formation. Many soils in developed countries are fertilized with two goals: first, to increase soil fertility, to support a high level of various soil processes such as soil microorganism activity, and second, to replace the nutrients removed by the crop to prevent soil "mining" and maintain soil fertility.

Soil pollution is discussed in detail in Chapter 17. Unfortunately, soil pollution is often one of the last issues addressed in developing countries. The overriding need for food often masks environmental issues. This is unfortunate, in that environ-

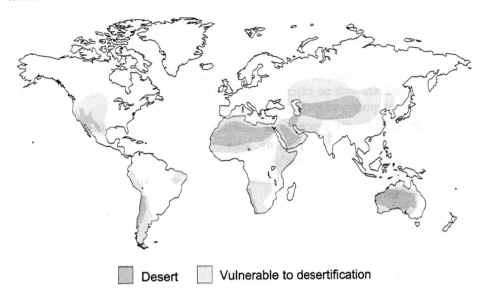

Desert Vulnerable to desertification

Figure 18-3 Areas of desert and areas subject to desertification in the world. (United Nations Conference on Desertification, 1977.)

mental issues must be addressed and included in development plans if serious and costly problems are to be avoided.

Desertification is the spread of desert into adjacent arid range and forest lands. Desertification of productive land is largely the result of overgrazing, deforestation, and low-input cultivation of crops. These activities remove the protective plant cover from the soil, lower the organic matter (humus) contents of the soils, and hence lower the soil's native fertility and increase its susceptibility to erosion. These processes are accelerated in periods of drought such as experienced in Africa's Sahel in the 1980s, but desertification is not solely the product of drought. Without the land degradation associated with overgrazing, poor farming practices, and indiscriminate harvesting of woodlands for timber and fuel, periods of drought usually do not result in desertification. Figure 18-3[3] shows the desert regions of the world and the arid adjacent regions that are susceptible to desertification.

STUDY QUESTIONS

1. If the world's population continues to double at an increasing rate, as it has up to the present time, what will the population be in 2050?

2. Discuss the factors that might result in the world's population leveling off at 10.5 billion in the year 2050 as predicted by many demographers.

[3]United Nations Conference on Desertification, 1977.

3. Compare the area of arable land in Africa (Table 18-3) with the area of soils with no inherent fertility limitations in Africa (Table 18-5). Explain the differences in these numbers.

4. What would be the consequences if all the 535.2 million hectares of soils in Africa with no inherent fertility limitations (Table 18-5) were used to produce temporary crops?

5. Discuss the concept and consequences of "mining" the soil of nutrients.

6. How does overgrazing, overcultivation, and deforestation prevent the recovery of a region's soils after the end of an extended drought?

7. Explain why low soil fertility often results in more soil erosion than a high level of soil fertility.

Glossary[1]

absorption, active Movement of ions and water into the plant root as a result of metabolic processes by the root, frequently against an electrochemical potential gradient.

absorption, passive Movement of ions and water into the plant root as a result of diffusion along an activity gradient.

acid soil Soil with a pH value < 7.0. *See also* reaction, soil.

acidic cations Hydrogen ions or cations that, on being added to water, undergo hydrolysis resulting in an acidic solution. Examples in soils are H^+, Al^{3+}, and Fe^{3+}.

acidulation The process of treating a fertilizer source with an acid. The most common process is treatment of phosphate rock with an acid (or mixture of acids) such as sulfuric, nitric, or phosphoric acid.

adsorption The process by which atoms, molecules, or ions are taken up and retained on the surfaces of solids by chemical or physical binding (e.g., the adsorption of cations by negatively charged minerals).

aerate To allow or promote exchange of soil gases with atmospheric gases.

aeration, soil The process by which air in the soil is replaced by air from the atmosphere. In a well-aerated soil, the soil air is very similar in composition

[1]This glossary was compiled and modified from the *Glossary of Soil Science Terms*, Soil Science Society of America, Madison, Wis., 1987.

to the atmosphere above the soil. Poorly aerated soils usually contain a much higher content of CO_2 and a lower content of O_2 than the atmosphere above the soil. The rate of aeration depends largely on the volume and continuity of air-filled pores within the soil.

aerobic (1) Having molecular oxygen as a part of the environment. (2) Growing only in the presence of molecular oxygen, as aerobic organisms. (3) Occurring only in the presence of molecular oxygen (said of certain chemical or biochemical processes such as aerobic decomposition).

aggregate A unit of soil structure, usually formed by natural processes in contrast with artificial processes, and generally < 10 mm in diameter.

aggregation The process whereby primary soil particles (sand, silt, clay) are bonded together, usually by natural forces and substances derived from root exudates and microbial activity.

air dry (1) The state of dryness at equilibrium with the water content in the surrounding atmosphere. The actual water content will depend upon the relative humidity and temperature of the surrounding atmosphere. (2) To allow to reach equilibrium in water content with the surrounding atmosphere.

air porosity The fraction of the bulk volume of soil that is filled with air at any given time or under a given condition, such as a specified soil-water content or soil-water matric potential.

albedo The ratio of the amount of solar radiation reflected by a body to the amount incident upon it, often expressed as a percentage, as: the albedo of the earth is 34%.

alkaline soil Any soil having a pH > 7.0. *See also* reaction, soil.

alluvial Pertaining to processes or materials associated with transportation or deposition by running water.

alluvium Sediments deposited by running water of streams and rivers. It may occur on terraces well above present streams or in the normally flooded bottomland of existing streams.

ammonia fixation Chemisorption of ammonia (NH_3), and possibly ammonium, by the organic fraction of the soil.

ammonia volatilization Mass transfer of nitrogen as ammonia gas from soil, plant, or liquid systems to the atmosphere.

ammonium fixation The process of converting exchangeable or soluble ammonium ions to those occupying positions similar to K^+ in the micas. They are counterions entrapped in the ditrigonal voids in the plane of basal oxygen atoms of some phyllosilicates as a result of contraction of the interlayer space. The fixation may occur spontaneously with some minerals in aqueous suspensions, or as a result of heating to remove interlayer water in others. Ammonium ions so adsorbed are exchangeable only after expansion of the interlayer space. *See also* potassium fixation.

amorphous material   Noncrystalline soil materials.

anaerobic   (1) The absence of molecular oxygen. (2) Growing in the absence of molecular oxygen (such as anaerobic bacteria). (3) Occurring in the absence of molecular oxygen (as a biochemical process).

anaerobic respiration   The metabolic process whereby electrons are transferred from an organic compound to an inorganic acceptor molecule other than oxygen. The most common acceptors are carbonate, sulfate, and nitrate. *See also* denitrification.

anion-exchange capacity   The sum total of exchangeable anions that a soil can adsorb. Expressed as centimoles of charge per kilogram of soil (or of other adsorbing material, such as clay).

anion exclusion   The exclusion or repulsion of anions from the vicinity of soil particle surfaces because of the negative potential associated with most soil particles.

autotroph   An organism capable of utilizing CO_2 or carbonates as a sole source of carbon and obtaining energy for carbon reduction and biosynthetic processes from radiant energy (photoautotroph or phototroph) or oxidation of inorganic substances (chemoautotroph or chemolithotroph).

available nutrients   (1) Nutrient ions or compounds in forms which plants can absorb and utilize in growth, and (2) contents of legally designated "available" nutrients in fertilizers determined by specified laboratory procedures which in most states constitute the legal basis for guarantees.

available water   The portion of water in a soil that can be absorbed by plant roots. It is the amount of water released between in situ field capacity and the permanent wiling point (usually estimated by water content at soil matric potential of -1.5 MPa).

B horizon   *See* soil horizon.

backslope   The slope component that is the steepest, straight then concave, or merely concave middle portion of an erosional slope.

badland   An area generally devoid of vegetation and broken by an intricate maze of narrow ravines, sharp crests, and pinnacles resulting from serious erosion of soft geologic materials. Most common in arid or semiarid regions.

banding   A method of fertilizer application. Refers to either placement of fertilizers close to the seed at planting or surface or subsurface applications of solids or fluids in strips before or after planting.

bar   A term used in a generic sense to include various types of submerged or exposed embankments of sand and gravel built on a sea or lake floor by waves and currents. Or a mass of sand, gravel, or alluvium deposited on the bed of a stream, sea, or lake, or at the mouth of a stream, forming an obstruction in navigation.

base saturation percentage   The extent to which the adsorption complex of a soil is saturated with alkali or alkaline earth cations expressed as a percentage of the cation-exchange capacity.

basic fertilizer   One that, after application to and reaction with soil, decreases residual acidity and increases soil pH.

basic slag   A by-product in the manufacture of steel, containing lime, phosphorus, and small amounts of other plant nutrients, such as sulfur, manganese, and iron.

bay   A swampy depression occupied by large shrubs and trees, many of which have broad, leathery, evergreen trees.

beaches   Sandy, gravelly, cobbly, or stony shores washed and reworked by waves. The areas may be partly covered by water during high tides or storms.

bedding   The mounding of soil into elevated strips performed with tillage tools such as sweep, shovel, disk, or moldboard.

bedrock   The solid rock underlying soils and the regolith in depths ranging from zero (where exposed by erosion) to several hundred centimeters.

bench terrace   *See* terrace.

bentonite   Layer silicates, largely composed of smectite minerals, produced by the alteration of volcanic ash in sites.

bioassay   A method for the quantitative determination of a substance by its effect on the growth of a suitable microorganism, plant, or animal under controlled conditions.

biodegradable   A substance capable of being decomposed by biological processes.

biological denitrification   *See* denitrification.

biological immobilization   *See* immobilization.

biomass   (1) The total mass of living microorganisms in a given volume or mass of soil. (2) The total weight of all microorganisms in a particular environment.

biome   A large, easily recognized community unit formed by the interaction of regional climates with regional biota and substrates. In a given biome the life form of the climatic climax vegetation in uniform. Thus the climax vegetation of the grassland biome is grass, although the dominant species of grass may vary in different parts of the biome.

biosequence   A sequence of related soils that differ, one from the other, primarily because of differences in kinds and numbers of plants and soil organisms as a soil-forming factor.

bisequum   One sola above another in the same profile.

blown-out   Areas from which all or almost all of the soil and soil material has been removed by wind erosion. Usually barren, shallow depressions with a flat or irregular floor consisting of a more resistant layer and/or accumulation of pebbles, or a wet zone immediately above a water table. Usually unfit for crop production.

BOD (biochemical oxygen demand)   The quantity of oxygen used in the biochemical oxidation of organic matter in a specified time, at a specified temperature, and under specified conditions. An indirect measure of the concentration of biologically degradable material present in organic wastes.

border strip irrigation   *See* irrigation methods.

bottomland   *See* floodplain.

breccia   A rock composed of coarse angular fragments cemented together.

broad-base terrace   *See* terrace.

broadcast   The application of fertilizer on the soil surface. Usually done prior to planting and normally incorporated with tillage but may be unincorporated in no-till systems.

buffer compounds, soil   The solid and solution phase components of soils that resist appreciable pH change in the soil solution (i.e., carbonates, phosphates, oxides, phyllosilicates, and some organic materials).

buffer power   The ability of ions associated with the solid phase to buffer changes in ion concentration in the solution phase.

bulk density, soil   The mass of dry soil per unit bulk volume. The bulk volume is determined before drying to constant weight at 105°C. The value is expressed in grams per cubic centimeter.

buried soil   Soil covered by an alluvial, loessal, or other depositional surface mantle of new material, usually to a depth greater than the thickness of the solum.

C horizon   *See* soil horizon.

calcareous soil   Soil containing sufficient free $CaCO_3$ and/or $MgCO_3$ to effervesce visibly when treated with cold 0.1 M HCl. These soils usually contain from as little as 10 to as much as 200 g kg^{-1} $CaCO_3$ equivalent.

calcitic lime   Limestone containing mostly $CaCO_3$.

caliche   (1) A zone near the surface, more or less cemented by secondary carbonates of Ca or Mg precipitated from the soil solution. It may occur as a soft thin soil horizon, as a hard thick bed, or as a surface layer exposed by erosion. (2) Alluvium cemented with $NaNO_3$, $NaCl$, and/or other soluble salts in the nitrate deposits of Chile and Peru.

capillary fringe   A zone in the soil just above the plane of zero gage pressure that remains saturated or almost saturated with water. The extent can be inferred from the retentivity curve and depends upon the size distribution of pores.

carbon cycle   The sequence of transformations whereby carbon dioxide is converted to organic forms by photosynthesis or chemosynthesis, recycled through the biosphere (with partial incorporation into sediments), and ultimately returned to its original state through respiration or combustion.

carbon/organic nitrogen ratio   The ratio of the mass of organic carbon to the mass of organic nitrogen in soil, organic material, plants, or the cells of microorganisms.

cat clay   Wet clay soils containing ferrous sulfide which become highly acidic when drained.

catena   (as used in the U.S.) A sequence of soils of about the same age, derived from similar parent material, and occurring under similar climatic conditions, but having different characteristics due to variation in relief and in drainage. *See also* toposequence.

cation exchange   The interchange between a cation in solution and another cation on the surface of any negatively charged material such as clay colloid or organic colloid.

cation-exchange capacity (CEC)   The sum of exchangeable cations that a soil, soil constituent, or other material can adsorb at a specific pH. It is usually expressed in centimoles of charge per kilogram of exchanger ($cmol_c$ kg^{-1}).

cemented   Indurated; having a hard brittle consistency because the particles are held together by cementing substances such as humus, $CaCO_3$, or the oxides of silicon, iron, and aluminum. The hardness and brittleness persist even when wet. *See also* consistence.

chelates   Certain organic chemicals, known as chelating agents, form ring compounds in which a metal is held between two or more atoms strongly enough to diminish the rate at which it becomes fixed by soil, thereby making it more available for plant uptake.

chemical weathering   The breakdown of rocks and minerals due to chemical activity, primarily due to the presence of water and components of the atmosphere. *See also* weathering.

chemigation   The process where fertilizers and pesticides are applied into irrigation water to fertilize crops and control pests.

chemodenitrification   Nonbiological processes leading to the production of gaseous forms of nitrogen (molecular nitrogen or an oxide of nitrogen).

chert   *See* coarse fragments.

chisel   A tillage implement with one or more shanks to which are attached chisel, spike, or narrow-shovel tools. When used to subsoil, and penetrate deeper than 0.4 m, the shanks are sturdier and spaced farther apart.

chlorite   A layer structured group of silicate minerals of the 2:1 type that has the interlayer filled with a positively charged metal-hydroxide octahedral sheet. There are both trioctahedral (e.g., $M = Fe^{2+}$, Mg^{2+}) and dioctahedral ($M = Al^{3+}$) varieties.

chroma   The relative purity, strength, or saturation of a color; directly related to the dominance of the determining wavelength of the light and inversely related to grayness; one of the three variables of color. *See also* Munsell color system, hue, *and* value, color.

chronosequence   A sequence of related soils that differ, one from the other, in certain properties primarily as a result of time as a soil-forming factor.

citrate-insoluble phosphorous   That portion of P in fertilizer remaining after water (*see* water-soluble phosphate) and ammonium citrate (*see* citrate-soluble phosphorus) extractions. Phosphorus content of fertilizers that is considered to be immediately unavailable to plants in the guaranteed analysis of the fertilizer.

citrate-soluble phosphorus   That part of the total P in fertilizer that is insoluble in water but soluble in neutral 0.33 M ammonium citrate and which, together with water-soluble P represents the readily available P content of the fertilizer.

clay   A soil separate consisting of particles < 0.002 mm in equivalent diameter.

clay films   Coatings of clay on the surfaces of soil peds and mineral grains and in soil pores. (Also called *clay skins, clay flows, illuviation cutans, argillans,* or *tonhautchen.*)

clay mineral   (1) Any crystalline inorganic substance of clay size (i.e., <2 μm equivalent spherical diameter). (2) Any phyllosilicate of clay size.

claypan   A dense, compact layer in the subsoil having a much higher clay content than the overlying material, from which it is separated by a sharply defined boundary; formed by downward movement of clay or by synthesis of clay in place during soil formation. Claypans are usually hard when dry, and plastic and sticky when wet. Also, they usually impede the movement of water and air, and the growth of plant roots.

climax The most advanced successional community of plants capable of development under, and in dynamic equilibrium with, the prevailing environment.

clod A compact coherent mass of soil varying in size, usually produced by plowing, digging, and so on, especially when these operations are performed on soils that are either too wet or too dry and usually formed by compression, or breaking off from a larger unit, as opposed to a building-up action as in aggregation.

coarse fragments Rock or mineral particles > 2.0 mm in diameter. The following names are used for coarse fragments in soils.

Shape[a]	*Material*	*Diameters < 7.6 cm*	*Diameters from 7.6 to 25 cm*	*Diameters > 25 cm*
Rounded or subrounded	All kinds of rock	Gravelly	Cobbly	Stony[b]
Irregular and angular	Chert	Cherty	Coarse cherty	Stony
	Other than chert	Angular gravelly	Angular cobbly	Stony
		Lengths up to 15.2 cm	Lengths from 15.2 to 38 cm	Lengths > 38 cm
Thin and flat	Limestone, sandstone, or schist	Channery	Flaggy	Stony
	Slate	Slaty	Flaggy	Stony
	Shale	Shaly	Flaggy	Stony

Source: Soil Survey Staff, *Soil Survey Manual*, USDA Handbook No. 18, p. 214, U.S. Government Printing Office, Washington, D.C., 1951.
[a]The adjectives describing fragments are also applied to lands and soils when they have significant amounts of such fragments.
[b]*Bouldery* is sometimes used when stones are larger than 61 cm.

coarse texture The texture exhibited by sands, loamy sands, and sandy loams except very fine sandy loam.

cobbly Containing appreciable quantities of cobblestones. (Said of soil and of land. The term *angular cobbly* is used when the fragments are less rounded.)

cohesion The force holding a solid or liquid together, owing to attraction between like molecules.

coliform A general term for a group of bacteria that inhabit the intestinal tract of humans and other animals. Their presence in water constitutes presumptive evidence for fecal contamination. Includes all aerobic and facultatively anaerobic, gram-negative rods that are nonspore forming and that ferment lactose with gas formation. *Escherichia coli* and *Enterobacter* are important members.

colluvium   A general term applied to deposits on a slope or at the foot of a slope or cliff that were moved there chiefly by gravity. Talus and cliff debris are included in such deposits.

color   *See* Munsell color system.

cometabolism   Transformation of a substrate by a microorganism without deriving energy, carbon, or nutrients from the substrate. The organism is able to transform the substrate into intermediate degradation products but fails to multiply at its expense.

community   All of the organisms that occupy a common habitat and that interact with one another.

compost   Organic residues, or a mixture of organic residues and soil, which have been mixed, piled, and moistened, with or without addition of fertilizer and lime, and generally allowed to undergo thermophilic decomposition until the original organic materials have been substantially altered or decomposed. The final product can be easily worked into the soil, or can be used in or as a potting mix. Sometimes called *artificial manure* or *synthetic manure*. In Europe, the term may refer to a potting mix for container-grown plants.

compressibility   The property of a soil pertaining to its susceptibility to decrease in bulk volume when subjected to a load.

concentrated flow   The flowing of a rather large accumulated body of water over a relatively narrow course. It often causes serious erosion and gullying.

concretion   A local concentration of a chemical compound, such as calcium carbonate or iron oxide, in the form of a grain or nodule of varying size, shape, hardness, and color.

cone penetrometer   An instrument in the form of a cylindrical rod with a cone-shaped tip designed for penetrating soil and for measuring the end-bearing component of penetration resistance. The resistance to penetration developed by the cone equals the vertical force applied to the cone divided by its horizontally projected area.

consistence   The attributes of soil material as expressed in its degree of cohesion and adhesion or in its resistance to deformation or rupture. Terms used in soil survey for describing consistence at various soil-water contents are:

wet soil   nonsticky, slightly sticky, sticky, very sticky, nonplastic, slightly plastic, plastic, and very plastic.
moist soil   loose, very friable, friable, firm, very firm, and extremely firm.
dry soil   loose, soft, slightly hard, hard, very hard, and extremely hard.

cementation   weakly cemented, strongly cemented, and indurated.

consistency   The manifestations of the forces of cohesion and adhesion acting within the soil at various water contents, as expressed by the relative ease with which a soil can be deformed or ruptured. Engineering descriptions include (1) the designation of five inplace categories (soft, firm, or medium, stiff, very stiff, and hard) as assessed by thumb and thumbnail penetrability and indentability; and (2) characterization by the Atterberg limits (i.e., liquid limit, plastic limit, and plasticity number). *See also* liquid limit, *and* plastic limit.

constant-charge surface   A mineral surface carrying a net electrical charge whose magnitude depends only on the structure and chemical composition of the mineral itself. Constant-charge surfaces usually arise from isomorphous substitution in mineral structures.

constant-potential surface   A solid surface carrying a net electrical charge which may be positive, negative, or zero, depending on the activity of one or more species of ion (called a potential-determining ion) in a solution phase contacting the surface. For minerals common in soils, the potential-determining ion is usually H^+ or OH^-, but any ion that forms a complex with the surface may be potential determining. *See also* pH-dependent charge.

coppice mound   A small mound of stabilized soil material around desert shrubs.

coprogenic material   Remains of fish excreta and similar materials that occur in some organic soils.

cradle knoll   A small knoll formed by earth that is raised and left by an uprooted tree.

creep   Slow mass movement of soil and soil material down relatively steep slopes primarily under the influence of gravity, but facilitated by saturation with water and by alternate freezing and thawing.

crest   The slope component that is commonly at the top of an erosional ridge, hill, mountain, and so on. *See also* summit.

critical nutrient concentration   The nutrient concentration in the plant, or specified plant part, below which the nutrient becomes deficient for optimum growth rate.

cross-slope bench   *See* terrace.

crushing strength   The force required to crush a mass of dry soil or, conversely, the resistance of the dry soil mass to crushing. Expressed in units of force per unit area (pressure).

crust   A soil-surface layer, ranging in thickness from a few millimeters to a few tens of millimeters, that is much more compact, hard, and brittle when dry than the material immediately beneath it.

cryogenic soil   Soil that has formed under the influence of cold soil temperatures.

crystal   A regular arrangement of atoms in space.

crystal structure   The orderly arrangement of atoms in a crystalline material.

crystalline rock   A rock consisting of various minerals that have crystallized in place from magma. *See also* igneous rock *and* sedimentary rock.

cultivation   A tillage operation used in preparing land for seeding or transplanting or later for weed control and for loosening the soil.

cyclic salt   Salt derived from the sea or salt lakes deposited on the landscape from wind or rainfall.

Darcy's law   A law describing the rate of flow of water through porous media. (Named for Henry Darcy of Paris, who formulated it in 1856 from extensive work on the flow of water through sand filter beds.)

deflation   The removal of fine soil particles from soil by wind.

deflocculate   (1) To separate the individual components of compound particles by chemical and/or physical means. (2) To cause the particles of the disperse phase of a colloidal system to become suspended in the dispersion medium. *See also* disperse.

degradation   The process whereby a compound is transformed into simpler compounds, although products more complex than the starting material may be formed.

denitrification   Reduction of nitrate or nitrite to molecular nitrogen or nitrogen oxides by microbial activity or by chemical reactions involving nitrite.

deposit   Material left in a new position by a natural transporting agent such as water, wind, ice, or gravity, or by the activity of humans.

desert crust   A hard layer, containing calcium carbonate, gypsum, or other binding material, exposed at the surface on desert regions.

desert pavement   The layer of gravel or stones left on the land surface in desert regions after the removal of the fine material by wind erosion.

desert varnish   A glossy sheen or coating on stones and gravel in arid regions.

desorption   The displacement of ions from the solid phase of the soil into solution by a displacing ion.

detoxification   Conversion of an inhibitory molecule into a nontoxic compound.

diatomaceous earth   A geologic deposit of fine, grayish siliceous material composed chiefly or wholly of the remains of diatoms. It may occur as a powder or as a porous, rigid material.

diatoms Algae having siliceous cell walls that persist as a skeleton after death. Any of the microscopic unicellular or colonial algae constituting the class Bacillariaceae. They are abundant in fresh and salt waters and their remains are widely distributed in soils.

diffuse double layer A heterogeneous system that consists of a solid surface having a net electrical charge, together with an ionic swarm under the influence of the solid and a solution phase that is in direct contact with the surface.

diffusion (nutrient) The movement of nutrients in soil that results from a concentration gradient.

dinitrogen fixation Conversion of molecular nitrogen (N_2) to ammonia and subsequently to organic combinations or to forms utilizable in biological processes.

dioctahedral An octahedral sheet or a mineral containing such a sheet that has two-thirds of the octahedral sites filled by trivalent ions such as aluminum or ferric iron.

disintegration *See* physical weathering.

disperse (1) To break up compound particles, such as aggregates, into the individual component particles. (2) To distribute or suspend fine particles, such as clay, in or throughout a dispersion medium, such as water.

dissection The partial erosional destruction of a land surface or landform by gully, arroyo, canyon, or valley cutting, leaving flattish remnants, or ridges, hills, or mountains separated by drainageways.

diversion dam A structure or barrier built to divert part or all of the water of a stream to a different course.

dolomitic lime A naturally occurring liming material composed chiefly of carbonates of Mg and Ca in approximately equimolar proportions.

drain, to (1) To provide channels, such as open ditches or drain tile, so that excess water can be removed by surface or by internal flow. (2) To lose water (from the soil) by percolation.

drainage class (natural) Has been used in humid areas: includes phrases such as well-drained, moderately well-drained, somewhat poorly, poorly, and very poorly drained.

drainage, surface Used to refer to surface movement of excess water: includes such terms as *ponded, flooded, slow*, and *rapid*.

drain tile Concrete, ceramic, or plastic pipe used to conduct water from the soil.

drip irrigation Irrigation whereby water is slowly applied to the soil surface through emitters having small orifices.

dryland farming The practice of crop production without irrigation (rainfed agriculture).

dry-mass content or ratio The ratio of the mass of any component (of a soil) to the oven-dry mass (ODwt) of the soil. *See also* oven-dry soil.

duff mull A type of forest humus that is transitional between mull and mor and usually contains four horizons: Oi(L), Oe(F), Oa(H), and A.

dune land Consists of ridges and the intervening troughs made up of sand-size particles that shift with the wind. It is devoid of vegetation.

dysic Low level of bases in soil material, specified at family level of classification.

EC$_e$ The electrolytic conductivity of an extract from saturated soil, normally expressed in units of siemens per meter at 25°C.

ecology The science that deals with the interrelations between organisms and between organisms and their environment.

ecosystem A community of organisms and the environment in which they live.

ectomycorrhiza A mycorrhizal association in which the fungal mycelia extend inward, between root cortical cells, to form a network (*Hartig net*) and outward into the surrounding soil. Usually the fungal hyphae also form a mantle on the surface of the roots.

edaphic (1) Of or pertaining to the soil. (2) Resulting from or influenced by factors inherent in the soil or other substrate, rather than by climatic factors.

edaphology The science that deals with the influence of soils on living things; particularly plants, including human use of land for plant growth.

effective precipitation That portion of the total precipitation which becomes available for plant growth.

Eh The potential that is generated between an oxidation or reduction half-reaction and the H electrode in the standard state.

eluvial horizon A soil horizon that has been formed by the process of eluviation. *See also* illuvial horizon.

eluviation The removal of soil material in suspension (or in solution) from a layer or layers of a soil. Usually, the loss of material in solution is described by the term *leaching*. *See also* illuviation *and* leaching.

endomycorrhiza A mycorrhizal association with intracellular penetration of the host root cortical cells by the fungus as well as outward extension into the surrounding soil.

eolian A term applied to materials deposited by wind such as sand dunes, sandsheets, and loess.

erodibility The state or condition of being erodible.

erodible Susceptible to erosion. (Expressed by terms such as *highly erodible, slightly erodible*, etc.)

erosion (1) The wearing away of the land surface by running water, wind, ice, or other geological agents, including such processes as gravitational creep. The following terms are used to describe different types of water erosion:

accelerated erosion Erosion much more rapid than normal, natural, geological erosion, primarily as a result of the influence of the activities of humans or, in some cases, of animals.

geological erosion The normal or natural erosion caused by geological processes acting over long geological periods. Synonymous with *natural erosion.*

gully erosion The erosion process whereby water accumulates in narrow channels and, over short periods, removes the soil from this narrow area to considerable depths, ranging from 0.5 m to as much as 25 to 30 m.

interrill erosion The removal of a fairly uniform layer of soil on a multitude of relatively small areas by splash due to raindrop impact and by film flow.

natural erosion Wearing away of the earth's surface by water, ice, or other natural agents under natural environmental conditions of climate, vegetation, and so on, undisturbed by humans. *See geological erosion.*

normal erosion The gradual erosion of land used by humans which does not greatly exceed natural erosion. *See natural erosion.*

rill erosion An erosion process in which numerous small channels of only several centimeters in depth are formed; occurs mainly on recently cultivated soils. *See* rill.

sheet erosion The removal of soil from the land surface by rainfall and surface runoff. Often interrupted to include rill and interrill erosion.

splash erosion The detachment and airborne movement of small soil particles caused by the impact of raindrops on soils.

erosion (2) Detachment and movement of soil or rock by water, wind, ice, or gravity. The following terms are used to describe different types by water and wind erosion:

saltation (1) The bouncing or jumping action of soil particles 0.1 to 0.5 mm in diameter by wind, usually at a height <15 cm above the soil surface, for relatively short distances. (2) The bouncing or jumping action of mineral particles, including gravel or stones, effected by the energy of flowing water. (3) The bouncing or jumping movement of material downslope in response to gravity.

surface creep (1) The rolling of dislodged soil particles 0.5 to 1.0 mm in diameter by wind along the soil surface. (2) The slow movement of soil and rock debris, which is usually not perceptible except through extended observation.

suspension The movement of soil particles usually <0.1 mm diameter through the air, usually at a height of > 15 cm above the soil surface, for relatively long distances.

universal soil-loss equation (USLE) An equation for predicting A, the average annual soil loss in mass per unit area per year, defined as A = RKLSPC, where R is the rainfall factor, K the soil-erodibility factor, L the length of slope, S the percent slope, P the conservation practice factor, and C the cropping and management factor.

wind-erosion equation An equation for predicting E, the average annual soil loss due to wind in mass per unit area per year, defined as E = IKCLV, where I is the soil-erodibility factor, K the soil-ridge-roughness factor, C the local climatic factor, L the field width, and V the vegetative factor.

erosion classes A grouping of erosion conditions based on the degree of erosion or on characteristic patterns. (Applied to accelerated erosion; not to normal, natural, or geological erosion.) Four erosion classes are recognized for water erosion and three for wind erosion. Specific definitions for each vary somewhat from one climatic zone, or major soil group, to another. (For details, see Soil Survey Staff, *Soil Survey Manual*, USDA Handbook No. 18, U.S. Government Printing Office, Washington, D.C., 1951.)

erosion pavement A layer of coarse fragments, such as sand or gravel, remaining on the surface of the ground after the removal of fine particles by erosion.

erosional surface A land surface shaped by the erosive action of ice, wind, water, but usually as the result of running water.

erosivity The potential ability of water, wind, gravity, and so on, to cause erosion.

essential chemical elements Elements required by plants to complete their life cycles.

euic High level of bases in soil material, specified at family level of classification.

eutrophic Having concentrations of nutrients optimal, or nearly so, for plant or animal growth. (Said of nutrient or soil solutions and bodies of water.) The term literally means "self-feeding."

evaporites Residue of salts (including gypsum and all more soluble species) precipitated by evaporation.

evapotranspiration The combined loss of water from a given area, and during a specified period of time, by evaporation from the soil surface and by transpiration from plants.

exchange capacity   The total ionic charge of the adsorption complex active in the adsorption of ions. *See also* anion-exchange capacity *and* cation-exchange capacity.

exchangeable anion   A negatively charged ion held on or near the surface of a solid particle by a positive surface charge and which may be replaced by other negatively charged ions.

exchangeable bases   *See* base saturation percentage.

exchangeable cation   A positively charged ion held on or near the surface of a solid particle by a negative surface charge of a colloid and which may be replaced by other positively charged ions in the soil solution. Often determined as the salt-extractable minus water-soluble cations in a saturation extract, and expressed in centimoles of charge per kilogram.

exchangeable cation percentage   The extent to which the adsorption complex of a soil is occupied by a particular cation. It is expressed as follows:

$$ECP = \frac{\text{exchangeable cation (cmol}_c\text{kg}^{-1}\text{ soil)}}{\text{cation exchange capacity (cmol}_c\text{kg}^{-1}\text{ soil)}} \times 100$$

exchangeable sodium fraction   The fraction of the cation exchange capacity of a soil occupied by sodium atoms.

exchangeable sodium percentage   Exchangeable sodium fraction expressed as a percentage.

exchangeable sodium ratio   The ratio of exchangeable sodium (cmol kg^{-1} soil) to all other exchangeable cations (cmol kg^{-1} soil).

exudate   Low-molecular-weight metabolites that enter the soil from plant roots.

facultative organism   An organism that is able to carry out both options of a mutually exclusive process (e.g., aerobic and anaerobic metabolism). May also be used in reference to other processes, such as photosynthesis (e.g., a facultative photosynthetic organism is one that can use either light or the oxidation of organic or inorganic compounds as a source of energy).

fan   A generic term for constructional landforms that are built of stratified alluvium with or without debris-flow deposits and that occur on the pediment slope, downslope from their source of alluvium.

ferrihydrite $Fe_5HO_8 \cdot 4H_2O$. A dark reddish-brown, poorly crystalline iron oxide mineral that forms in wet soils. Occurs in concretions and placic horizons and often can be found in ditches and pipes that drain wet soils.

fertigation   Application of plant nutrients in irrigation water to accomplish fertilization.

fertility, soil   The ability of a soil to supply the nutrients essential to plant growth.

fertilization, foliar Application of a dilute solution of fertilizer nutrients to plant foliage, usually made to supplement nutrient absorbed by plant roots.

fertilizer Any organic or inorganic material of natural or synthetic origin (other than liming materials) that is added to a soil to supply one or more elements essential to the growth of plants.

fertilizer, acid forming Fertilizer that, after application to and reaction with soil, increases residual acidity and decreases soil pH.

fertilizer analysis The percent composition of a fertilizer as determined in a laboratory and expressed as total N, available phosphoric acid (P_2O_5), and water-soluble potash (K_2O).

fertilizer, blended A mechanical mixture of different fertilizer materials.

fertilizer, bulk-blended A physical mixture of dry granular fertilizer materials to produce specific fertilizer ratios and grades. Individual granules in the bulk-blended fertilizer do not have the same ratio and content of plant food as does the mixture as a whole.

fertilizer, complete A chemical compound or a blend of compounds used for its plant nutrient content containing significant quantities of N, P, and K. It may contain other plant nutrients.

fertilizer, compound A fertilizer formulated with two or more plant nutrients.

fertilizer, controlled-release A fertilizer term used interchangeably with *delayed release, slow release, controlled availability, slow acting,* and *metered release* to designate a controlled dissolution of fertilizer at a lower rate than that of conventional water-soluble fertilizers. Controlled-release properties may result from coatings on water-soluble fertilizers or from low dissolution and/or mineralization rates of fertilizer materials in soil.

fertilizer, fluid Fertilizer wholly or partially in solution that can be handled as a liquid, including clear liquids and liquids containing solids in suspension.

fertilizer, granular Fertilizer in the form of particles sized between an upper and lower limit or between two screen sizes, usually within the range of 1 to 4 mm and often more closely sized. The desired size may be obtained by agglomerating smaller particles, crushing and screening larger particles, controlling size in crystallization processes, or prilling.

fertilizer, injected Placement of fluid fertilizer or anhydrous ammonia into the soil either through use of pressure or nonpressure systems.

fertilizer, inorganic A fertilizer material in which carbon is not an essential component of its basic chemical structure. Urea is often considered an inorganic fertilizer because of its rapid hydrolysis to form ammonium ions in soil.

fertilizer, mixed Two or more fertilizer materials blended or granulated together into individual mixes. The term includes dry mix powders, granulated, clear liquid, suspension, and slurry mixtures.

fertilizer, organic A material containing carbon and one or more plant nutrients in addition to hydrogen and/or oxygen. Urea is often considered an inorganic fertilizer because of its rapid hydrolysis to form ammonium ions in soil.

fertilizer, pop-up Fertilizer placed in small amounts in direct contact with the seed.

fertilizer ratio The relative proportions of primary nutrients in a fertilizer grade divided by the highest common denominator for that grade (e.g., grades 10–6–4 and 20–12–8 have a ratio 5–3–2).

fertilizer requirement The quantity of certain plant nutrients needed, in addition to the amount supplied by the soil, to increase plant growth to a designated level.

fertilizer, salt index The ratio of the decrease in osmotic potential of a solution containing a fertilizer compound or mixture to that produced by the same weight of $NaNO_3 \times 100$.

fertilizer, sidedressed Application made to the side of crop rows after plant emergence.

fertilizer, slow-release *See* fertilizer, controlled-release.

fertilizer, starter A fertilizer applied in relatively small amounts with or near the seed for the purpose of accelerating early growth of the crop plants.

fertilizer, suspension A fertilizer containing dissolved and undissolved plant nutrients. The undissolved plant nutrients are kept in suspension with a suspending agent, usually a swelling-type clay. The suspension must be flowable enough to be mixed, pumped, agitated, and applied to the soil in a homogeneous mixture.

fertilizer, top-dressed A surface application of fertilizer to a soil after the crop has been established.

fibric material Mostly undecomposed plant remains that contain large amounts of fibers that are well preserved and are recognizable. Bulk density is usually very low.

field capacity, in situ (field water capacity) The content of water, on a mass or volume basis, remaining in a soil 2 or 3 days after having been wetted with water and after free drainage is negligible. *See also* available water.

film water A thin layer of water, in close proximity to soil-particle surfaces, that varies in thickness from 1 or 2 to perhaps 100 or more molecular layers.

fine texture   Consisting of or containing large quantities of the fine fractions, particularly of silt and clay. (Includes clay loam, sandy clay loam, silty clay loam, sandy clay, silty clay, and clay textural classes. Sometimes subdivided into clayey texture and moderately fine texture.)

firm   A term describing the consistency of moist soil that offers distinctly noticeable resistance to crushing but can be crushed with moderate pressure between the thumb and forefinger. *See also* consistence.

first bottom   The lowest and most frequently flooded part of the floodplain of a stream.

fixation   The process by which available plant nutrients are rendered less available or unavailable in the soil.

fixed ammonium   The ammonium in soil that cannot by replaced by a neutral potassium salt solution (e.g., 1 *M* KCl). *See also* ammonium fixation.

flagstone   A relatively thin fragment, 15.2 to 38 cm long, of sandstone, limestone, slate, shale, or rarely, of schist. *See also* coarse fragments.

flexible cropping   A nonsystematic sequence of growing adapted crops with cropping and fallow decisions at each prospective date of planting based on available water in the soil plus expected growing season precipitation.

floodplain   The land bordering a stream, built up of sediments from overflow of the stream and subject to inundation when the stream is at flood stage. *See also* first bottom.

fluvioglacial   *See* glaciofluvial deposits.

foliar diagnosis   An estimation of mineral nutrient deficiencies (excesses) of plants based on examination of the chemical composition of selected plant parts, and the color and growth characteristics of the foliage of the plants.

footslope   The relatively gently sloping, slightly concave to straight slope component of an erosional slope that is at the base of the backslope component.

forest floor   All dead vegetable or organic matter, including litter and unincorporated humus, on the mineral soil surface under forest vegetation.

free iron oxides   A general term for those iron oxides that can be reduced and dissolved by a dithionite treatment. Generally includes goethite, hematite, ferrihydrite, lepidocrocite, and maghemite, but not magnetite. *See also* iron oxides.

friable   A consistency term pertaining to the ease of crumbling of soils. *See also* consistence.

friction cone penetrometer   A cone penetrometer with the additional capacity of measuring the local side-friction component of penetration resistance. The resistance to penetration developed by the friction sleeve equals the vertical force applied to the sleeve divided by its surface area.

fritted trace elements   Sintered silicates having total guaranteed analyses of micronutrients with controlled (relatively slow) release characteristics.

frost heaving   Lifting or lateral movement of soil as caused by freezing processes; in association with the formation of ice lenses or ice needles.

fulvic acid   (1) The mixture of organic substances remaining in solution upon acidification of a dilute alkali extract from soil, in which case the expression *fulvic acid fraction* is often used, and (2) the *colored* material that remains in solution after removal of humic acid by acidification.

furrow irrigation   *See* irrigation methods.

genetic   Resulting from, or produced by, soil-forming processes: for example, a genetic soil profile or a genetic horizon.

geological erosion   *See* erosion.

geomorphic surface   A portion of the landscape specifically defined in space and time that has determinable geographic boundaries and is formed by one or more agencies during a given time period.

gilgai   The microrelief of soils produced by expansion and contraction with changes in water content. Found in soils that contain large amounts of clay, which swells and shrinks considerably with wetting and drying. Usually a succession of microbasins and microknolls in nearly level areas or of microvalleys and microridges parallel to the direction of the slope. *See also* microrelief.

glacial drift   Rock debris that has been transported by glaciers and deposited, either directly from the ice or from the meltwater. The debris may or may not be heterogeneous.

glacial till   *See* till (1).

glaciers   Large masses of ice that formed, in part, on land by the compaction and recrystallization of snow. They may be moving downslope or outward in all directions because of the stress of their own weight or they may be retreating or be stagnant.

glaciofluvial deposits   Material moved by glaciers and subsequently sorted and deposited by streams flowing from the melting ice. The deposits are stratified and may occur in the form of outwash plains, deltas, kames, eskers, and kame terraces. *See also* glacial drift *and* till (1).

granule   A natural soil aggregate or ped of relatively low porosity.

gravelly Containing appreciable or significant amounts of gravel. (Used to describe soils or sands.) *See also* coarse fragments.

gravitational potential The amount of work that must be done per unit quantity of pure water in order to transport reversibly and isothermally and infinitesimal quantity of water, identical in composition to the soil water, from a pool at a specified elevation and at atmospheric pressure, to a similar pool at the elevation of the point under question.

green manure Plant material incorporated into soil while green or at maturity, for soil improvement.

groundwater That portion of the water below the surface of the ground at a pressure equal to or greater that atmospheric. *See also* water table.

guano The decomposed dried excrement of birds and bats, used for fertilizer purposes.

gullied land Areas where all diagnostic soil horizons have been removed by water, resulting in a network of V- or U-shaped channels. Some areas resemble miniature badlands. Generally, gullies are so deep that extensive reshaping is necessary for most uses.

gully A channel resulting from erosion and caused by the concentrated but intermittent flow of water usually during and immediately following heavy rains. Deep enough to interfere with, and not to be obliterated by, normal tillage operations.

gully erosion *See* erosion (1).

gypsum The common name for calcium sulphate ($CaSO_4 \cdot 2H_2O$), used to supply carbon and sulfur and to ameliorate sodic soils.

gypsum requirement The quantity of gypsum or its equivalent required to reduce the exchangeable sodium content of a given amount of soil to an acceptable level.

habitat The place where a given organism lives.

halophytic vegetation Vegetation requiring or tolerating a saline environment.

hardpan A hardened soil layer, in the lower A or in the B horizon, caused by cementation of soil particles with organic matter or with materials such as silica, sesquioxides, or calcium carbonate. The hardness does not change appreciably with changes in water content and pieces of the hard layer do not slake in water. *See also* caliche *and* claypan.

harvest index The quantity of biomass produced per unit of input of a plant nutrient.

heavy metals Those metals which have densities > 5.0 Mg(megagram) m^{-3}. In agronomic usage these include the metallic elements Cu, Fe, Mn, Mo, Co, Zn, Cd, Hg, Ni, and Pb.

hemic material An intermediate degree of decomposition; as much as two-thirds of the material cannot be recognized. Bulk density is very low.

heterotroph An organism capable of deriving carbon and energy for growth and cell synthesis by the utilization of organic compounds.

horizon *See* soil horizon.

hue One of the three variables of color. It is caused by light of certain wavelengths and changes with the wavelength. *See also* Munsell color system, chroma, *and* value, color.

humic acid The dark-colored organic material that can be extracted from soil by various reagents (e.g., dilute alkali) and that is precipitated by acidification to pH 1 to 2.

humic substances A series of relatively high-molecular-weight, brown to black substances formed by secondary synthesis reactions. The term is used in a generic sense to describe the colored material or its fractions obtained on the basis of solubility characteristics (e.g., humic acid, fulvic acid).

humification The process whereby the carbon of organic residues is transformed and converted to humic substances through biochemical and/or chemical processes.

humin The fraction of the soil organic matter that is not dissolved upon extraction of the soil with dilute alkali.

humus (1) Total of the organic compounds in soil exclusive of undecayed plant and animal tissues, their "partial decomposition" products, and the soil biomass. The term is often used synonymously with *soil organic matter*. (2) Organic layers of the forest floor. *See also* duff mull, mor, *and* mull.

hydric soils Soils that are periodically wet long enough to produce anaerobic conditions, thereby influencing the growth of plants.

hydrogen bond The chemical bond between a hydrogen atom of one molecule and two unshared electrons of another molecule.

hydrogenic soil Soil developed under the influence of water standing within the profile for considerable periods; formed mainly in cold, humid regions.

hydrologic cycle The fate of water from the time of precipitation until the water has been returned to the atmosphere by evaporation and is again ready to be precipitated.

hydrophobic soils   Soils that are water repellent, often due to dense fungal mycelial mats or hydrophobic substances vaporized and reprecipitated during fire.

hydroxy-aluminum interlayers   Polymers of general composition $[\text{Al(OH)}_{3-x}]_m^{+nx}$ which adsorbed on interlayer cation-exchange sites. Although not exchangeable by unbuffered salt solutions, they are responsible for a considerable portion of the titratable acidity (and pH-dependent charge) in soils.

hydroxy-interlayered vermiculite   A vermiculite with partially filled interlayers of hydroxyaluminum groups. It is normally dioctahedral in both the interlayer and the vermiculite layer. It is common in the coarse clay fraction of acid surface soil horizons. It has intermediate cation-exchange properties between vermiculite and chlorite. Synonyms are *chlorite-vermiculite intergrade* and *vermiculite-chlorite intergrade*.

hymatomelanic acid   The alcohol-soluble fraction of humic acid.

igneous rock   Rock formed from the cooling and solidification of magma, and that has not been changed appreciable since its formation.

illite   The mica component of a structurally mixed fine grained mica and smectite or vermiculite. Sometimes the entire mixture is referred to as illite. Illites are dioctahedral.

illuvial horizon   A soil layer or horizon in which material carried from an overlying layer has been precipitated from solution or deposited from suspension. The layer of accumulation. *See also* eluvial horizon.

illuviation   The process of deposition of soil material removed from one horizon to another in the soil; usually from an upper to a lower horizon in the soil profile. *See also* eluviation.

immobilization   The conversion of an element from the inorganic to the organic form in microbial tissues or in plant tissues.

impeded drainage   A condition that hinders the movement of water through soils under the influence of gravity.

impervious   Resistant to penetration by fluids or by roots.

infiltration   The downward entry of water into the soil through the soil surface.

inoculate   To treat (usually seeds) with microorganisms for the purpose of creating a favorable response. Most often refers to the treatment of legume seeds with *Rhizobium* to stimulate dinitrogen fixation.

interception   *See* precipitation interception.

interstratification   Mixing of silicate layers in a given stack. Interstratification may be regular or random.

intrinsic permeability   The property of a porous material that expresses the ease with which gases or liquids flow through it.

ion selectivity   The relative adsorption of an ion by the solid phase in relation to the adsorption of other ions. The relative adsorption of an ion by a root in relation to absorption of other ions.

ions   Atoms, groups of atoms, or compounds, which are electrically charged as a result of the loss of electrons (cations) or the gain of electrons (anions).

iron oxides   Group name for the oxides and hydroxides of iron. Includes the minerals goethite, hematite, lepidocrocite, ferrihydrite, maghemite, and magnetite. Sometimes referred to as *free iron oxides, sesquioxides,* or *hydrous oxides.*

iron-pan   An indurated soil horizon in which iron oxide is the principal cementing agent. *See also* plinthite.

ironstone   Hardened plinthite materials often occurring as nodules and concretions.

irrigation   The intentional application of water to the soil.

irrigation efficiency   The ratio of the water actually consumed by crops on an irrigated area to the amount of water applied to the area.

irrigation methods   The manner in which water is intentionally applied to an area. The methods and the manner of applying the water are as follows:

   border strip   The water is applied at the upper end of a strip with earth borders to confine the water to the strip.
   check basin   The water is applied rapidly to relatively level plots surrounded by levees. The basin is a small check.
   corrugation   The water is applied to small, closely spaced furrows, frequently in grain and forage crops, to confine the flow of irrigation water to one direction.
   flooding   The water is released from field ditches and allowed to flood over the land.
   furrow   The water is applied to row crops in ditches made by tillage implements.
   sprinkler   The water is sprayed over the soil surface through nozzles from a pressure system.
   subirrigation   The water is applied in open ditches or tile lines until the water table is raised sufficiently to wet the soil.
   trickle   Water applied under low pressure from small openings.
   wild flooding   The water is released at high points in the field and distribution is uncontrolled.

isomorphous substitution   The replacement of one atom by another of similar size (but not necessarily of the same valence) in a crystal structure without disrupting or seriously changing the structure.

jarosite A pale yellow potassium iron sulfate mineral.

kame An irregular ridge or hill of stratified glacial drift.

kaolin A subgroup name of aluminum silicates with a 1:1 layer structure. Kaolinite is the most common clay mineral in the subgroup. Also, a soft, usually white rock composed largely of kaolinite.

kaolinite A clay mineral of the kaolin subgroup. Its general formula is $Al_2Si_2O_5(OH)_4$. It has a 1:1 layer structure composed of shared sheets of Si–O tetrahedrons and Al–(O,OH) octahedrons.

labile A substance that is readily transformed by microorganisms or is readily available to plants.

labile pool The sum of an element in the soil solution and the amount of that element readily solubilized or exchanged when the soil is equilibrated with a salt solution.

lacustrine deposit Material deposited in lake water and later exposed either by lowering of the water level or by the elevation of the land.

landform A three-dimensional part of the land surface, formed of soil, sediment, or rock that is distinctive because of its shape, that is significant for land use or to landscape genesis, that repeats in various landscapes, and that also has a fairly consistent position relative to surrounding landforms.

landscape All the natural features such as fields, hills, forests, water, and so on, which distinguish one part of the earth's surface from another part. Usually, that portion of land or territory which the eye can comprehend in a single view, including all its natural characteristics.

landslide (1) A mass of material that has slipped downhill under the influence of gravity, frequently assisted by water (i.e., when the material is saturated). (2) Rapid movement downslope of a mass of soil, rock, or debris.

land, wild Uncultivated land; it may or may not be maintained by the owner for its productive vegetative cover or for wood, forage production, recreation, or wildlife.

lava flows Areas that are covered by lava. Most have a sharp jagged surface, crevices, and angular blocks characteristic of lava. Others are relatively smooth and have a ropey, glazed surface. Some soil material may be in a few cracks and sheltered pockets, but the flows are virtually devoid of plants except for lichens.

layer charge Magnitude of charge per formula unit balanced by ions of opposite charge external to the unit layer.

layer silicate minerals Minerals with the sheet silicate structures of the phyllosilicates.

leaching   The removal of materials in solution from the soil. *See also* eluviation.

leaching fraction   The fraction of infiltrated irrigation water that percolates below the root zone.

leaching requirement   The leaching fraction necessary to keep soil salinity, chloride, or sodium (the choice being that which is most demanding) from exceeding a tolerance level of the crop in question. It applies to steady-state or long-term average conditions.

lime, agricultural   A soil amendment containing calcium carbonate, magnesium carbonate, and other materials used to neutralize soil acidity and furnish calcium and magnesium for plant growth. Classification including calcium carbonate equivalent and limits in lime particle size is usually prescribed by law or regulation.

lime requirement   The amount of liming material required to change the soil to a specified state with respect to pH or soluble Al content.

limnic material   One of the common components of organic soils and includes both organic and inorganic materials that were either (1) deposited in water by precipitation or through the action of aquatic organisms, or (2) derived from underwater and floating aquatic plants and aquatic animals.

liquid limit   The minimum water mass content at which a small sample of soil will barely flow under a standard treatment. Synonymous with *upper plastic limit*.

lithic contact   A boundary between soil and continuous, coherent, underlying material. The underlying material must be sufficiently coherent to make hand-digging with a spade impractical. If mineral, it must have a hardness of 3 or more (Mohs scale), and gravel size chunks that can be broken out do not disperse with 15 hours' shaking in water or sodium hexametaphosphate solution.

lithosequence   A group of related soils that differ, one from the other, in certain properties primarily as a result of differences in the parent rock as a soil-forming factor.

litter   The surface layer of the forest floor consisting of freshly fallen leaves, needles, twigs, stems, bark, and fruits.

loess   Material transported and deposited by wind and consisting of predominantly silt-sized particles.

loose   A soil consistence term. *See also* consistence.

lowland vs. upland soils   Terms commonly used in connection with rice (*Oryza sativa* L.) culture to denote flooded (paddy) vs. unflooded conditions.

luxury uptake   The absorption by plants of nutrients in excess of their need for growth. Luxury concentrations during early growth may be utilized in later growth.

lysimeter (1) A device for measuring percolation and leaching losses from a column of soil under controlled conditions. (2) A device for measuring grains (irrigation, precipitation, and condensation) and losses (evapotranspiration) by a column of soil.

macronutrient A plant nutrient usually attaining a concentration of >500 mg kg^{-1} in mature plants. Usually refers to N, P, K, Ca, Mg, and S.

maintenance application Application of fertilizer materials in amounts and at intervals to maintain available soil nutrients at levels necessary to produce a desired yield.

manure The excreta of animals, with or without an admixture of bedding or litter, fresh or at various stages of further decomposition or composting. In some countries may denote any fertilizer material.

marl Soft and unconsolidated calcium carbonate, usually mixed with varying amounts of clay or other impurities.

marsh Periodically wet or continually flooded areas with the surface not deeply submerged. Covered dominantly with sedges, cattails, rushes, or other hydrophytic plants. Subclasses include freshwater and saltwater marshes. *See also* swamp.

mass flow (nutrient) The movement of solutes associated with net movement of water.

matric potential The amount of work that must be done per unit quantity of pure water in order to transport reversibly and isothermally an infinitesimal quantity of water, identical in composition to the soil water, from a pool at the elevation and the external gas pressure of the point under consideration, to the soil water.

medium-texture Intermediate between fine-textured and coarse-textured (soils). (It includes the following textural classes: very fine sandy loam, loam, silt loam, and silt.)

metamorphic rock Rock derived from preexisting rocks but that differ from them in physical, chemical, and mineralogical properties as a result of natural geological processes, principally heat and pressure, originating within the earth. The preexisting rocks may have been igneous, sedimentary, or another form of metamorphic rock.

mica A layer-structured aluminosilicate mineral group of the 2:1 type that is characterized by its high layer charge, which is usually satisfied by potassium. The major types are muscovite, biotite, and phlogopite.

microclimate (1) The climatic condition of a small area resulting from the modification of the general climatic conditions by local differences in elevation or exposure. (2) The sequence of atmospheric changes within a very small region.

micronutrient A chemical element necessary for plant growth found in small amounts, usually < 100 mg kg^{-1} in the plant. These elements consist of B, Cl, Cu, Fe, Mn, Mo, and Zn.

microrelief (1) Small scale, local differences in topography, including mounds, swales, or pits that are usually < 1 m in diameter and with elevation differences of up to 2 m. (2) Differences in topography altered by tillage operation, generally over an area of about 1 m^2 with elevation differences of a few centimeters or less.

mineralization The conversion of an element from an organic form to an inorganic state as a result of microbial activity.

mineral soil A soil consisting predominantly of, and having its properties determined predominantly by, mineral matter. Usually contains < 200 g kg^{-1} organic carbon (< 120 to 180 g kg^{-1} if saturated with water), but may contain an organic surface layer up to 30 cm thick.

minor elements *See* micronutrients.

montmorillonite An aluminum silicate (smectite) with a layer structure composed of two silica tetrahedral sheets and a shared aluminum and magnesium octahedral sheet. Montmorillonite has a permanent negative charge that attracts interlayer cations that exist in various degrees of hydration thus causing expansion and collapse of the structure (i.e., shrink-swell). Its general formula is $Si_4Al_{1.5}Mg_{0.5}O_{10}(OH)_2Ca_{0.25}$. The calcium is exchangeable.

mor A type of forest humus in which the Oa horizon is present and in which there is practically no mixing of surface organic matter with mineral soil; that is, the transition from the Oa to A horizon is abrupt. (Sometimes differentiated into thick mor, thin mor, granular mor, greasy mor, or felty mor.)

mottles Spots or blotches of different color or shades of color interspersed with the dominant color.

muck soil (1) A soil containing between 200 and 500 g kg^{-1} of organic matter. (2) An organic soil in which the plant residues have been altered beyond recognition.

mulch (1) Any material such as straw, sawdust, leaves, plastic film, loose soil, and so on, that is spread upon the surface of the soil to protect the soil and plant roots from the effects of raindrops, soil crusting, freezing, evaporation, and so on. (2) To apply mulch to the soil surface.

mulch farming A system of tillage and planting operations resulting in minimum incorporation of plant residues or other mulch into the soil surface.

mull A type of forest humus in which the Oe horizon may or may not be present and in which there is no Oa horizon. The A horizon consists of an intimate mixture of organic matter and mineral soil with gradual transition between the

A horizon and the horizon beneath. (Sometimes differentiated into firm mull, sand mull, coarse mull, medium mull, and fine mull.)

Munsell color system A color designation system that specifies the relative degrees of the three simple variables of color: hue, value, and chroma. For example: 10YR 6/4 is a color (of soil) with a hue = 10YR, value = 6, and chroma = 4. These notations can be translated into several different systems of color names as desired. *See also* chroma, hue, *and* value, color.

mycorrhiza Literally "fungus root." The association, usually symbiotic, of specific fungi with the roots of higher plants. *See also* endomycorrhiza *and* ectomycorrhiza.

neutral soil A soil in which the surface layer, at least in the tillage zone, is in the pH range 6.6 to 7.3. *See also* acid soil, alkaline soil, pH, soil *and* reaction, soil.

niche (1) The particular role that a given species plays in the ecosystem; (2) the physical space occupied by an organism.

nitrate reduction (biological) The process whereby nitrate is reduced by plants and microorganisms to ammonium for cell synthesis (nitrate assimilation, assimilatory nitrate reduction) or to various lower oxidation states (N_2, N_2O, NO, NO_2^-) by bacteria using nitrate as the terminal electron acceptor in anaerobic respiration (respiratory nitrate reduction, dissimilatory nitrate reduction, denitrification).

nitrification Biological oxidation of ammonium to nitrite and nitrate, or a biologically induced increase in the oxidation state of nitrogen.

nitrogen cycle The sequence of biochemical changes undergone by nitrogen wherein it is used by a living organism, transformed upon the death and decomposition of the organism, and converted ultimately to its original state of oxidation.

nitrogen fixation *See* dinitrogen fixation.

nodules (1) Specialized tissue enlargements, or swellings, on the roots or leaves of plants, such as are caused by nitrogen-fixing microorganisms, (2) glaebules with an undifferentiated fabric; in the context undifferentiated fabric includes recognizable rock and soil fabrics.

nutrient antagonism The depressing effect caused by one or more plant nutrients on the uptake and availability of another.

nutrient balance An as yet undefined ratio among concentrations of nutrients necessary for maximum growth rate and yield. An imbalance results when one or more nutrients are present in either deficient or excess supply.

O horizon *See* soil horizon.

organic soil A soil that contains a high percentage (> 200 g kg^{-1}, or > 120-180 g kg^{-1} if saturated with water) of organic carbon throughout the solum.

orthophosphate A salt of orthophosphoric acid, such as NH_4HPO_4, $CaHPO_4$, or K_2HPO_4.

osmotic potential The amount of work that must be done per unit quantity of pure water in order to transport reversibly and isothermally an infinitesimal quantity of water from a pool of pure water, at a specified elevation and at atmospheric pressure, to a pool of water identical in composition to the equilibrium soil solution (at the point under consideration), but in all other respects being identical to the reference pool.

osmotic pressure The pressure to which a pool of water, identical in composition to the soil water, must be subjected in order to be in equilibrium, through a semipermeable membrane, with a pool of pure water (*semipermeable* means permeable only to water). May be identified with the osmotic potential defined above.

outwash Stratified glacial drift deposited by meltwater streams beyond active glacier ice.

oven-dry soil Soil that has been dried at 105°C until it reaches constant weight (ODwt).

overburden (1) Material recently deposited by a transportation mode, that occurs immediately superjacent to the surface horizon of a contemporaneous soil. (2) A term used to designated disturbed or undisturbed material of any nature, consolidated or unconsolidated, that overlies a deposit of useful materials, ores, lignites, or coals, especially those deposits mined from the surface by open cuts.

paleosol, buried A soil formed on a landscape during the geological past and subsequently buried by sedimentation.

paleosol, exhumed A formerly buried paleosol that has been exposed on the landscape by the erosive stripping of an overlying mantle of sediment.

pan, genetic A natural subsurface soil layer of low or very low permeability, with a high concentration of small particles, and differing in certain physical and chemical properties from the soil immediately above or below the pan. *See also* claypan *and* hardpan.

pan, pressure or induced A subsurface horizon or soil layer having a higher bulk density and a lower total porosity than the soil directly above or below it, as a result of pressure that has been applied by normal tillage operations or by other artificial means. Frequently referred to as *plowpan, plowsole*, or *traffic pan*.

pans   Horizons or layers, in soils, that are strongly compacted, indurated, or very high in clay content. *See also* caliche, claypan, *and* hardpan.

paralithic contact   Similar to a lithic contact except that it is softer, can be dug with difficulty with a spade, if a single mineral has a hardness < 3 (Mohs scale), and gravel size chunks that can be broken out will partially disperse within 15 hours' shaking in water or sodium hexametaphosphate solution.

parasitism   An association whereby one organism (parasite) lives in or on another organism (host) and benefits at the expense of the host.

parent material   The unconsolidated and more or less chemically weathered mineral or organic matter from which the solum of soils is developed by pedogenic processes.

particle density   The density of the soil particles, the dry mass of the particles being divided by the solid (not bulk) volume of the particles, in contrast with bulk density. Units are g cm^{-3}.

particle size   The effective diameter of a particle measured by sedimentation, sieving, or micrometric methods.

particle-size analysis   Determination of the various amounts of the different separates in a soil sample, usually by sedimentation, sieving, micrometry, or combinations of these methods.

particle-size distribution   The fractions of the various soil separated in a soil sample, often expressed as mass percentages.

peat   Unconsolidated soil material consisting largely of undecomposed, or only slightly decomposed, organic matter accumulated under conditions of excessive moisture.

peat soil   An organic soil containing > 500 g kg^{-1} organic matter. Used in the United States to refer to the stage of decomposition of the organic matter, "peat" referring to the slightly decomposed materials. *See also* peat *and* muck soil.

ped   A unit of soil structure such as an aggregate, crumb, prism, block, or granule, formed by natural processes (in contrast with a clod, which is formed artificially).

pediment   The footslope component of an erosional slope; geomorphologically an erosional surface that lies at the foot of a receded slope, with underlying rocks or sediments that also underlie the upland, which is barren of, or mantled with sediment, and which normally has a concave upward profile.

pediplain   A geomorphic term for an outwash plain landform.

pedon   A three-dimensional body of soil with lateral dimensions large enough to permit the study of horizon shapes and relations. Its area ranges from 1 to 10 m^2. Where horizons are intermittent or cyclic, and recur at linear intervals of 2 to 7 m, the pedon includes one-half of the cycle. Where the cycle is < 2 m, or all horizons are continuous and of uniform thickness, the pedon has an area of approximately 1 m^2. If the horizons are cyclic but recur at intervals > 7 m, the pedon reverts to the 1-m^2 size, and more than one soil will usually be represented in each cycle.

peneplain   A once high, rugged area that has been reduced by erosion to a low, gently rolling surface resembling a plain.

penetrability   The ease with which a probe can be pushed into the soil. (May be expressed in units of distance, speed, force, or work, depending on the type of penetrometer used.)

percolation, soil water   The downward movement of water through soil. Especially, the downward flow of water in saturated or nearly saturated soil at hydraulic gradients of the order of 1.0 or less.

permafrost   (1) Permanently frozen material underlying the solum. (2) A perennially frozen soil horizon.

permanent wilting point   The largest water content of a soil at which indicator plants, growing in that soil, wilt and fail to recover when placed in a humid chamber. Often estimated by the water content at -1.5-bar soil matric potential.

permeability, soil   (1) The ease with which gases, liquids, or plant roots penetrate or pass through a bulk mass of soil or a layer of soil. Since different soil horizons vary in permeability, the particular horizon under question should be designated. (2) The property of a porous medium itself that expresses the ease with which gases, liquids, or other substances can flow through it, and is the same as intrinsic permeability k.

petroferric contact   A boundary between soil and a continuous layer of indurated soil in which iron is an important cement. Contains little or no organic matter.

pH-dependent charge   The portion of the cation- or anion-exchange capacity that varies with pH.

pH, soil   The negative logarithm of the hydrogen ion activity of a soil. The degree of acidity (or alkalinity) of a soil as determined by means of a glass, quinhydrone, or other suitable electrode or indicator at a specified moisture content or soil-water ratio, and expressed in terms of the pH scale.

phosphate   In fertilizer trade terminology, phosphate is used to express the sum of the water-soluble and the citrate-soluble phosphoric acid (P_2O_5); also referred to as the available phosphoric acid (P_2O_5).

phosphate rock   A porous, lower-density, microcrystalline, calcium fluorophosphate of sedimentary of igneous origin. It is usually concentrated and solubilized to be used directly or concentrated in manufacture of commercial phosphate fertilizers.

phosphoric acid   In commercial fertilizer manufacturing, it is used to designate orthophosphorphoric acid, H_3PO_4. In fertilizer labeling, it is the common term used to represent the phosphate content in terms of available P, expressed as percent P_2O_5.

P_2O_5   Phosphorous pentoxide; designation on the fertilizer label that denotes the percentage of available phosphate.

physical weathering   The breakdown of rock and mineral particles into smaller particles by physical forces such as frost action. *See also* weathering.

phytoliths   Inorganic bodies derived from replacement of plant cells; they are usually opaline.

plain   A flat, undulating, or even rolling area, larger or smaller, that includes few prominent hills or valleys, that usually is at low elevation in reference to surrounding areas, and that may have considerable overall slope and local relief.

plant analysis   Analytical procedures to determine the nutrient content of plants or plant parts.

plant nutrient   An element that is absorbed by plants and is necessary for completion of the life cycle.

plastic limit   The minimum water mass content at which a small sample of soil material can be deformed without rupture. Synonymous with *lower plastic limit*.

plastic soil   A soil capable of being molded or deformed continuously and permanently, by relatively moderate pressure, into various shapes. *See also* consistence.

platy   Consisting of soil aggregates that are developed predominantly along the horizontal axes; laminated; flaky.

playa   An ephemerally flooded, vegetatively barren area on a basin floor that is veneered with fine-textured sediment and acts as a temporary or as the final sink for drainage water.

plinthite   A nonindurated mixture of iron and aluminum oxides, clay, quartz, and other diluents that commonly occurs as red soil mottles usually arranged in platy, polygonal, or reticulate patterns. Plinthite changes irreversibly to ironstone hardpans or irregular aggregates on exposure to repeated wetting and drying.

plow pan   *See* pan, pressure *or* induced.

plowsole   *See* pan, pressure *or* induced.

pocosin   A swamp, usually containing organic soil, and partly or completely enclosed by a sandy rim (e.g., the Carolina Bays of the southeastern United States).

pore-size distribution   The volume fractions of the various size ranges of pores in a soil, expressed as percentages of the soil bulk volume (soil particles plus pores).

pore space   The portion of soil bulk volume occupied by soil pores.

pore volume   *See* pore space.

porosity   The volume of pores in a soil sample (nonsolid volume) divided by the bulk volume of the sample.

potash   Term used to refer to potassium or potassium fertilizers and usually designated as K_2O.

potassium fixation   The process of converting exchangeable or water-soluble potassium to that occupying the position of K^+ in the micas. They are counterions entrapped in the ditrigonal voids in the plane of basal oxygen atoms of some phyllosilicates as a result of contraction of the interlayer space. The fixation may occur spontaneously with some minerals in aqueous suspensions or as a result of heating to remove interlayer water in others. Fixed K^+ ions are exchangeable only after expansion of the interlayer space. *See also* ammonium fixation.

precipitation interception   The stopping, interrupting, and temporary holding of precipitation in any form by a vegetative canopy or vegetation residue.

predation   A relationship between two organisms whereby one organism (predator) engulfs and digests the second organism (prey).

primary mineral   A mineral that has not been altered chemically since deposition and crystallization from molten lava. *See also* secondary mineral.

primary nutrients   Refers to N, P, and K in fertilizers. *See also* macronutrients.

prismatic soil structure   A soil structure type with prismlike aggregates that have a vertical axis much longer that the horizontal axes.

profile, soil   A vertical section of the soil through all its horizons and extending into the C horizon.

pyroclastics   A general term applied to detrital volcanic material that have been explosively or aerially ejected from a volcanic vent.

pyrophosphate A class of phosphorus compounds produced by the reaction of either anhydrous ammonia or potassium hydroxide with pyrophosphoric acid ($H_4P_2O_7$). Pyrophosphoric acid is a condensation product of two molecules of orthophosphoric acid (H_3PO_4). The main polyphosphate species in polyphosphate fertilizers.

R layer *See* soil horizon.

rainfall erosion index A measure of the erosive potential of a specific rainfall event. In the universal soil-loss equation it is defined as the product of two rainstorm characteristics: total kinetic energy of the storm × its maximum 30-min intensity.

rainfall interception *See* precipitation interception.

reaction, soil The degree of acidity or alkalinity of a soil, usually expressed as a pH value. Descriptive terms commonly associated with certain ranges in pH are: *extremely acid,* < 4.5; *very strongly acid,* 4.5–5.0; *strongly acid,* 5.1–5.5; *moderately acid,* 5.6–6.0; *slightly acid,* 6.1–6.5; *neutral,* 6.6–7.3; *slightly alkaline,* 7.4–7.8; *moderately alkaline,* 7.9–8.4; *strongly alkaline,* 8.5–9.0; and *very strongly alkaline,* >9.1.

regolith The unconsolidated mantle of weathered rock and soil material on the earth's surface; loose earth materials above solid rock. (Approximately equivalent to the term *soil* as used by many engineers.)

remote sensing In the broadest sense, the measurement or acquisition of information of some property of an object or phenomenon, by a recording device that is not in physical or intimate contact with the object or phenomenon under study [e.g., the utilization at a distance (as from aircraft, spacecraft, or ship) of any device and its attendant display for gathering information pertinent to the environment, such as measurements of force fields, electromagnetic radiation, or acoustic energy]. The technique employs such devices as the camera, lasers, radio-frequency receivers, radar systems, sonar, seismographs, gravimeters, magnetometers, and scintillation counters.

residual fertility The available nutrient content of a soil carried over to subsequent crops.

residual material Unconsolidated and partly weathered mineral materials accumulated by disintegration of consolidated rock in place.

retardation factor The capability of a soil for slowing or retarding the movement of a solute, and is defined for solutes subject to equilibrium reactions with the soil matrix.

reticulate mottling A network of streaks of different color; most commonly found in the deeper profiles of Lateritic soils containing plinthite.

rhizobia Bacteria capable of living symbiotically in roots of leguminous plants, from which they receive energy and often utilize molecular nitrogen. Collective common name for the genus *Rhizobium*.

rhizocylinder The plant root plus the adjacent soil that is influenced by the root. *See also* rhizosphere.

rhizosphere The zone of soil immediately adjacent to plant roots in which the kinds, numbers, or activities of microorganisms differ from that of the bulk soil.

rill A small, intermittent water course with steep sides; usually only several centimeters deep, and, hence, no obstacle to tillage operations.

rill erosion *See* erosion (2).

river wash Barren alluvial areas, usually coarse-textured, exposed along streams at low water and subject to shifting during normal high water. A miscellaneous area.

rock outcrop Consists of exposures of base bedrock, other than lava flows and rock-lined pits. Most rock outcrops are of hard rock, but some are soft rock.

runoff That portion of the precipitation on an area which is discharged from the area through stream channels. That which is lost without entering the soil is called *surface runoff* and that which enters the soil before reaching the stream is called *groundwater runoff* or *seepage flow* from groundwater. (In soil science *runoff* usually refers to the water lost by surface flow; in geology and hydraulics *runoff* usually includes both surface and subsurface flow.)

saline soil A nonsodic soil containing sufficient soluble salt to adversely affect the growth of most crop plants. The lower limit of saturation extract electrical conductivity of such soils is conventionally set at 0.4 siemens per meter. Actually, sensitive plants are affected at half this salinity and highly tolerant ones at about twice this salinity.

salination The process whereby soluble salts accumulate in the soil.

saline seep Intermittent or continuous saline water discharge at or near the soil surface under dryland conditions which reduces or eliminates crop growth. It is differentiated from other saline soil conditions by recent and local origin, shallow water table, saturated root zone, and sensitivity to cropping systems and precipitation.

saline-sodic soil A soil containing both sufficient soluble salt and exchangeable Na^+ to affect crop production adversely under most soil and crop conditions. The electrical conductivity and sodium adsorption ratio of the saturation extract are at least 0.4 siemens per meter and 13, respectively.

salt-affected soil   Soil that has been adversely modified for the growth of most crop plants by the presence of soluble salts, exchangeable sodium, or both. *See also* saline-sodic soil *and* sodic soil.

salt balance   The quantity of soluble salt removed from an irrigated area in the drainage water minus that delivered in the irrigation water.

salt flats   Undrained areas in closed basins in arid regions. In these areas, 10 to 75 cm of crystalline salt overlie stratified, very strongly saline sediments. The water table may be within 20 cm of the surface at some period during the year.

salt tolerance   The ability of plants to resist the adverse, nonspecific effects of excessive soluble salts in the root medium. Salt tolerance is distinguished from tolerances to specific ions (e.g., Na^+ and Cl^-) or solutes (e.g., H_3BO_3) and from nutritional imbalances.

sand   (1) A soil particle between 0.05 and 2.0 mm in diameter. (2) Any one of five soil separates: very coarse sand, coarse sand, medium sand, fine sand, and very fine sand.

sapric material   One of the components of organic soils with highly decomposed plant remains. Material is not recognizable and bulk density is low.

saturate   (1) To fill all the voids between soil particles with a liquid. (2) To form the most concentrated solution possible under a given set of physical conditions in the presence of an excess of the solute. (3) To fill to capacity, as the adsorption complex with a cation species (e.g., H^+-saturated, etc.).

saturated soil paste   A particular mixture of soil and water. At saturation, the soil paste glistens as it reflects light, flows slightly when the container is tipped, and the paste slides freely and cleanly from a spatula.

saturation content   The water content of a saturated soil paste, expressed as a fraction of the dry soil mass.

saturation extract   The solution extracted from a soil at its saturation water content.

second bottom   The first terrace above the normal floodplain of a stream.

secondary mineral   A mineral resulting from the decomposition of a primary mineral or from the reprecipitaion of the products of decomposition of a primary mineral. *See also* primary mineral.

secondary nutrients   Refers to Ca, Mg, and S in fertilizers. *See also* macronutrients.

sediment   Transported and deposited particles derived from rocks, soil, or biological material.

sedimentary rock   A rock formed from materials deposited from suspension or precipitated from solution and usually being more or less consolidated. The principal sedimentary rocks are sandstones, shales, limestones, and conglomerates.

sedimentation   The process of sediment deposition.

self-mulching soil   A soil in which the surface layer becomes so well aggregated that it does not crust and seal under the impact of rain but instead serves as a surface mulch upon drying.

sesquioxides   A general term for oxides and hydroxides of iron and aluminum.

shaly   (1) Containing a large amount of shale fragments, as a soil. (2) A soil phase, as for example, the shale phase. *See also* coarse fragments.

shear   Force, as with a tillage tool, acting at a right angle to the direction of movement of the tillage implement.

shear strength   The maximum resistance of a soil to shearing stresses.

sheet erosion   *See* erosion (2).

shoulder   The convex slope component at the top of an erosional sideslope.

silt   A soil separate consisting of particles between 0.05 and 0.002 mm in equivalent diameter.

silting   The deposition of waterborne sediments in stream channels, lakes, reservoirs, or on floodplains, usually resulting from a decrease in the velocity of the water.

sinkhole (karst)   Term refers to geomorphic component in a karst topography, and so on.

site   (1) In ecology, an area described or defined by its biotic, climatic, and soil conditions as related to its capacity to produce vegetation. (2) An area sufficiently uniform in biotic, climatic, and soil conditions to produce a particular climax vegetation.

site index   (1) A quantitative evaluation of the productivity of a soil for forest growth under the existing or specified environment. (2) The height in meters of the dominant forest vegetation taken at or calculated to an index age, commonly 25, 50, or 100 years.

slaty   Containing a considerable quantity of slate fragments. (Used to modify soil texture class names, such as "slaty clay loam," etc.) *See also* coarse fragments.

slickensides   Polished and grooved surfaces produced by one mass sliding past another. Slickensides are common in Vertisols.

slickspots   Areas of sodic soils.

slow release A fertilizer term used interchangeably with *delayed release, controlled release, controlled availability, slow acting,* and *metered release* to designate a rate of dissolution (usually in water) much less than is obtained for completely water-soluble compounds. Slow release may involve either compounds that dissolve slowly or soluble compounds coated with substances relatively impermeable to water.

smectite A group of 2:1 layer structured silicates with a high cation-exchange capacity and variable interlayer spacing. Formerly called the montmorillonite group. The group includes di- and trioctahedral members.

sodic soil A nonsaline soil containing sufficient exchangeable sodium to adversely affect crop production and soil structure under most conditions of soil and plant type. The lower limit of the saturation extract SAR of such soils is conventionally set at 13.

sodium adsorption ratio (SAR) A relation between soluble sodium and soluble divalent cations which can be used to predict the exchangeable sodium percentage of soil equilibrated with a given solution. It is defined as follows:

$$SAR = \frac{(sodium)}{\sqrt{\frac{(calcium) + (magnesium)}{2}}}$$

where concentrations, denoted by parentheses, are expressed in moles per liter.

soil (1) The unconsolidated mineral or material on the immediate surface of the earth that serves as a natural medium for the growth of land plants. (2) The unconsolidated mineral or organic matter on the surface of the earth that has been subjected to and influenced by genetic and environmental factors of: parent material, climate (including water and temperature effects), macro- and microorganisms, and topography, all acting over a period of time and producing a product—soil—that differs from the material from which it is derived in many physical, chemical, biological, and morphological properties and characteristics.

soil air The soil atmosphere; the gaseous phase of the soil, being that volume not occupied by solid or liquid.

soil amendment Any material such as lime, gypsum, sawdust, compost, animal manures, crop residue, or synthetic soil conditioners, that is worked into the soil or applied on the surface to enhance plant growth. Amendments may contain important fertilizer elements, but the term commonly refers to added materials other than those used primarily as fertilizers.

soil association A kind of map unit used in soil surveys comprised of delineations, each of which shows the size, shape, and location of a landscape unit composed of two or more kinds of component soils or component soils and miscellaneous areas, plus allowable inclusions in either case. The bodies of component soils and miscellaneous areas are large enough to be delineated individually at the scale of 1:24,000. Several to numerous bodies of each kind of component soil or miscellaneous area are apt to occur in each delineation and they occur in a fairly repetitive and describable pattern. The proportions of the components may vary appreciable from one delineation to another and all of the components need not occur in every delineation though they will be present in most delineations.

soil auger A tool for boring into the soil and withdrawing a small sample for field or laboratory observation. Soil augers may be classified into several types as follows: (1) those with worm-type bits, uninclosed; (2) those with worm-type bits inclosed in a hollow cylinder; and (3) those with a hollow cylinder with a cutting edge at the lower end.

soil compaction Increasing the soil bulk density, and concomitantly decreasing the soil porosity, by the application of mechanical forces to the soil.

soil conditioner A material that will measurably improve the physical characteristics of the soil as a plant growth medium. Examples include sawdust, peat, compost, and various inert materials.

soil conservation (1) Protection of the soil against physical loss by erosion or against chemical deterioration; that is, excessive loss of fertility by either natural or artificial means. (2) A combination of all management and land use methods that safeguard the soil against depletion or deterioration by natural or by human-induced factors. (3) The branch of soil science that deals with soil conservation (1) and (2).

soil creep *See* creep.

soil extract The solution separated from a soil suspension or from a soil by filtration, centrifugation, suction, or pressure. (May or may not be heated prior to separation.)

soil formation factors The variable, usually interrelated natural agencies that are active in and responsible for the formation of soil. The factors are usually grouped into five major categories: parent rock, climate, organisms, topography, and time.

soil genesis (1) The mode of origin of the soil with special reference to the processes or soil-forming factors responsible for development of the solum, or true soil, from unconsolidated parent material. (2) The branch of soil science that deals with soil genesis (1).

soil horizon A layer of soil or soil material approximately parallel to the land surface and differing from adjacent genetically related layers in physical, chemical, and biological properties or characteristics such as color, structure, texture, consistency, kinds and number of organisms present, degree of acidity or alkalinity, and so on.

soil loss tolerance (1) The maximum average annual soil loss that will allow continuous cropping and maintain soil productivity without requiring additional management inputs. (2) The maximum soil erosion loss that is offset by the theoretical maximum rate of soil development which will maintain an equilibrium between soil losses and gains.

soil management (1) The sum total of all tillage and planting operations, cropping practices, fertilizer, lime, irrigation, herbicide and insecticide application, and other treatments conducted on or applied to a soil for the production of plants. (2) The branch of soil science that deals with the items listed in (1).

soil mineral (1) Any mineral that occurs as a part of or in the soil. (2) A natural inorganic compound with definite physical, chemical, and crystalline properties (within the limits of isomorphism) that occurs in the soil. *See also* clay mineral.

soil monolith A vertical section of a soil profile removed from the soil and mounted for display or study.

soil morphology (1) The physical constitution, particularly the structural properties, of a soil profile as exhibited by the kinds, thickness, and arrangement of the horizons in the profile, and by the texture, structure, consistency, and porosity of each horizon. (2) The structural characteristics of the soil or any of its parts.

soil organic matter The organic fraction of the soil exclusive of undecayed plant and animal residues. Often used synonymously with *humus*.

soil organic residue Animal and vegetative materials added to the soil and which are recognizable as to their origin.

soil pores That part of the bulk volume of soil not occupied by soil particles. Soil pores have also been referred to as *interstices* or *voids*.

soil porosity *See* porosity.

soil productivity The capacity of a soil to produce a certain yield of crops or other plants with optimum management.

soil salinity The amount of soluble salts in a soil. The conventional measure of soil salinity is the electrical conductivity of a saturation extract.

soil separates Mineral particles, < 2.0 mm in equivalent diameter, ranging between specified size limits. The names and size limits of separates recognized in the United States are: *very coarse sand,* 2.0 to 1.0 mm; *coarse sand,* 1.0 to 0.5 mm; *medium sand,* 0.5 to 0.25 mm; *fine sand,* 0.25 to 0.10 mm; *very fine sand,* 0.10 to 0.05 mm; *silt,* 0.05 to 0.002 mm; and *clay,* <0.002 mm. The separates recognized by the International Society of Soil Science are (I) *coarse sand,* 2.0 to 0.2 mm; (II) *fine sand,* 0.2 to 0.02 mm; (III) *silt,* 0.02 to 0.002 mm; and (IV) *clay,* < 0.002 mm.

soil solution   The aqueous liquid phase of the soil and its solutes.

soil structure   The combination or arrangement of primary soil particles into secondary particles, units, or peds. These secondary units may be, but usually are not, arranged in the profile in such a manner as to give a distinctive characteristic pattern. The secondary units are characterized and classified on the basis of size, shape, and degree of distinctness into classes, types, and grades, respectively. *See also* soil structure grades.

soil structure grades   A grouping or classification of soil structure on the basis of inter- and intra-aggregate adhesion, cohesion, or stability within the profile. Four grades of structure designated from 0 to 3 are recognized as follows:

   0. Structureless   No observable aggregation or no definite and orderly arrangement of natural lines of weakness. *Massive,* if coherent; *single-grain,* if noncoherent.

   1. Weak   Poorly formed indistinct peds, barely observable in place.

   2. Moderate   Well-formed distinct peds, moderately durable and evident, but not distinct in undisturbed soil.

   3. Strong   Durable peds that are quite evident in undisturbed soil, adhere weakly to one another, withstand displacement, and become separated when the soil is disturbed.

soil survey   The systematic examination, description, classification, and mapping of soils in an area. Soil surveys are classified according to the kind and intensity of field examination. Also, the program of the National Cooperative Soil Survey that includes developing and implementing standards for describing, classifying, mapping, writing, and publishing information about soils of a specific area.

soil test   A chemical, physical, or biological procedure that estimates a property of the soil pertinent to the suitability of the soil to support plant growth. (Sometimes used as an adjective to define fractions of soil components, e.g., *soil test phosphorus.*)

soil texture   The relative proportions of the various soil separates in a soil as described by the classes of soil texture. The textural classes may be modified by the addition of suitable adjectives when coarse fragments are present in substantial amounts; for example, *stony silt loam* or *silt loam, stony phase.* (For other modifications *see* coarse fragments.) The sand, loamy sand, and sandy loam are further subdivided on the basis of the proportions of the various sand separates present.

soil water characteristic or characteristic curve   The relationship between the soil-water content (by mass or volume) and the soil-water matric potential.

solum (plural: **sola**) The upper and most weathered part of the soil profile; the A, E, and B horizons.

spatial variability The variation in soil properties (1) laterally across the landscape, at a given depth, or with a given horizon, or (2) vertically downward through the soil.

splash erosion *See* erosion (2).

spoil bank Rock waste, banks, and dump depositions resulting from the excavation of ditches and strip mines.

static penetrometer A penetrometer that is pushed into the soil at a constant and slow rate of penetration.

stemflow That portion of precipitation or irrigation water that is intercepted by plants and then flows down the stem to the ground.

stones Rock fragments > 25 cm in diameter if rounded, and > 38 cm along the greater axis if flat. *See also* coarse fragments.

stony Containing sufficient stones to interfere with or to prevent tillage. To be classified as stony, > 0.01% of the surface of the soil must be covered with stones. Used to modify soil class, such as *stony clay loam* or *clay loam, stony phase*. *See also* coarse fragments.

stratified Arranged in or composed of strata or layers.

strip cropping The practice of growing crops that require different types of tillage, such as row and sod, in alternate strips along contours or across the prevailing direction of wind.

strip planting A method of simultaneous tillage and planting in isolated bands of varying width separated by bands of erect residues essentially undisturbed by tillage.

structural charge The charge (usually negative) on a mineral caused by isomorphous substitution within the mineral layer. (Expressed as moles of charge per kilogram of clay.)

stubble mulch The stubble of crops or crop residues left essentially in place on the land as a surface cover before and during the preparation of the seedbed and at least partly during the growing of a succeeding crop.

subsoiling Any treatment to loosen soil below the tillage zone without inversion and with a minimum of mixing with the tilled zone.

substrate (1) That which is laid or spread under; an underlying layer, such as the subsoil. (2) The substance, base, or nutrient on which an organism grows. (3) Compounds or substances that are acted upon by enzymes or catalysts and changed to other compounds in the chemical reaction.

substratum   Any layer lying beneath the soil solum, either conforming or unconforming.

subsurface tillage   Tillage beneath the surface, cutting plant roots and loosening the soil, without inverting it and without appreciably incorporation surface residues.

sulfidic material   Waterlogged material or organic material that contains 7.5 g kg^{-1} or more of sulfur.

sulfur cycle   The sequence of transformations undergone by sulfur wherein it is used by living organisms, transformed upon death and decomposition of the organism, and converted ultimately to its original state of oxidation.

summit   The highest point of any landform remanant, hill, or mountain.

superphosphate   A product obtained when phosphate rock is treated with H_2SO_4, H_3PO_4, or a mixture of those acids.

superphosphate, ammoniated   A product obtained when superphosphate is treated with NH_3 or with solutions containing NH_3 and/or other NH_4–N containing compounds.

superphosphate, concentrated   Also called *triple* or *treble superphosphate*, made with phosphoric acid. This includes any grade between 10 and 19% P (44 to 48% P_2O_5).

superphosphate, normal   Also called *ordinary* or *single superphosphate*. Superphosphate made by reaction of phosphate rock with sulfuric acid, usually containing 7 to 10% P (16 to 22% P_2O_5).

superphosphoric acid   The acid form of polyphosphates, consisting of a mixture of orthophosphoric and polyphosphoric acids. Species distribution varies with concentration, which is typically 30 to 36% P (68 to 83% P_2O_5).

surface area   The area of the solid particles in a given quantity of soil or porous medium.

surface sealing   The orientation and packing of dispersed soil particles in the immediate surface layer of the soil, thus greatly reducing it permeability.

surface soil   The uppermost part of the soil, ordinarily moved in tillage, or its equivalent in uncultivated soils and ranging in depth from 7 to 20 cm. Frequently designated as the *surface layer*, *Ap layer*, or *Ap horizon*.

surfactant   A substance that lowers the surface tension of a liquid.

swamp   An area saturated with water throughout much of the year but with the surface of the soil usually not deeply submerged. Usually characterized by tree or shrub vegetation. *See also* marsh.

talc   A magnesium silicate mineral with a 2:1 layer structure but without isomorphous substitution. May occur in soils as an inherited mineral. It is trioctahedral.

talud   A short, steep slope formed gradually at the downslope margin of a field by deposition against a hedge, a stone wall, or similar barrier.

talus   Fragments of rock and other soil material accumulated by gravity at the foot of cliffs or steep slopes.

tensiometer   A device for measuring the soil-water matric potential (or tension, or suction) of water in soil in situ; a porous, permeable ceramic cup connected through a water-filled tube to a manometer, vacuum gage, pressure transducer, or other pressure-measuring device.

terrace   (1) A level, usually narrow, plain bordering a river, lake, or the sea. Rivers sometimes are bordered by terraces at different levels. (2) A raised, more or less level or horizontal strip of earth usually constructed on or nearly on a contour and supported on the downslope side by rocks or other similar barrier and designed to make the land suitable for tillage and to prevent accelerated erosion. For example, the ancient terraces built by the Incas in the Andes. (3) An embankment with the uphill side sloping toward and into a channel for conducting water, and the downhill side having a relatively sharp decline; constructed across the direction of the slope for the purpose of conducting water from the area above the terrace at a regulated rate of flow and to prevent the accumulation of large volumes of water on the downslope side of cultivated fields. The depth of the channel, the width of the terrace ridge, and the spacings of the terraces on a field are varied with soil types, cropping systems, climatic conditions, and other factors.

thermosequence   A sequence of related soils that differ, one from the other, primarily as a result of temperature as a soil-formation factor.

threshold moisture content   (biology)The minimum moisture condition, measured either in terms of moisture content or moisture stress, at which biological activity just becomes measurable.

throughfall   That portion of precipitation that falls through or drips off of a plant canopy.

tidal flats   Areas of nearly flat, barren mud periodically covered by tidal waters. Normally, these materials have an excess of soluble salt. A miscellaneous area.

tile drain   Concrete or ceramic pipe placed at suitable depths and spacings in the soil or subsoil to provide water outlets from the soil.

till   (1) Unstratified glacial drift deposited by ice and consisting of clay, silt, sand, gravel, and boulders, intermingled in any proportion. (2) To prepare the soil for seeding; to seed or cultivate the soil.

tillage terminology

> *tillage* The manipulation, generally mechanical, of soil properties for any purpose; but in agriculture it is usually restricted to modifying soil conditions for crop production.
>
> *tillage action* The specific form or forms of soil manipulation performed by the application of mechanical forces to the soil with a tillage tool, such as cutting, shattering, inversion, or mixing.
>
> *tillage equipment (tools)* Field tools and machinery which are designed to lift, invert, stir, and pack soil, reduce the size of clods and uproot weeds (i.e., plows, harrows, disks, cultivators, and rollers).
>
> *tillage objective* A desired soil, soil surface, or soil residue cover condition that is to be produced by one or more tillage actions.
>
> *tillage operation* Act of applying one or more tillage actions in a distinct mechanical application of force to all or part of the soil mass.
>
> *tillage requirements* The soil physical conditions, which after a complete evaluation of fundamental utilitarian and economic requirements, is deemed necessary. Currently, often based on practical experience and judgment.

tilth The physical condition of soil as related to its ease of tillage, fitness as a seedbed, and its impedance to seedling emergence and root penetration.

top dressing An application of fertilizer to a soil surface, without incorporation, after the crop stand has been established.

toposequence A sequence of related soils that differ, one from the other, primarily because of topography as a soil-formation factor.

topsoil (1) The layer of soil moved in cultivation. *See also* surface soil. (2) The A horizon. (3) The Al horizon. (4) Presumably fertile soil material used to top-dress roadbanks, gardens, and lawns.

tortuosity The nonstraight nature of soil pores.

traffic pan *See* pan, pressure *or* induced.

trioctahedral An octahedral sheet that has all of the sites filled, usually by divalent ions such as magnesium or ferrous iron.

truncated Having lost all or part of the upper soil horizon or horizons.

tuff Volcanic ash usually more or less stratified and in various states of consolidation.

tundra A level or undulating treeless plain characteristic of arctic regions.

underground runoff (seepage) Water that seeps toward stream channels after infiltration into the ground.

universal soil loss equation   *See* erosion (2).

unsaturated flow   The movement of water in soil in which the pores are not completely filled with water.

value, color   The relative lightness or intensity of color and approximately a function of the square root of the total amount of light. One of the three variables of color. *See also* Munsell color system, hue, *and* chroma.

varnish   *See* desert varnish.

vermiculite   A highly charged layer-structured silicate of the 2:1 type that is formed from mica. It is characterized by adsorption preference for potassium and cesium over smaller exchange cations. It may be di- or trioctahedral.

void ratio   The ratio of the volume of soil pore (or void) space to the solid particle volume.

volumetric water content   The soil-water content expressed as the volume of water per unit bulk volume of soil.

vughs   Relatively large voids, usually irregular and not normally interconnected with other voids of comparable size; at the magnifications at which they are recognized they appear as discrete entities.

water content   The water lost from the soil upon drying to constant mass at 105°C; expressed either as the mass of water per unit mass of dry soil or as the volume of water per unit bulk volume of soil.

waterlogged   Saturated or nearly saturated with water.

water mass content   The soil-water content expressed as the mass of water per unit mass of oven-dry soil. *See also* dry-mass content or ratio.

water-soluble phosphate   That part of the phosphorus in a fertilizer that is soluble in water as determined by prescribed chemical tests.

water-stable aggregate   A soil aggregate that is stable to the action of water, such as falling drops, or agitation as in wet-sieving analysis.

water table   The upper surface of groundwater or that level in the ground where the water is at atmospheric pressure.

water table, perched   The water table of a saturated layer of soil which is separated from an underlying saturated layer by an unsaturated layer (vadose water).

water use efficiency   Dry matter or harvested portion of crop produced per unit of water consumed.

weathering   All physical and chemical changes produced in rocks, at or near the earth's surface, by atmospheric agents. *See also* chemical *and* physical weathering.

wetland   An area of land that has hydric soils and hydrophytic vegetation.

wilting point   *See* permanent wilting point.

windbreak   A planting of trees, shrubs, or other vegetation, usually perpendicular or nearly so to the principal wind direction, to protect soil, crops, homesteads, roads, and so on, against the effects of winds, such as wind erosion and the drifting of soil and snow.

windthrow mound   *See* cradle knoll *and* microrelief.

xenobiotic   A compound foreign to biological systems. Often refers to human-made (anthropogenic) compounds that are resistant or recalcitrant to biodegradation and/or decomposition.

xerophytes   Plants that grow in or on extremely dry soils or soil materials.

yield   The amount of a specified substance produced (e.g., grain, straw, total dry matter) per unit area.

Appendix

Conversion Factors for SI and Non-SI Units

To convert column 1 into column 2, multiply by:	Column 1 SI unit	Column 2 non-SI unit	To convert column 2 into column 1, multiply by:
		Length	
0.621	kilometer, km (10^3 m)	mile, mi	1.609
1.904	meter, m	yard, yd	0.914
3.28	meter, m	foot, ft	0.304
1.0	micrometer, μm (10^{-6} m)	micron, μ	1.0
3.94×10^{-2}	millimeter, mm (10^{-3} m)	inch, in.	25.4
10	nanometer, nm (10^{-9} m)	Angstrom, Å	0.1
		Area	
2.47	hectare, ha	acre	0.405
247	sq. kilometer, km^2 (10^3 m)^2	acre	4.05×10^{-3}
0.386	sq. kilometer, km^2 (10^3 m)^2	sq. mile, mi^2	2.590
2.47×10^{-4}	sq. meter, m^2	acre	4.05×10^3
10.76	sq. meter, m^2	sq. foot, ft^2	9.29×10^{-2}
1.55×10^{-3}	sq. millimeter, mm^2 (10^{-6} m)^2	sq. inch, in.^{2}	645

To convert column 1 into column 2, multiply by:	Column 1 SI unit	Column 2 non-SI unit	To convert column 2 into column 1, multiply by:
		Volume	
9.73×10^{-3}	cubic meter, m^3	acre–inch	102.8
35.3	cubic meter, m^3	cubic foot, ft^3	2.83×10^{-2}
6.10×10^4	liter, L (10^{-3} m^3)	cubic inch, $in.^3$	1.64×10^{-5}
2.84×10^{-2}	liter, L (10^{-3} m^3)	bushel, bu	35.24
1.057	liter, L (10^{-3} m^3)	quart (liquid), qt	0.946
3.53×10^{-2}	liter, L (10^{-3} m^3)	cubic foot, ft^3	28.3
0.265	liter, L (10^{-3} m^3)	gallon	3.78
33.78	liter, L (10^{-3} m^3)	ounce (fluid), oz	2.96×10^{-2}
2.11	liter, L (10^{-3} m^3)	pint (fluid), pt	0.473
		Temperature	
$1.00(K - 273)$	Kelvin, K	Celsius, °C	$1.00(°C + 273)$
$(9/5 °C) + 32$	Celsius, °C	Fahrenheit, °F	$5/9(°F - 32)$
		Mass	
2.20×10^{-3}	gram, g (10^{-3} kg)	pound, lb	454
3.52×10^{-2}	gram, g (10^{-3} kg)	ounce (avdp), oz	28.4
2.205	kilogram, kg	pound, lb	0.454
10^{-2}	kilogram, kg	quintal (metric), q	10^2
1.10×10^{-3}	kilogram, kg	ton (2000 lb), ton	907
1.102	megagram, Mg (tonne)	ton (U.S.), ton	0.907
1.102	tonne, t (mt)	ton (U.S.), ton	0.907
		Yield and Rate	
0.893	kilogram per hectare, kg ha^{-1}	pound per acre, lb $acre^{-1}$	1.12
7.77×10^{-2}	kilogram per cubic meter, kg m^{-3}	pound per bushel, lb bu^{-1}	12.87
1.49×10^{-2}	kilogram per hectare, kg ha^{-1}	bushel per acre, 60 lb	67.19
1.59×10^{-2}	kilogram per hectare, kg ha^{-1}	bushel per acre, 56 lb	62.71
1.86×10^{-2}	kilogram per hectare, kg ha^{-1}	bushel per acre, 48 lb	53.75
0.107	liter per hectare, L ha^{-1}	gallon per acre	9.35
893	tonnes per hectare, t ha^{-1}	pound per acre, lb $acre^{-1}$	1.12×10^{-3}
893	megagram per hectare, Mg ha^{-1}	pound per acre, lb $acre^{-1}$	1.12×10^{-3}
0.446	megagram per hectare, Mg ha^{-1}	ton (2000 lb) per acre, ton $acre^{-1}$	2.24
2.24	meter per second, m s^{-1}	mile per hour	0.447

To convert column 1 into column 2, multiply by:	Column 1 SI unit	Column 2 non-SI unit	To convert column 2 into column 1, multiply by:
		Pressure	
9.90	megapascal, MPa (10^6 Pa)	atmosphere, atm	0.101
10	megapascal, MPa (10^6 Pa)	bar	0.1
1.00	megagram per cubic meter, Mg m^{-3}	gram per cubic centimeter, g cm^{-1}	1.00
2.09×10^{-2}	pascal, Pa	pound per square foot, lb ft^{-2}	47.9
1.45×10^{-4}	pascal, Pa	pound per square inch, lb in.$^{-2}$	6.90×10^3
		Water Measurement	
9.73×10^{-3}	cubic meter, m^3	acre-inches, acre-in.	102.8
9.81×10^{-3}	cubic meter per hour, m^3 h^{-1}	cubic feet per second, ft^3 s^{-1}	101.9
4.40	cubic meter per hour, m^3 h^{-1}	U.S. gallons per minute, gal min^{-1}	0.227
8.11	gram per kilogram, g kg^{-1}	acre-feet, acre-ft	0.123
97.28	megagram per cubic meter, Mg m^{-3}	acre-inches, acre-in.	1.03×10^{-2}
8.1×10^{-2}	milligram per kilogram, mg kg^{-1}	acre-feet, acre-ft	12.33
		Energy, Work, Quantity of Heat	
9.52×10^{-4}	joule, J	British thermal unit, Btu	1.05×10^3
0.239	joule, J	calorie, cal	4.19
10^7	joule, J	erg	10^{-7}
0.735	joule, J	foot-pound	1.36
2.387×10^{-5}	joule per sq. meter, J m^{-2}	calorie per sq. centimeter (langley)	4.19×10^4
10^5	newton, N	dyne	10^{-5}
1.43×10^{-1}	watt per sq. meter, W m^{-2}	cal cm^{-2} min^{-1}	698

Source: Soil Science Society of America, 1987.

Index

KEY TO SOIL MAP ON INSIDE BACK COVER

A Alfisols Soils with subsurface horizons of clay accumulation and medium-to-high basic cation supply; either usually moist or moist for 90 consecutive days during a period when temperature is suitable for plant growth

A1   BORALFS cool
        A1a—with Histosols, cryic temperature regimes common
        A1b—with Spodosols, cryic temperature regimes

A2   UDALFS temperate to hot, usually moist
        A2a—with Aqualfs
        A2b—with Aquolls
        A2c—with Hapludults
        A2d—with Ochrepts
        A2e—with Troporthents
        A2f—with Udorthents

A3   USTALFS temperate to hot, dry more than 90 cumulative days during periods when temperature is suitable for plant growth
        A3a—with Tropepts
        A3b—with Troporthents
        A3c—with Tropudults
        A3d—with Usterts
        A3e—with Ustochrepts
        A3f—with Ustolls
        A3g—with Ustorthents
        A3h—with Ustox
        A3j—Plinthustalfs with Ustorthents

A4   XERALFS temperate or warm, moist in winter and dry more than 60 consecutive days in summer
        A4a—with Xerochrepts
        A4b—with Xerorthents
        A4c—with Xerults

D Aridisols Soils with pedogenic horizons, usually dry in all horizons and are never moist as long as 90 consecutive days during a period when temperature is suitable for plant growth

D1   ARIDISOLS undifferentiated
        D1a—with Orthents
        D1b—with Psamments
        D1c—with Ustalfs

D2   ARGIDS with horizons of clay accumulation
        D2a—with Fluvents
        D2b—with Torriorthents

E Entisols Soils without pedogenic horizons; either usually wet, usually moist, or usually dry

E1   AQUENTS seasonally dry or perennially wet
        E1a—Haplaquents with Udifluvents
        E1b—Psammaquents with Haplaquents
        E1c—Tropaquents with Hydraquents

E2   ORTHENTS loamy or clayey textures, many shallow to rock
        E2a—Cryorthents

E2b—Cryorthents with Orthods
E2c—Torriorthents with Aridisols
E2d—Tirriorthents with Ustalfs
E2e—Xerorthents with Xeralf

E3   PSAMMENTS sand or loamy sand textures
E3a—with Aridisols
E3b—with Orthox
E3c—with Torriorthents
E3d—with Ustalfs
E3e—with Ustox
E3f—with shifting sands
E3g—Ustipsamments with Ustolls

H Histosols Organic soils

H1   HISTOSOLS undifferentiated
H1a—with Aquods
H1b—with Boralfs
H1c—with Cryaquepts

I Inceptisols Soils with pedogenic horizons of alteration or concentration but without accumulations of translocated materials other than carbonates or silica; usually moist or moist for 90 consecutive days during a period when temperature is suitable for plant growth

I1   ANDEPTS amorphous clay or vitric volcanic ash or pumice
I1a—Dystrandepts with Ochrepts

I2   AQUEPTS seasonally wet
I2a—Cryaquepts with Orthents
I2b—Halaquepts with Salorthids
I2c—Haplaquepts with Humaquepts
I2d—Haplaquepts with Ochraqualfs
I2e—Humaquepts with Psamments
I2f—Tropaquepts with Hydraquents
I2g—Tropaquepts with Plinthaquults
I2h—Tropaquepts with Tropaquents
I2j—Tropaquepts with Tropudults

I3   OCHREPTS thin, light-colored surface horizons and little organic matter
I3a—Dystrochrepts with Fragiochrepts
I3b—Dystrochrepts with Orthox
I3c—Xerochrepts with Xerolls

I4   TROPEPTS continuously warm or hot
I4a—with Ustalfs
I4b—with Tropudults
I4c—with Ustox

I5   UMBREPTS dark-colored surface horizons with medium-to-low basic cation supply
I5a—with Aqualfs

M Mollisols Soils with nearly black, organic-rich surface horizons and high basic cation supply; either usually moist or usually dry

M1   ALBOLLS light gray subsurface horizon over slowly permeable horizon, seasonally wet
M1a—with Aquepts

M2   BOROLLS cool or cold
M2a—with Aquolls

            M2b—with Orthids
            M2c—with Torriorthents

M3   RENDOLLS subsurface horizons have much calcuim carbonate but no accumulation of clay
            M3a—with Usterts

M4   UDOLLS temperate or warm, usually moist
            M4a—with Aquolls
            M4b—with Eutrochrepts
            M4c—with Humaquepts

M5   USTOLLS temperate to hot, dry more than 90 cumulative days in the year
            M5a—with Argialbolls
            M5b—with Ustalfs
            M5c—with Usterts
            M5d—with Ustochrepts

M6   XEROLLS cool to warm, moist in winter and dry more than 60 consecutive days in summer
            M6a—with Xerorthents

O Oxisols Soils with pedogenic horizons that are mixtures principally of kaolinite, hydrated oxides, and quartz, and are low in weatherable minerals

O1   ORTHOX hot, nearly always moist
            O1a—with Plinthaquults
            O1b—with Tropudults

O2   USTOX warm or hot, dry for long periods but moist more than 90 consecutive days in the year
            O2a—with Plinthaquults
            O2b—with Tropudults
            O2c—with Ustalfs

S Spodosols Soils with accumulation of amorphous materials in subsurface horizons usually moist or wet

S1   SPODOSOLS undifferentiated
            S1a—cryic temperature regimes; with Boralfs
            S1b—cryic temperature regimes; with Histosols

S2   AQUODS seasonally wet
            S2a—Haplaquods with Quartzipsamments

S3   HUMODS with accumulations of organic matter in subsurface horizons
            S3a—with Hapludalfs

S4   ORTHODS with accumulation of organic matter, iron, and aluminum in subsurface horizons
            S4a—Haplothods with Boralfs

U Ultisols Soils with subsurface horizons of clay accumulation and low basic cation supply, usually moist or moist for 90 consecutive days during a period with temperature is suitable for plant growth

U1   AQUULTS seasonably wet
            U1a—Ochraquults with Udults
            U1b—Plinthaquults with Orthox
            U1c—Plinthaquults with Plinthaquox
            U1d—Plinthaquults with Tropaquepts

U2   HUMULTS temperate or warm and moist all of year, high content of organic matter
         U2a—with Umbrepts

U3   UDULTS temperate to hot, never dry more than 90 consecutive days in the year
         U3a—with Andepts
         U3b—with Dystrochrepts
         U3c—with Udalfs
         U3d—Hapludults with Dystrochrepts
         U3e—Rhodudults with Udalfs
         U3f—Tropudults with Aquults
         U3g—Tropudults with Hydraquents
         U3h—Tropudults with Orthox
         U3j—Tropudults with Tropepts
         U3k—Tropudults with Tropudalfs

U4   USTULTS warm or hot, dry more than 90 cumulative days in the year
         U4a—with Ustochrepts
         U4b—Plinthustults with Ustorthents
         U4c—Rhodustults with Ustalfs
         U4d—Tropostults with Tropaquepts
         U4e—Tropustults with Ustalfs

V Vertisols   Soils with high content of swelling clays; deep, wide cracks develop during dry
periods

V1   UDERTS usually moist in some part in most years, cracks open less than 90 cumula-
         tive days in the year
         V1a—with Usterts

V2   USTERTS cracks open more than 90 consecutive days in the year
         V2a—with Tropaquepts
         V2b—with Tropofluvents
         V2c—with Ustalfs

X Soils in areas with mountains   Soils with various moisture and temperature regimes;
many steep slopes, relief and total elevation vary greatly from place to place. Soils vary
greatly within short distances and with changes in altitude; vertical zonation common

X1   Cryic great groups of Entisols, Inceptisols, and Spodosols

X2   Boralfs and cryic great groups of Entisols and Inceptisols

X3   Udic great groups of Alfisols, Entisols, Inceptisols, and Ultisols

X4   Ustic great groups of Alfisols, Inceptisols, Mollisols, and Ultisols

X5   Xeric great groups of Alfisols, Entisols, Inceptisols, Mollisols, and Ultisols

X6   Aridisols, torric great groups of Entisols

X7   Ustic and cryic great groups of Alfisols, Entisols, Inceptisols, and Mollisols; ustic
         great groups of Ultisols, cryic great groups of Spodisols

X8   Aridisols, torric and cryic great groups of Entisols, and cryic great groups of Spodisols
         and Inceptisols

Z Miscellaneous

Z1   Icefields

Z2   Rugged Mountains—mostly devoid of soil (includes glaciers , permanent snow fields,
         and, in some places, small areas of soil)

JOHN TAGGART HINCKLEY LIBRARY
NORTHWEST COLLEGE
POWELL, WY 82435

WITHDRAWN